67 Springer Series in Solid-State Sciences

Edited by Klaus von Klitzing

Springer Series in Solid-State Sciences

Editors: M. Cardona   P. Fulde   K. von Klitzing   H.-J. Queisser

Two-Dimensional Systems: Physics and New Devices

Proceedings of the International Winter School
Mauterndorf, Austria, February 24–28, 1986

Editors:
G. Bauer, F. Kuchar, and H. Heinrich

With 201 Figures

Springer-Verlag Berlin Heidelberg GmbH

Professor Dr. Günther Bauer

Institut für Physik, Montanuniversität Leoben, Franz-Josef-Straße 18
A-8700 Leoben, Austria

Professor Dr. Friedemar Kuchar

Institut für Festkörperphysik, Universität Wien, A-1090 Wien, Austria

Professor Dr. Helmut Heinrich

Institut für Experimentalphysik, Universität Linz, A-4040 Linz, Austria

Series Editors:
Professor Dr., Dr. h. c. Manuel Cardona
Professor Dr., Dr. h. c. Peter Fulde
Professor Dr. Klaus von Klitzing
Professor Dr. Hans-Joachim Queisser

Max-Planck-Institut für Festkörperforschung, Heisenbergstrasse 1
D-7000 Stuttgart 80, Fed. Rep. of Germany

ISBN 978-3-662-02472-0   ISBN 978-3-662-02470-6 (eBook)
DOI 10.1007/978-3-662-02470-6

© Springer-Verlag Berlin Heidelberg 1986
Originally published by Springer-Verlag Berlin Heidelberg New York in 1986.
Softcover reprint of the hardcover 1st edition 1986

Offset printing: Beltz Offsetdruck, 6944 Hemsbach/Bergstr.
2153/3150-543210

Preface

In the series of International Winter Schools on New Developments in Solid State Physics, the fourth one was devoted to the subject: "Two-Dimensional Systems: Physics and Devices". For the second time the proceedings of one of these Winter Schools appear as a volume in the Springer Series in Solid-State Sciences (the earlier proceedings were published as Vol. 53). The school was held in the castle of Mauterndorf/Salzburg (Austria) February 24–28, 1986. These proceedings contain contributions based on the thirty invited lectures. The school was attended by 179 registered participants (40% students), who came from western European countries, the United States of America, Japan, the People's Republic of China and Poland.

As far as the subjects are concerned, several papers deal with the growth and characterization of heterostructures. Dynamical RHEED techniques are described as a tool for in situ studies of MBE growth mechanisms. Various growth techniques, including MBE, MOMBE, MOCVD and modifications of these, are discussed. The limiting factors for the carrier mobilities and the influence of the spacer thickness in single heterostructures of GaAs/GaAlAs seem to be understood and are no longer a matter of controversy. In addition, the growth of two fascinating systems, Si/SiGe and $Hg_{1-x}Cd_x Te/CdTe$, is discussed in detail.

A whole part is devoted to the subject of band edge discontinuities, which were not treated at all in the previous volume. Both experimental and theoretical contributions are presented. For GaAs/GaAlAs (for x = 0.3) a tendency towards a 60:40 ratio (ratio of the conduction to valence band discontinuity) is manifested.

Doping and compositional superlattices are treated, with regard to both their basic physics and their potential applications. These topics include nipi and the new sawtooth superlattices. The optical and electronic properties of Si/SiGe superlattices and possible optoelectronic applications of these structures are discussed.

The recent developments in bound states in quantum wells and two-dimensional impurity layers are reviewed. Of course, reviews on the present status of understanding of the integer and fractional quantum Hall effects are also presented. Experimental papers deal with the observation of the integer quantum Hall effect at microwave frequencies and with the fractional quantum Hall effect. The significance of the observed fractions with even denominators is discussed. New determinations of the density of states of Landau levels in the quantum Hall effect regime from activated transport, capacitance, specific heat and magnetization measurements are presented.

Beside the device-related papers on growth and characterization, several papers deal with transport properties of two-dimensional electron gas devices, e.g. tunnelling devices, hot-carrier transport in submicron MOS-FETs, microwave FETs (TEGFET, HEMT).

Whereas in the proceedings of the Winter School which was held two years ago the tutorial aspect was emphasized by many speakers, the new volume contains thorough reviews of the most recent developments in the topics mentioned above. We have also included papers on time-resolved spectroscopy of two-dimensional systems, a rapidly developing field of research, and the fascinating subject of electronic excitations in microstructured MIS and heterostructures. These two volumes now cover a broad spectrum of our present-day knowledge of the physics of two-dimensional systems.

The organizers are grateful to the authors, attendants and competitors in the final ski-race, who all helped to contribute to a successful meeting.

It is a pleasure to acknowledge the generous financial support received from

Bundesministerium für Wissenschaft und Forschung, Austria,
European Office of Aerospace Research and Development,
Österreichische Forschungsgemeinschaft,
Österreichische Physikalische Gesellschaft,
Office of Naval Research,
Salzburger Landesregierung,
US Army Research Development and Standardisation Group,

and from the following companies: Balzers, Bomem Europe, Bruker, EG and G. IBM Österreich, Instruments S.A. (Riber), Odelga Physik, Oxford Instruments and VG Instruments.

Mauterndorf, Austria   *G. Bauer F. Kuchar*
March 1986   *H. Heinrich*

Contents

Part VI   New Structures and Devices

Part VII   High Field Transport and Optical Excitation

Part I

Epitaxial Growth:
Methods and Characterization

New Epitaxial Growth Methods and Their Application to Quantum Wells and 2DEG Structures

Y. Horikoshi, N. Kobayashi, and H. Sugiura

NTT Research Laboratories, Musashino-shi, Tokyo 180, Japan

New thin film epitaxial growth methods have been developed which provide the possibility of accurately controlling thicknesses and doping profiles. These methods can be applied to 2DEG structures having a modulation doped single heterojunction, and those having an asymmetric quantum well doped with an atomic layer profile.

1. Introduction

Both metal-organic chemical vapor desposition (MOCVD) and molecular beam epitaxy (MBE) have been shown to be useful methods of producing ultrathin semiconductor layers with reasonable quality for device applications. Even for these methods, however, it has seemed to be almost impossible to control the growing thickness to an accuracy of one atomic layer. Moreover, these methods still present difficulties in growing GaAs and AlGaAs. MOCVD requires relatively high substrate temperatures. Because of this restriction, it is difficult to obtain sharp heterojunction interfaces and steep impurity concentration profiles. It is rather difficult to grow GaAs and AlGaAs by MBE with precise control and rapid changes in arsenic beam intensity. These factors are very important to obtain high quality layers and heterojunctions.

Two new epitaxial growth methods reported here may solve the above-mentioned problems: One is based on MOCVD and using an alternate supply of source gases. We have termed this method "Flow-rate Modulation Epitaxy (FME)". This method effectively lowers substrate temperature, and improves thickness controllability. The other is based on MBE utilizing arsenic molecule transport by purified hydrogen gas. It is termed "Vapor Transport Epitaxy (VTE)". With this method, arsenic intensity can be precisely regulated. In addition, it was found that the hydrogen gas flow through the chamber is effective in purifying the grown layers. By applying the FME method, we fabricated single heterojunction modulation-doped 2DEG structures and a new 2DEG structure based on an asymmetric quantum well doped with an atomic layer profile. Reasonably high 2DEG mobilities were attained using this method.

2. Flow-rate Modulation Epitaxy

An epitaxial growth method based on an alternate supply of source gases has recently been reported and applied to the growth of II-VI compound semiconductors/1/. Much attention has been paid to this method because it provides the possibility of controlling thickness to an accuracy of one atomic layer. This method was applied to GaAs growth by Nishizawa et al. /2/ and has been proved to be useful for obtaining well-thickness controlled layers. However, the GaAs layer showed heavy p-type conductivity, suggesting an incorporation of p-type impurities.

This phenomenon is probably caused by the use of trimethyl gallium (TMG) for the Ga source. This is because a large amount of carbon atoms are incorporated in the crystal when TMG is used for the Ga source in the conventional MOCVD growth. This problem almost disappears when triethyl gallium (TEG) is employed instead of TMG. For the MOCVD growth of AlGaAs, the combination of triethyl aluminium (TEA) and TEG yields layers with higher purity than that of trimethyl aluminium (TMA) and TMG. Another cause of the heavy p-type problem may be due to the formation of arsenic vacancies near the growing surface during the Ga adsorption period followed by a reaction with impurity atoms such as carbon.

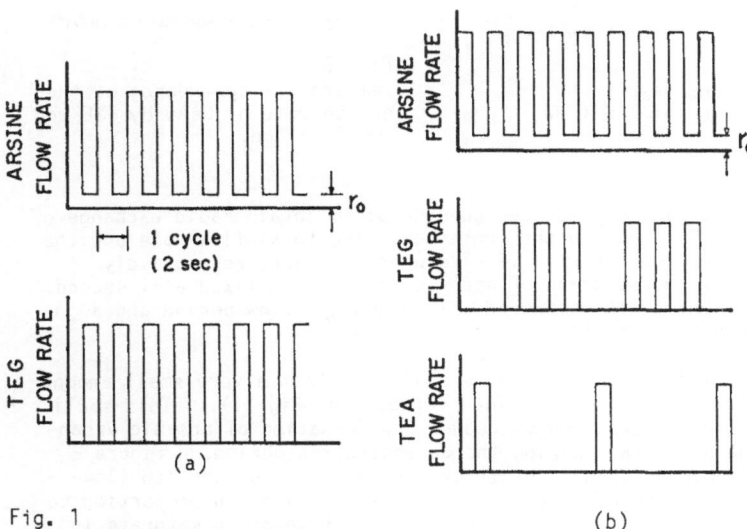

Fig. 1
Gas flow sequence for (a) GaAs, and (b) $Al_{0.25}Ga_{0.75}As$ growth by Flow-rate Modulation Epitaxy.

The FME method/3/ is based on the alternating gas flow of triethyl compounds (TEG and/or TEA) and arsine by using hydrogen gas, as shown in Fig. 1. The alternating gas flow of metallorganic compounds and arsine inherently has a very important advantage in addition to the possibility of improving thickness controllability. In the conventional MOCVD growth, additional compounds are formed between metallorganic compounds and arsine during growth. These compounds degrade the purity of layers when the growth temperature is lowered. In contrast, the formation of such compounds will be much reduced using the present method.

One of the practical advantages of this method is that both the layer thickness and alloy composition are determined by the number of gas flow pulses with fixed peak flow rates as shown in Fig. 1. Therefore, the growth process can be easily controlled by a computer. The growth appara-

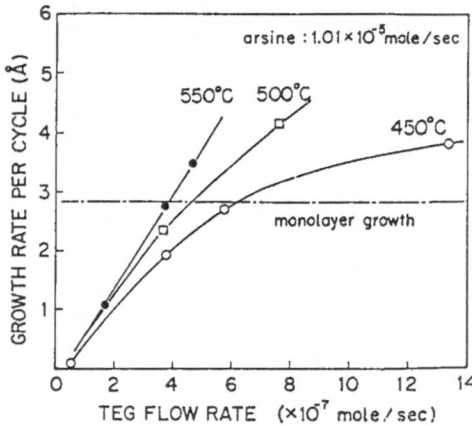

Fig. 2
Growth rate of GaAs layers per
cycle as a function of TEG flow
rate.

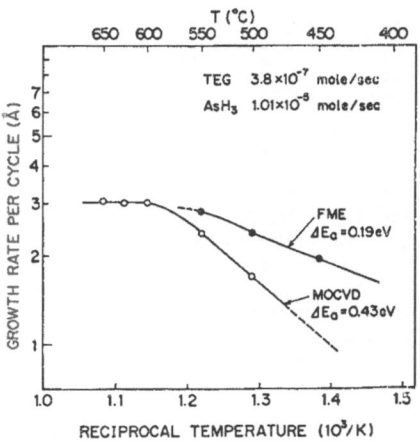

Fig. 3
Temperature dependence of the
growth rate of GaAs by FME and
MOCVD method.

tus used in this experiment was designed so as to obtain rapid exchange of
gas composition (<0.1 sec) on the substrate. The total flow-rate and the
pressure inside the reactor were 10 sl/min and 90 Torr, respectively.
The period for the introduction of each source gas was fixed at 1 second.
For GaAs growth, one cycle comprises a 1 second TEG flow period and a
succeeding 1 second arsine flow period.

Another characteristic point of this method is that a very small amount
of arsine is added during the TEG flow period (see Fig. 1). This small
amount of arsine was introduced to reduce the formation of arsenic vacancies in the growing surface during the Ga adsorption period. Figure 2
shows the dependence of GaAs layer thickness per cycle on the TEG flow-
rate. At 450°C, the layer thickness per cycle increases in proportion to
the TEG flow-rate in the low flow rate region, but tends to saturate in
the higher flow rate region. This phenomenon may indicate the onset of
"atomic layer growth". Although no such evidence is found at the higher
growth temperatures shown in Fig. 2, it is interesting to note that if the
growth rate per unit cycle is chosen to be equal or less than one atomic
layer thickness (0.28 nm) high-quality layers with improved lateral uni-
formity were obtained. Therefore, the flow-rate of TEG was chosen to yield
a growth of one monolayer per cycle. A typical growth duration was 7200
cycles (4 h) which produces layers totaling about 2 microns in thickness.

The FME method produces uniform GaAs and AlGaAs layers with high crystal
quality even when the substrate temperature is as low as 450°C. In the
conventional MOCVD method, however, good quality AlGaAs layers can be
grown when the substrate temperature is higher than about 700°C. This
remarkable reduction in the growth temperature implies that a different
growth mechanism is operative in FME. To see this effect more clearly,
we measured the temperature dependence of the growth rate as shown in Fig.
3. The activation energy of FME estimated at lower growth temperatures
is much less than that of MOCVD. This difference can be accounted for by
the adsorption-decomposition mechanism with the FME method/3/.

4

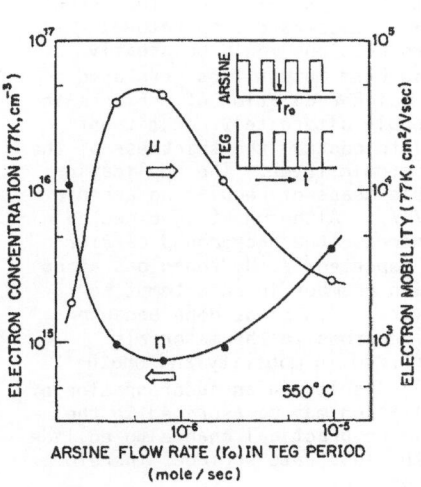

Fig. 4
Electron mobility and carrier
concentration of FME grown GaAs
as a function of r_0.

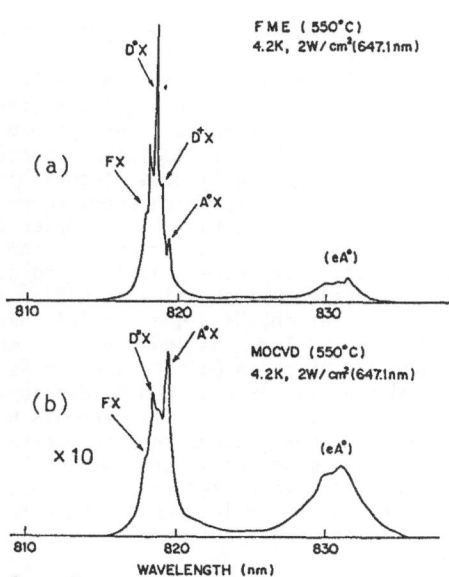

Fig. 5
Photoluminescence from GaAs grown
by: (a) FME and (b) MOCVD at 550°C.

Next, we will describe the effect of a very small amount of arsine
during the TEG flow period. Even without this small amount of arsine,
the resulting layers grown at 550°C exhibit n-type conductivity. However,
the mobilities at 77 K are less than 3000 cm²/Vsec. An important effect
of this small amount of arsine is to enhance the electron mobility in the
GaAs layers. Figure 4 demonstrates the electron mobility and residual
carrier concentration of GaAs layers grown at 550°C as a function of the
arsine flow rate (r_0) during the TEG period. Electron mobility at 77 K
drastically increases from 2000 to 50000 cm²/Vsec, and electron concentration decreases from 10^{16} to less than 10^{15} /cm³ when r_0 is increased from 0 to
5×10^{-6} mole/sec. This r_0 value corresponds to an arsine flow of only 4
% during the arsine flow period.

Photoluminescence measurements also suggest that the quality of GaAs
layers grown by the FME method is superior to that of layers grown by the
MOCVD method. Figure 5 compares 4.2 K photoluminescence spectra of
the GaAs layers grown by the FME and MOCVD method at 550°C. The sample
was excited by a Kr-ion laser (647.1 nm) with an intensity of about 2
W/cm². The GaAs layers grown by the FME method exhibit a lower carrier
concentration than those grown by the conventional MOCVD method. Nevertheless, photoluminescence intensity was about 10 times that for the
MOCVD-grown sample. In addition, relative intensities of acceptor related
emissions (band-to-acceptor and excitons bound to acceptors) are much
reduced in the FME-grown sample.

3. Vapor Transport Epitaxy

Quick, precise regulation of arsenic beam intensity is difficult during
MBE growth of GaAs and AlGaAs. One fixed arsenic beam intensity has been

used during the growth of GaAs/AlGaAs heterojunctions because of this difficulty. This is in spite of the fact that optimum values for arsenic beam intensity for GaAs and AlGaAs at a given substrate temperature differ slightly. The quality of the heterojunctions would be greatly improved if the prospective optimum arsenic beam intensities were used during growth. In addition, precise and quick regulation of the arsenic beam intensity will make it possible to supply alternately constituent source materials, which seems to be useful to control the thickness of the layers to an accuracy on the order of one atomic layer. The application of arsine instead of arsenic is one possible means of regulating arsenic beam intensity more precisely and quickly/4/. Although it is expected that residual impurities in arsine will increase the background carrier concentration, high quality GaAs has been reported/5/. Hydrogen gas at up to 5 x 10^{-6} Torr was injected into the growth chamber in an attempt to purify the grown GaAs/6/ and AlGaAs/7/ layers. This was done because hydrogen gas is expected to deoxidize and decarbonize the materials. Although discernible improvements were reported in mobility and photo-luminescence measurement, this method still results in an incorporation of a large amount of impurities. This limitation may be ascribed to the fact that evaporated arsenic molecules have no practical chance to collide with hydrogen molecules on their trip to the substrate surface. Therefore, purification by hydrogen gas is not so effective.

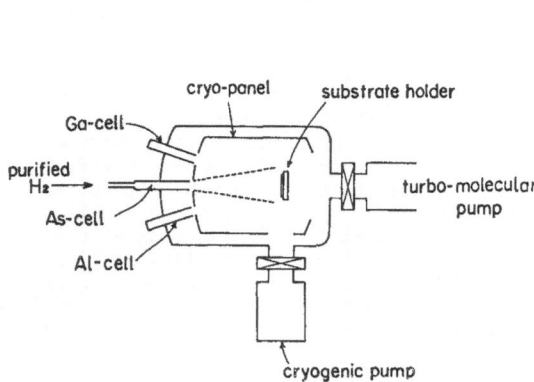

Fig. 6
High-vacuum growth chamber for the VTE growth. Purified hydrogen gas is introduced through the arsenic effusion cell.

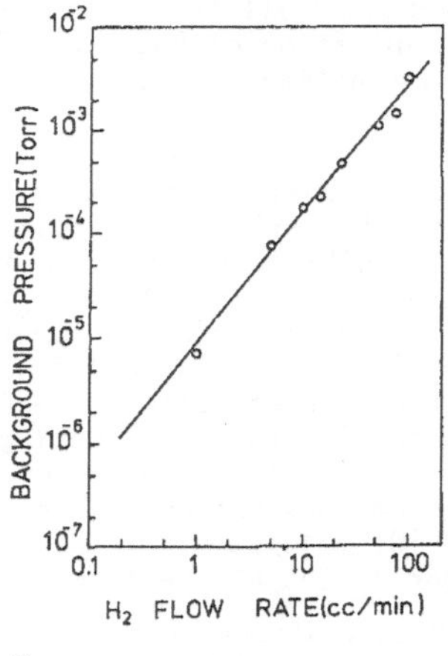

Fig. 7
Plotting of background hydrogen gas pressure as a function of hydrogen gas flow rate through the arsenic cell.

In the present experiment, GaAs layers were grown in a high-vacuum MBE chamber using arsenic molecules transported by purified hydrogen carrier gas. This was done to demonstrate the possibility of quick, precise arsenic beam intensity regulation as well as the purification effect due to hydrogen gas. Figure 6 shows the apparatus used in this study. A conventional MBE system was modified by adding a turbo-molecular pump with a capacity of 1800 1/sec to evacuate a large amount of hydrogen gas. The minimum background pressure of the system (2 x 10^{-10} Torr) was attained using a cryogenic pump (1500 l/sec). The hydrogen gas flow rate through the heated arsenic cell varied over a range of 0.3 to 100 cc/min. This corresponds to a background hydrogen gas pressure range of 1×10^{-6} to 1×10^{-3} Torr, as demonstrated in Fig. 7.

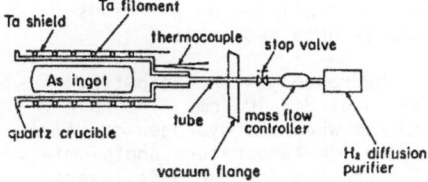

Fig. 8
Structure of the arsenic cell designed for the VTE study.

As shown in Fig. 8, the arsenic effusion cell has a quartz crucible with a thin quartz tube at its bottom. The other end of this tube was connected with a stainless steel tube to introduce palladium-diffused hydrogen gas. Since a large amount of hydrogen molecules are introduced through the heated arsenic cell, the arsenic molecules over the arsenic ingot will frequently collide with hydrogen molecules. This is because the mixture of hydrogen and arsenic molecules in the arsenic crucible has characteristics of a viscous fluid. Thus, we expect an effective purification of arsenic by hydrogen gas. The transport of arsenic vapor from the heated arsenic ingot to the substrate surface by hydrogen gas flow was directly observed by measuring the thickness of the adsorbed arsenic on the cooled (< 0°C) substrate surface. Figure 9 represents the observed arsenic deposition rate as a function of the hydrogen gas flow rate. In this experiment, the arsenic beam was impinged for 3 hours, and the arsenic cell temperature was maintained at a value to attain an arsenic beam-equivalent-pressure of 7×10^{-6} Torr (without hydrogen gas flow). A distinct increase in the amount of adsorbed arsenic was observed as the flow rate increased. This result clearly indicates that the arsenic beam intensity can be regulated by the flow rate of hydrogen gas.

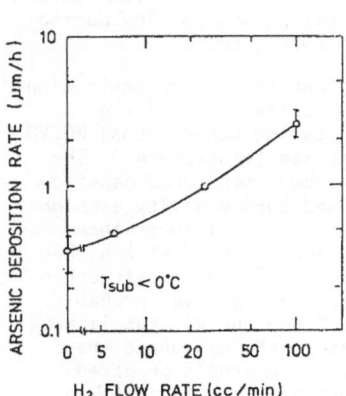

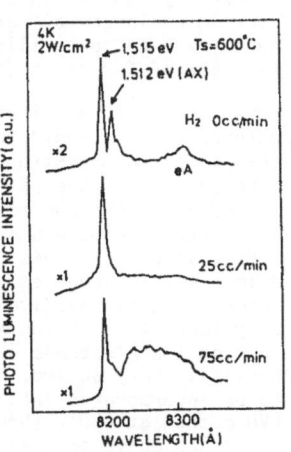

Fig. 9
Arsenic deposition rate on cooled GaAs substrate as a function of the hydrogen gas flow rate.

Fig. 10
Photoluminescence spectra of GaAs layers grown by the VTE method.

The surface morphology of grown GaAs layers was examined to see the effect of the arsenic transported by hydrogen gas. In our apparatus, the arsenic stabilized surface condition cannot be established with an arsenic beam-equivalent-pressure of 5×10^{-7} Torr when the substrate temperature is 600°C. Under this condition, the GaAs layers exhibit rough surfaces. By introducing a flow of hydrogen gas through the arsenic cell at a rate of 50 cc/min, an arsenic stabilized condition is established by the transported arsenic, and a specular grown surface is obtained.

Undoped GaAs layers grown without flowing hydrogen gas showed p-type conductivity with carrier concentrations of about 2×10^{14}/cm. The hole concentrations were reduced to about 5×10^{13}/cm when the hydrogen gas flow was applied. This result is consistent with low-temperature photoluminescence measurements shown in Fig. 10. The spectra for the GaAs layers grown at 600°C with hydrogen gas flow rates of 25 and 75 cc/min are compared with the results for a zero-flow-rate sample. It is interesting to note that the intensities of acceptor related luminescence (AX and eA) are much reduced by introducing a flow of hydrogen gas. However, a broad emission band appears at an increased flow rate(75 cc /min). This is probably caused by the defect-induced-bound excitons. This phenomenon may be explained as follows: At a higher hydrogen gas flow rate, arsenic molecules over the arsenic ingot will be supercooled by the collision with hydrogen molecules. Thus, the arsenic vapor tends to form clusters composed of a larger number of arsenic atoms. This may enhance emission due to the defect induced excitons.

4. Application to 2DEG Structures

By applying the Flow-rate Modulation Epitaxy method, we have grown single heterojunction modulation-doped structures at 550°C /8/. An undoped GaAs layer was grown on a [001] oriented semi-insulating GaAs substrate using the gas flow rate program shown in Fig. 1(a). The succeeding $Al_{0.25}Ga_{0.75}As$ layer was grown using the program shown in Fig. 1(b). Thus, the $Al_{0.25}Ga_{0.75}As$ layer is formed by repeating the growth of three GaAs monolayers followed by the growth of one AlAs monolayer. Therefore, the $Al Ga_{0.75}As$ layer is no longer a "random alloy", but has an "ordered" structure. Si impurity was added by allowing silane gas pulses to flow during the TEG gas flow period for the middle GaAs monolayer growth.

Figure 11 shows the measured electron mobility and carrier concentration for this structure. Also shown for comparison is the result for a modulation doped single heterostructure prepared by the conventional MOCVD method using the same apparatus at a higher substrate temperature. The structure prepared using the FME method with the above-mentioned gas-flow program at a substrate temperature of 550°C showed 2DEG mobility as high as 8×10^4 cm^2/ Vsec at 5 K. This value is, however, still lower than that obtained in a sample prepared by the conventional method at a higher growth temperature. This relatively low mobility in FME grown structure may be caused by impurities in the AlGaAs "ordered" alloy layer probably incorporated during the Al-atomic layer growth. Since the Al-stabilized surface is chemically very active, impurity atoms in the gas phase are anticipated to be incorporated into the surface. The sample prepared using the MOCVD method at 680°C showed much higher electron mobility as indicated by a dashed curve in this figure (as high as 4.5×10^5 cm^2/Vsec at 5 K). When the growth temperature is decreased, however, the observed 2DEG mobility drops very rapidly. At 550°C, there is no evidence for 2DEG formation. These experimental results indicate that the crystal

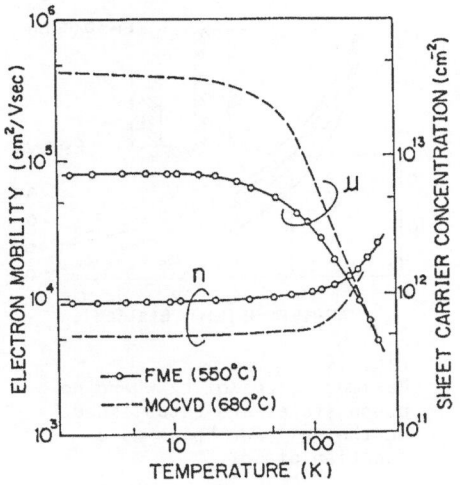

Fig. 11
Electron mobility and carrier
concentration in a FME grown
modulation-doped single heterostructure grown at 550°C. The
dashed curve represents the
results for a MOCVD grown wafer
at a higher temperature(680°C).

quality of GaAs and AlGaAs is much deteriorated by decreasing the substrate temperature in the conventional MOCVD growth, and that the FME method
is useful to reduce the substrate temperature   maintaining a reasonable
quality of the grown crystals.

In the following, we propose a new 2DEG (or 2DHG) structure which utilizes characteristics of an asymmetric quantum well.   An asymmetric
quantum well need not have any quantized states in the well if the well
width Lz is smaller than some critical value (see Fig. 12). If donor (or
acceptor) impurities are doped in such a quantum well, no electrons (or
holes) from these donors (or acceptors) will be confined in the well.
Instead, they form 2DEG (or 2DHG) at the neighboring heterojunction, as
shown in Fig. 12 (b).   The AlGaAs barrier layers are undoped in this
structure.   Therefore, 2DEG characteristics are expected to be free from
the undesirable effects of DX centers.

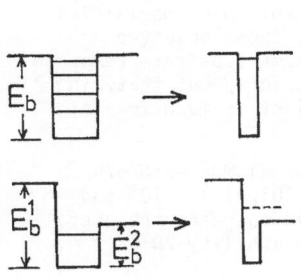

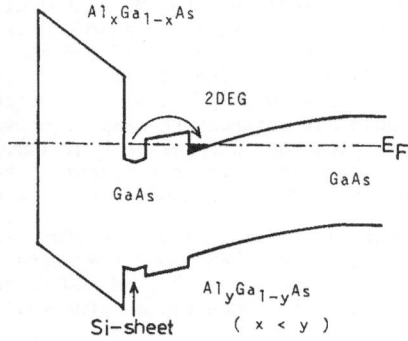

Fig. 12 (a)
Comparison between symmetric and
asymmetric quantum wells when the
well thickness is decreased.

Fig. 12 (b)
Schematic 2DEG structure having
an asymmetric quantum well. The
well width is so narrow that we
applied an atomic layer doping.

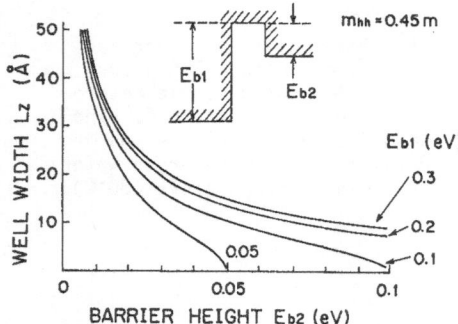 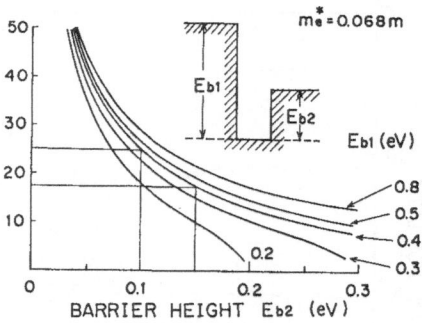

Fig. 13
Maximum values of Lz where no
bound states are established
in the well for electrons as a
function of Eb2. The Eb1 values
are given in the figure.

Fig. 14
Maximum values of Lz where no
bound states are established
in the well for holes as a
function of Eb2.

When a square well is bounded by two different potential barriers of Eb1
and Eb2 (Eb1>Eb2), the quantum well has no bound state if

$$Lz < \hbar(2m\ Eb2)^{-1/2} \cdot arctan\{[(Eb1-Eb2)/Eb2]\}^{1/2}.$$

Figures 13 and 14 demonstrate the maximum values of Lz given by this equation as a function of Eb2 for several respective Eb1 values for electrons
and holes.   If $Al_{0.4}Ga_{0.6}As$ and $Al_{0.1}Ga_{0.9}As$ are chosen for the barrier
layers, the resulting barrier heights for electrons, Eb1 and Eb2, are 0.4
eV and 0.1 eV, respectively, when the conduction band edge discontinuity
$\Delta Ec/\Delta Eg$ is assumed to be 0.8.   The maximum Lz corresponding to these
barrier heights is about 2.5 nm.   If a much lower $\Delta Ec/\Delta Eg$ value of 0.6
is used, the resulting Lz will be 2.9 nm.

    Therefore, we have grown a structure with a well width of about 2.2 nm
(8 monolayers) by the FME method at 550°C.   The well is so narrow that we
doped Si impurity in an atomic layer profile at the center of the well, as
shown in Fig. 12(b).   If the above principle works, we may have high
mobility 2DEG.   Indeed, this structure showed reasonably high mobilities
of $4 - 5 \times 10^4 cm^2/Vsec$ at 77 K which were similar to those observed in
single heterostructure samples grown by FME at the same substrate temperature (see open circles in Fig. 11).   It should be pointed out that, in
this structure, the persistent photoconductivity effect is much smaller
than in modulation doped single heterostructures.

    Wafers with a similar structure were grown by a normal MBE method/9/.
The resulting 2DEG mobilities were as high as 3.5×10^5, 1.1×10^5 and
$7000\ cm^2/Vsec$ at 5, 77 and 300 K, respectively.   This suggests the useful-
ness of the asymmetric quantum well in inducing high mobility 2DEG.

5. Summary

    Two new thin film epitaxal growth methods have been presented.   Flow-
rate Modulation Epitaxy effectively reduced the substrate temperature by
more than 100 degrees compared with the conventional MOCVD method.   The

electron mobility in GaAs layers grown at low temperatures was found to be enhanced by the small amount of arsine added during the TEG (or TEA) period. In the Vapor Transport Epitaxy, arsenic molecules are transported from the heated arsenic ingot to the growing surface. This suggests the possibility of regulating arsenic beam intensity using hydrogen gas. The hydrogen gas flow through the chamber was found to be effective in reducing acceptor type impurities in the grown GaAs layers. The FME method was applied to grow 2DEG structures. Experimental results suggest that the FME method is useful in growing high-mobility 2DEG structures at relatively low substrate temperatures. A 2DEG structure having an asymmetric quantum well has been proposed. Our preliminary results indicate that this structure is useful in inducing high-mobility 2DEGs.

Acknowledgments: The authors would like to thank Drs. Yoshinori Kato and Tatsuya Kimura for their encouragement throughout this work. They also wish to thank Dr. Klaus Ploog and Prof. Herbert Kroemer for useful discussions on the 2DEG structure having an asymmetric quantum well.

References

/1/ T. Suntola: Extended Abstract of 16th Conf. Solid. State Devices & Materials (Kobe, Japan) p.647.
/2/ J. Nishizawa, H. Abe and T. Kurabayashi: J. Electrochem. Soc. 132 (1985) 1197.
/3/ N. Kobayashi, T. Makimoto and Y. Horikoshi: Jpn. J. Appl. Phys. 24 (1985) L962.
/4/ M. B. Panish: J. Electrochem. Soc. 127 (1980) 2729.
/5/ B. J. Skromme, G. L. Stillman, A. R. Calawa and G. M. Metze: Appl. Phys. Lett. 44 (1984) 240.
/6/ A. R. Calawa: Appl. Phys. Lett. 33 (1978) 1020.
/7/ K. Kondo, S. Muto, K. Nanbu, T. Ishikawa, S. Hiyamizu, and H. Hashimoto: Jpn. J. Appl. Phys. 22 (1983) L121.
/8/ T. Makimoto, N. Kobayashi and Y. Horikoshi: to be published in Jpn. J. Appl Phys.
/9/ Y. Horikoshi, A. Fischer and K. Ploog: to be published in Jpn. J. Appl. Phys.

Metalorganic MBE – A New Technique for the Growth of III-V Semiconductor Layers

H. Lüth

II. Physikalisches Institut der Rheinisch-Westfälischen Technischen Hochschule Aachen, D-5100 Aachen, Fed. Rep. of Germany

Abstract

A new method for the controlled growth of thin III-V semiconductor films is described. The technique called metalorganic MBE (MOMBE) combines advantages of MOCVD and MBE. As in MBE the process is performed in an UHV system; as sources molecular beams of hydrides like AsH_3, PH_3 and/or metalorganic alkyls like $Ga(CH_3)_3$ [TMG], $Ga(C_2H_5)_3$ [TEG] , $In(C_2H_5)_3$ [TEIn] , $As(CH_3)_3$ [TMAs] etc. are used. In the present review, different experimental approaches to MOMBE are briefly discussed. The main emphasis is on the epitaxial growth of GaAs from AsH_3, TMG and TEG. The same approach is used to realize p- and n-type doping by carbon and silicon. The feasibility of the doping techniques is demonstrated by the growth of modulation-doped multilayer structures (nipi structures). Furthermore, results of selective growth studies are presented.

1. Introduction

The quest for faster devices and new structures has caused continued interest in the improvement of techniques for the epitaxial growth of III-V semiconductor multilayer structures [1]. In the field of fundamental research the quantum physics of two-dimensional (2D) electron gases and of superlattices in general is of central interest and requires further development of such structures. With regard to microelectronics, there is a strong demand for the production of semiconductor LASERS and for very fast integrated circuits on the basis of III-V compound layer structures. The production of integrated circuits, in particular, requires the availability of large area wafers with epitaxial overlayers of extremely low defect density. Furthermore, high precision is necessary for the control of layer thickness, doping concentration and composition. For the production of sophisticated circuits and devices as e.g. HEMTs (high electron mobility transistors) [1] sharp doping and compositional profiles with abrupt changes within one atomic layer must be achieved. For industrial production these requirements must be met with a high degree of reproducibility.

Three experimental approaches are widely used at present to meet these requirements. In the metal organic chemical vapour deposition (MOCVD) technique a cold wall flow reactor is used to grow III-V epitaxial layers at atmospheric or at low pressure (LP-MOCVD) from gas sources, e.g. arsine (AsH_3) and trimethyl gallium [TMG, $Ga(CH_3)_3$].

In the molecular beam epitaxy (MBE) the growth process runs in an ultrahigh vacuum (UHV) system, where molecular beams as sources are generated by effusion from ovens (effusion cells) containing the corresponding solid elements (Ga, In, GaAs, Sb etc.). The obvious advantages of MOCVD, continuous performance even for large numbers of wafers and easier control of deposition rates by gas flow valves, are contrasted in MBE by the possibility of easily producing extremely sharp doping and composition profiles. Since no gas volume has to be changed, as in a flow reactor, switching times between different source materials are below the growth time for a monolayer. Furthermore, MBE as an UHV technique is compatible with in-situ control methods like RHEED (reflection high-energy electron diffraction) and, in particular, with technological processes running in high vacuum, as ion implantation, ion etching etc.. This could likely save steps in a production line. On the other hand, the necessary refilling of the effusion cells with source materials in MBE causes irreproducibilities in the properties of the grown layers.

The third technique, metalorganic MBE (MOMBE) [2] was developed in order to combine the advantages of both techniques MOCVD and MBE. In MOMBE, an UHV system is used as growth chamber and like in MOCVD gas sources (AsH$_3$, PH$_3$, TMG, TEG etc.) supply the growth material. Their flow can easily be controlled by valves and a continuous performance is possible. As in MBE short switching times are achieved and UHV analysis techniques and equipment for technological processes can be added to a MOMBE system by UHV transfer units. Beside the real MOMBE technique, where both the III- and the V components as well as the doping material are supplied from gas lines, there are several intermediate approaches in which standard UHV systems for MBE are equipped with one or two gas line sources for the supply of a metal alkyl or a group V hydride. These combinations of gas line sources with effusion cells for the other components are briefly considered in the next chapter and then the main emphasis is laid on the discussion of MOMBE in connection with results from our own work.

2. Different Experimental Approaches to MOMBE

Mainly, the consideration that effusion cells for the group V elements in MBE are rapidly depleted and cause irreproducibilities and inhomogeneities of the layers lead PANISH [3] to the use of gas sources of AsH$_3$ and PH$_3$ for the growth of GaAs and InP. The group III elements were supplied in his conventional MBE apparatus from standard thermally heated boron nitride crucibles. The element V hydride was decomposed by heating the end part of the alumina tube being used as the gas inlet. In this first work the possibility of epitaxial growth of GaAs and InP was demonstrated, and later on with an improved equipment GaInAsP/InP heterostructure LASERS [4] and GaInAs(P)/InP quantum well structures [5] were produced.

CALAWA [6] replaced the effusion cell for arsenic in his MBE apparatus by an AsH$_3$ molecular beam, in order to decrease the number of point defects in his GaAs layers. Elemental arsenic sources produce As$_4$ which dissociates at the growing surface. Incomplete decomposition is thought to be responsible for formation of defects [7]. On the other hand, the use of AsH$_3$, if predecomposed in the inlet capillary, might result in a substantial supply of the monomer As$_1$ to the surface. Thus 77 K mobilities as high as 110 000 cm^2/Vs could be achieved in GaAs by CALAWA [6]. Also defect-related photoluminescence bands characteristic for MBE grown material could be suppressed by using a molecular beam of AsH$_3$ with precracking facilities [8]. With respect to the growth of InP the handling of elemental red P in MBE is dangerous. Replacing the effusion cell by a gas source for PH$_3$ removes most of the difficulties [9]. A further advantage of using molecular beams of AsH$_3$ and PH$_3$ seems to be the production of atomic hydrogen during the growth process. In conventional MBE atomic H was shown to reduce unintentional background doping [10].

One severe problem with MBE grown layers of GaAs is the relatively high density of surface defects, in particular of so called oval defects. As one possible origin contaminations, e.g. oxides, in the gallium melt are discussed [11]. These considerations lead to the use of gaseous element III sources. Consequently some Japanese groups use molecular beams of TMG, TEG and TEIn as Ga and In source, respectively, whereas the group V component is supplied by conventional effusion cells [12-14]. As in MOMBE with two gaseous sources (see next chapter) the use of TMG always causes high carbon p-type dopings in the 10^{19}cm^{-3} range. With TEG much lower p-type doping at lower growth temperature was achieved [13]. Also the introduction of atomic hydrogen into the system decreased the background p-type doping level by reducing the incorporation of carbon into the layers [13].

An interesting experimental approach to gas phase sources in MBE has been introduced recently by TSANG [15-17]. He supplies both the III and the V component as metalorganics (TMIn, TEIn, TMG, TEG, TMAs, TEP). By this method the use of the extremely dangerous hydrids AsH$_3$ and PH$_3$ is avoided. To improve the quality of the grown layers H$_2$ is mixed in into the gas lines. The density of oval defects is considerably reduced [16] and by use of TEG in the growth of GaAs the background p-type doping is decreased into the range of 10^{16}cm^{-3} [17].

The method which is directly derived from MOCVD and which is meanwhile called MOMBE uses gas sources, metalorganics for the group III and hydrides as AsH$_3$, PH$_3$ etc. for the V component. The general possibility of this growth technique was first shown by our group in 1981 [2]; somewhat later similar results were obtained by VODJDANI et al. [18]. The discussion of this technique is the main topic of the present paper. In particular, growth and doping of GaAs will be considered.

3. Growth of GaAs from AsH$_3$, TMG and TEG

The experimental set-up for MOMBE which is used by our group [19] consists of a standard UHV chamber for MBE equipped with a load lock chamber and pumped by a turbomolecular pump (500 l/s) and a cryopump (pumping capacity for H$_2$: 9 bar x l). An additional Ti-sublimation pump is only used for pumping down. The substrate holder and the capillaries being used as gas sources are surrounded by LN$_2$ cooled cryoshields. Two gas inlet quartz capillaries for the supply of metalorganics (TMG, TEG etc.) are kept slightly above room temperature (heated by a Ta filament) to avoid condensation of the material. Two other capillaries for injecting AsH$_3$ and for SiH$_4$ as a doping material can be heated up to temperatures of 1000 K by a Ta filament. This Ta filament causes catalytic decomposition of the AsH$_3$ in the capillary [6,9] thus leading to a strong H$_2$/H background pressure between 10^{-3} and 10^{-4}Pa during growth. Before mounting on the substrate holder the GaAs wafers are chemically cleaned in boiling organic solvents and HCl and subsequently etched in H$_2$SO$_4$/H$_2$O$_2$/H$_2$O (3:1:1). After treatment with boiling deionized H$_2$O the samples are loaded into the chamber, where an initial pressure in the 10^{-8}Pa range is established. Prior to growth the wafers are heat treated at about 915 K in an As flux to obtain a clean surface. In order to change composition and doping rapidly during growth, mechanical shutters in front of the capillaries are used in addition to leak valve control.

Growth of GaAs can only be achieved by precracking the AsH$_3$ in the injection capillary [19]. Layers could be grown both by using TMG and TEG for the III component. According to the higher stability of TMG there is a remarkably different behaviour of the growth rate for both metalorganics (Figs. 1 and 2).

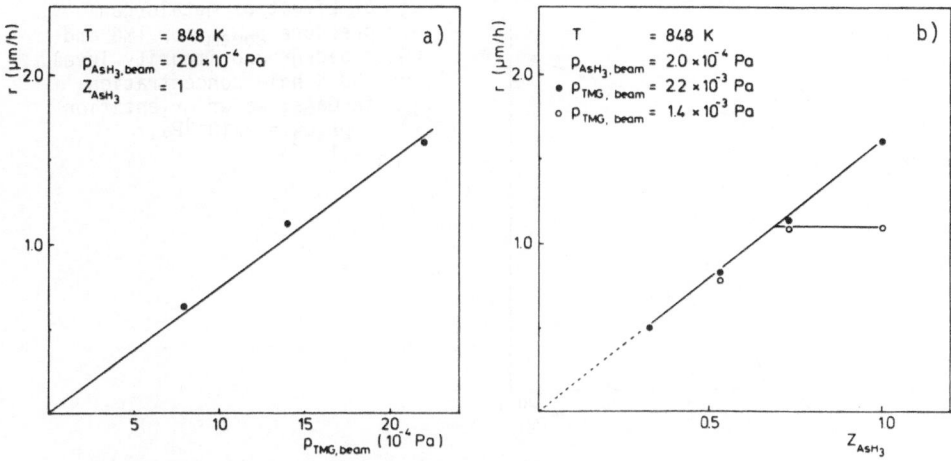

Fig. 1 Dependence of growth rate r of GaAs(100) for TMG on
a) TMG beam pressure $p_{TMG,beam}$
b) AsH$_3$ cracking efficiency Z_{AsH_3}

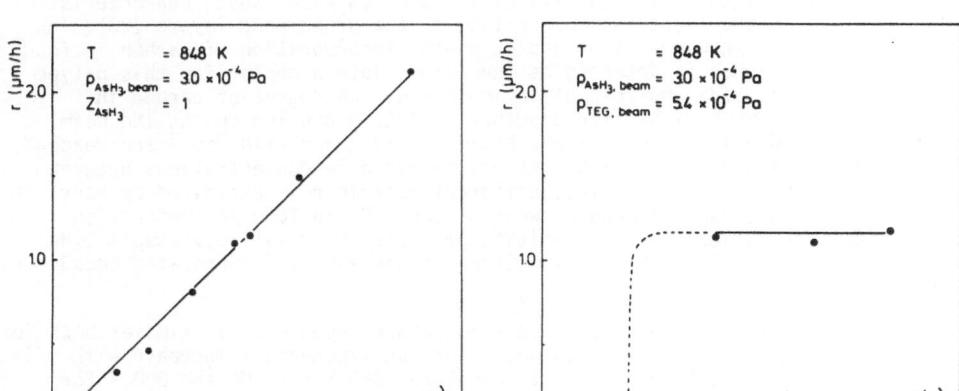

Fig. 2 Dependence of growth rate r of GaAs(100) for TEG on
a) TEG beam pressure $p_{TEG,beam}$
b) AsH$_3$ cracking efficiency Z_{AsH_3}

With both TMG and TEG a linear dependence of the growth rate on the TMG and
TEG beam pressure is given, if enough arsenic is supplied to the growing surface by sufficient precracking of AsH$_3$. On the other hand, with enough Ga
supply the influence of AsH$_3$ precracking is significantly different in both
cases; with TMG there is a linear increase of the growth rate with increasing
arsenic supply. With TEG the growth rate is essentially independent on
arsenic supply above a certain critical minimum value where lack of arsenic

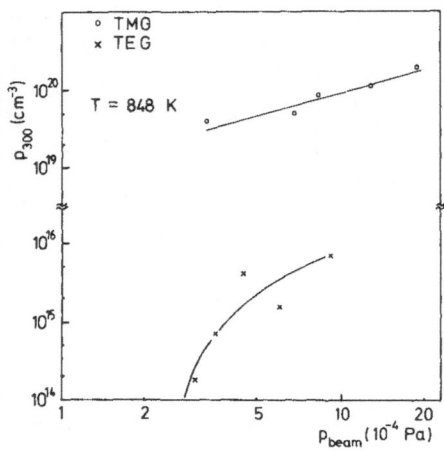

Fig. 3 Effect of metalorganic beam pressure p_{beam} for TMG and TEG on background impurity level, i.e. 300 K hole concentration p_{300} in GaAs; wafer orientation (100), $p_{AsH_3} = 3 \times 10^{-4} Pa$, $z_{AsH_3} = 1$

sense the use of TEG allows sufficient Ga supply on the surface, and the growth kinetics becomes similar to that in MBE, where excess Ga controls the sticking probability for arsenic.

The different thermal stability of TMG and TEG also causes characteristic differences in the electrical properties of the grown GaAs layers [20]. In both cases a background p-type doping due to incorporation of carbon is found as in MOCVD. Carbon is detected as the responsible acceptor for this p-type conductivity in SIMS and in photoluminescence. The degree of carbon incorporation is much higher for growth with TMG. Depending on the TMG beam pressure, 300 K hole concentrations between 2×10^{19} and $2 \times 10^{20} cm^{-3}$ are reached, whereas a variation of the TEG pressure yields hole concentrations between 10^{14} and $10^{16} cm^{-3}$ (Fig. 3). This different behaviour is explained by easy breaking of the chemical bond between Ga and C_2H_5 in TEG. In contrast to TMG with its smaller CH_x decomposition products, the relatively stable C_2H_5 complex resulting from TEG decomposition is not easily incorporated because of steric hindrances.

The growth rate versus substrate temperature dependence is similar both for growth with TMG and TEG [20]. In both cases an exponential increase with 1/T is found above a certain minimum temperature (850 K for TMG and 800 K for TEG). Both with TMG and TEG the activation energy amounts to 1.0-1.1 eV [20], a value which is also found in MOCVD.

4. Doping of GaAs in the MOMBE system

The observation of the different degree of carbon incorporation into a growing GaAs film by using TMG or TEG opens the possibility for intentional p-type doping by carbon. TEG can be used as the basic Ga source, and controlled amounts of TMG are added to achieve the desired doping level [21]. Fig. 4 shows the obtained 330 K hole concentration as a function of the TMG supply. At a fixed TEG beam pressure of $3.6 \times 10^{-4} Pa$ the TMG pressure is varied. The p-type doping, i.e. the carbon incorporation, is sublinear in the low-pressure regime and becomes superlinear for higher pressures. The reason for this change is due to the special property of TMG as a dopant, to supply simultaneously the acceptor and Ga as a basic constituent for GaAs. Beside a variation of the doping concentration also the V/III ratio is

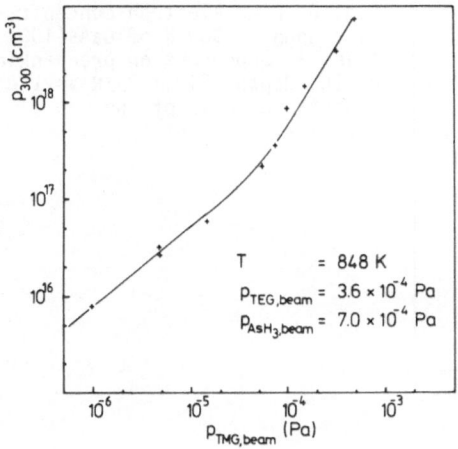

Fig. 4 Room temperature hole concentration p_{300} of GaAs(100) versus TMG pressure at fixed TEG beam pressure $p_{TEG,beam}$

$T = 848$ K
$p_{TEG,beam} = 3.6 \times 10^{-4}$ Pa
$p_{AsH_3,beam} = 7.0 \times 10^{-4}$ Pa

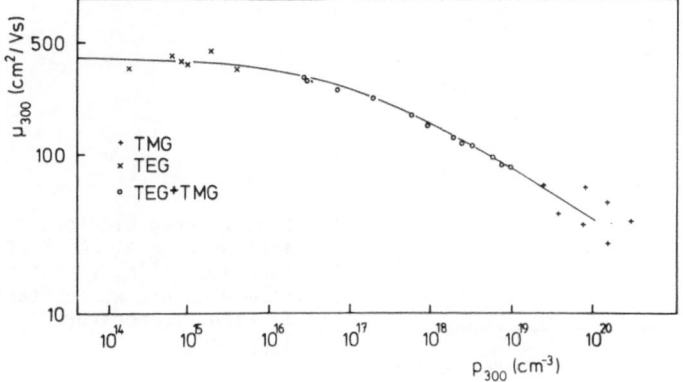

+ TMG
× TEG
○ TEG+TMG

Fig. 5 Hall mobilities μ_{300} at 300 K of intentionally doped GaAs layers [orientation (100)] by use of TMG only (+), TEG only (×), and mixtures of both alkyles (o). Solid line: best mobilities from literature [22]

modified on the growing surface. Thus the growth rate as a function of TMG pressure increases considerably for beam pressures higher than 2×10^{-5} Pa. The plot of the 300 K Hall mobilities contains samples with different hole concentrations (Fig. 5) obtained by growth with pure TEG, pure TMG and mixtures of both alkyls. The solid line represents a theory based on the best values reported in literature [22]. The comparison indicates the high quality of the C-doped layers. The high quality of the films is also documented by extremely sharp exciton lines (between 1.51 and 1.52 eV) in the photoluminescence spectra [21]. With this approach a new p-type dopant is available,which may replace the toxic element beryllium in MBE.

n-type conductivity in GaAs may be obtained by incorporation of Si or S. Doping by Si with a molecular beam of SiH_4 (5 % SiH_4 in H_2) has been investigated so far [23]. In principle,doping is possible with undecomposed SiH_4, but higher doping levels can only be achieved by cracking the SiH_4

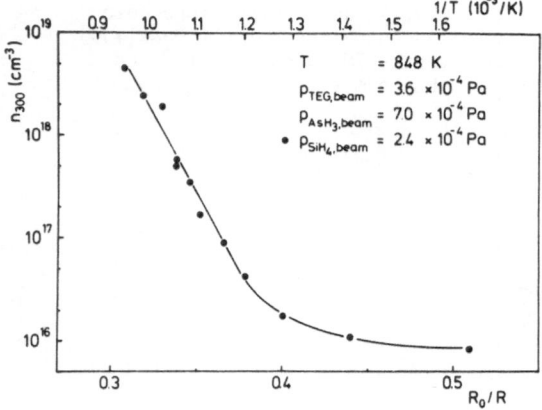

Fig. 6 Free electron concentration n_{300} at 300 K of GaAs(100) films as dependent on precracking of the dopant SiH₄; R_0/R is the resistance ratio of the capillary filament

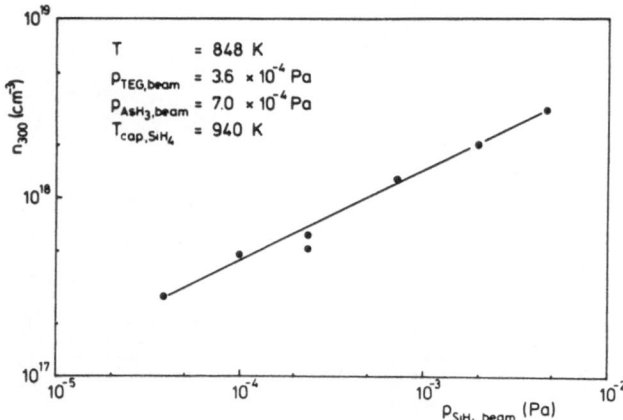

Fig. 7 Free electron concentration n_{300} at 300 K of GaAs(100) films versus SiH₄ beam pressure at constant filament temperature T_{cap,SiH_4}

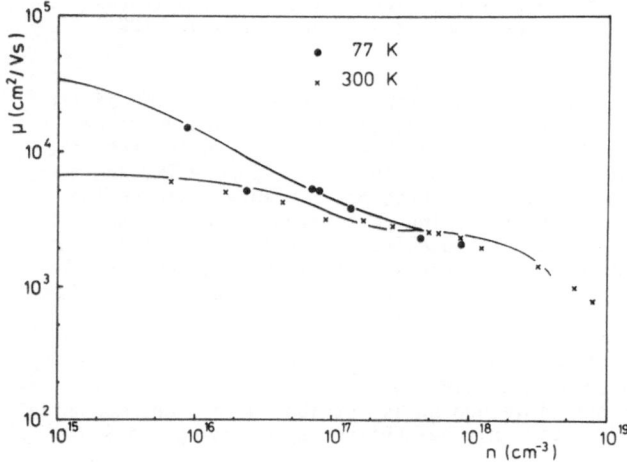

Fig. 8 Hall mobilities μ at 77 and 300 K of n-doped GaAs(100) films versus electron concentration n. SiH₄ was used as dopant. Solid line represents literature values from various techniques [24].

in the injection capillary (temperatures around 850 K). At higher cracking rates 300 K electron concentrations slightly below $10^{19}cm^{-3}$ can easily be obtained (Fig. 6). The doping level can, of course, also be varied by changing the SiH_4 beam pressure at constant cracking efficiency, i.e. constant filament temperature in the capillary (Fig. 7). The free electron concentration, i.e. also the density of incorporated Si atoms, is proportional to $(p_{SiH4,beam})^{1/2}$ (Fig. 7). This dependence might be understood in terms of an equilibrium between adsorbed Si atoms and built-in acceptor centers ($Si \rightleftharpoons Si^+ + e$).

The quality of this n-doping technique in MOMBE can be seen from the obtained mobilities [23]. Fig. 8 shows 300 K and 70 K Hall mobilities measured on layers with different doping levels. The best 77 K mobility amounts to 15000 cm^2/Vs at $n_{77} = 8\times10^{15}cm^{-3}$, the highest obtained doping level lies at $n_{300} \approx 8\times10^{18}cm^{-3}$. The curves in full line describe literature values for n-doped layers from various other techniques [24]. The comparison suggests that there is essentially no compensation.

5. Doping Superlattices (nipi structures)

The feasibility of applying MOMBE to device structures was demonstrated by growing modulation doped multilayer structures of GaAs [23]. In these structures the growing film is doped layerwise, in turn n- and p-type, with small intrinsic regions in between. From this doping sequence the superlattices are called sometimes nipi structures. In the present example a nipi structure was grown in MOMBE with TEG and AsH_4 as the basic materials; p- and n-type doping (nominal concentration $\sim 10^{18}cm^{-3}$) was achieved each time over a thickness of 400 Å by adding TMG and SiH_4 (precracked), respectively, in separate injection capillaries, 10 periods were grown. Fig. 9 shows the SIMS pattern of Si versus sputter depth measured on this structure [23]. The Si concentration decreases within a length of about 10 nm by an order of magnitude. The background and also the sharpness of the profile is determined by the analysis technique rather than by the real nipi structure. The complementary carbon profile of the p-type regions could not be determined because of experimental limitations of the used SIMS equipment.

The sequence of n- and p-type regions causes a spatial modulation of the GaAs band structure (inset in Fig. 9); free electrons and holes are spatially separated and between conduction band minimum and valence band maximum there

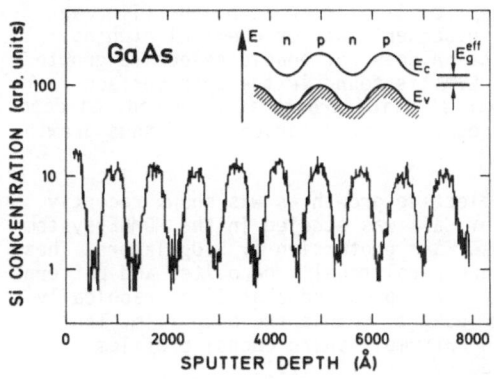

Fig. 9 Spatial variation of Si concentration measured by SIMS on a GaAs(100) nipi structure consisting of 10 periods (800 Å thick); n-doping by Si, p-doping by C with nominal concentrations of about $10^{18}cm^{-3}$. Inset: qualitative band structure of a nipi structure.

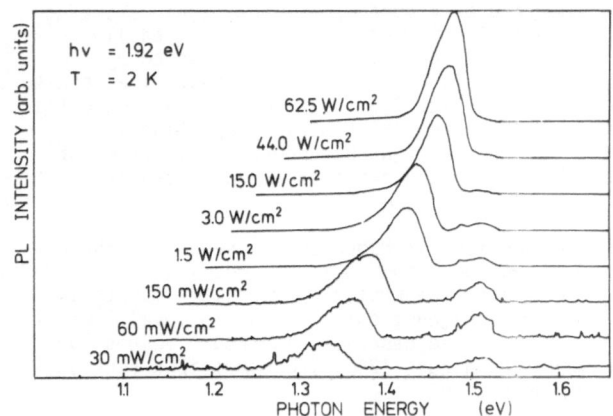

Fig. 10 Photolumin-
escence (PL) spectra
of the GaAs(100) nipi
structure of fig. 9
measured with different
LASER excitation
intensities

is an effective forbidden gap E_g^{eff} which is smaller than that of GaAs. Light-induced generation of electron hole pairs in such a structure decreases the space charge and thus smooths out the band modulation. The effective gap increases. This is exactly found in the photoluminescence (PL) spectra of Fig. 10 [23]. The PL band due to recombination of non-equilibrium carriers occurs at about 1.33 eV, i.e. at an effective gap energy which is signifi-cantly below that of GaAs ($\sim$1.5 eV at 2 K). For higher illumination in-tensities the luminescence band shifts to higher photon energies, thus indi-cating a flattening of the bands. In the limit of extremely high intensity the band gap energy of GaAs is approached. The results thus obtained with MOMBE agree well with those which have been found on GaAs nipi structures grown by standard MBE [25].

6. Selective Growth

For the production of devices and integrated circuits it is desirable to have a method for lateral structuration of the growing surface, in particu-lar, to generate lateral geometrical and doping patterns. This can be achieved by selective growth in certain spatial regions and suppression of growth in others. For this purpose glassy protective layers might be depo-sited on the clean wafer and patterns, i.e. limited areas of the film, can be removed photolithographically by subsequent ion- or chemical etching. Such a wafer structured by a SiO_2 pattern does not enable selective growth in standard MBE. Epitaxial growth of GaAs is found in the open surface areas, whereas on the SiO_2 film polycrystalline growth is observed. Ga deposi-tion on the SiO_2 mask obviously induces sticking of arsenic, and thus growth is initiated.

On the other hand MOMBE enables selective growth, as was shown recently [26]. In our group selective growth of GaAs was studied in the MOMBE system using TMG, TEG and AsH_3 sources and surface protection by SiO_2 layers. These layers, approximately 200 nm thick, were pyrolytically deposited and patterns with lateral dimensions down to 0.8 µm were produced photolithographically. Chemical or reactive ion etching was used to remove the SiO_2 film. The latter technique, of course, leads to extremely sharp edges; profiles in the submicron range have been produced recently [26].

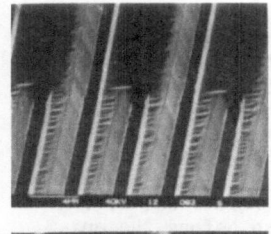
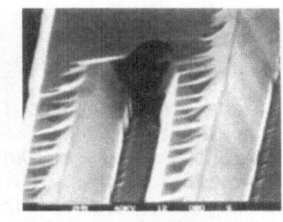
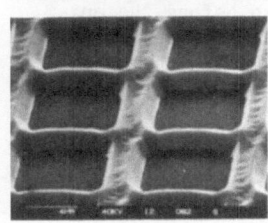

Fig. 11 Selective growth of GaAs(100) at two different substrate temperatures; The open areas above (893 K) and the areas with polycrystalline growth below (848 K) were covered with a 200 nm thick SiO2 film. The SiO2 pattern was produced by wet chemical etching and growth was performed from AsH3 and TEG

T = 893 K

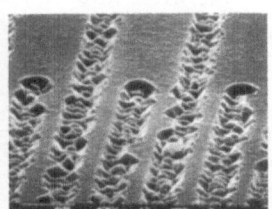

T = 848 K

If TMG is used as the metalorganic source, selective growth is observed in the whole range of temperatures (800-900 K) where GaAs growth is possible [25]. In contrast, with TEG selective growth is observed only at temperatures above 880 K (Fig. 11). For lower temperatures polycrystalline growth of GaAs is found in the covered areas, i.e. on the surface of the SiO2 film (Fig. 11). This effect is similar to what is seen in MBE. The reason that TEG in the MOMBE system leads to similar effects at lower temperatures is due to the relatively high decomposition rate of the alkyl even on the SiO2 surface. Thus enough elemental Ga is deposited, and with the additional arsenic polycrystalline growth of GaAs can start. The more stable TMG in the other case obviously needs the clean GaAs surface for catalytic decomposition and therefore no growth of GaAs occurs on the SiO2 masks.

Another approach to selective growth has been applied by TSANG [17]. Using Si shadow masks with structures in the 2-5 μm range to shadow certain areas of the wafer against the metalorganic beams, he could demonstrate selective growth both for GaAs and InGaAs. But because of the finite divergence of the gas beams there appear problems in producing sharp edges of the grown layers. These problems will probably become more important if one tries to apply this mask technique to large area wafers where even more dispergent molecular beams must be used.

7.   Conclusions

The present review shows that MOMBE is a promising technique to grow epitaxial III-V semiconductor films and layered structures. Both growth and doping can be achieved "in one package", consistently by gas line sources. The technique also has the potential to be applied to a wider field of applications: Epitaxial metal films like e.g. Fe on GaAs have already been deposited recently by means of an $Fe(CO)_5$ gas source [27]. In particular, in comparison with MBE, it should be emphasized that more complex surface reactions, e.g. a catalytic surface decomposition of alkyls etc. take place before the epitaxial growth is initiated. The technique thus also presents the possibility to control these intermediate surface reactions by external parameters, like light or electron irradiation. In the future it might be possible to control and to enhance growth by such probes in spatially limited surface areas and to "write" two-dimensional patterns during the growth process itself. First promising results in the LPMOCVD system are already reported [28].

Acknowledgement

I would like to thank P. Balk, H. Heinecke, N. Pütz, M. Weyers and K. Werner for their pleasant cooperation in this field of MOMBE and for many interesting discussions.

References

1   L.L. Chang and K. Ploog (Editors): "Molecular Beam Epitaxy and Hetero-
       structures", Martinus Nijhoff Publishers, Dordrecht (1985)
2   E. Veuhoff, W. Pletschen, P. Balk and H. Lüth: J. Cryst. Growth 55,
       30 (1981)
3   M.B. Panish: J. Electrochem. Soc. 127, 2730 (1980)
4   M.B. Panish and H. Tempkin: Appl. Phys. Lett. 44, 785 (1983)
5   H. Tempkin, M.B. Panish, P.M. Petroff, R.A. Hamm, J.M. Vandenberg
       and S. Sumski: Appl. Phys. Lett. 47, 394 (1985)
6   A.R. Calawa: Appl. Phys. Lett. 38, 701 (1981)
7   J.H. Neave, P. Blood and B.A. Joyce: Appl. Phys. Lett. 36, 311 (1980)
8   B.J. Skromme, G.E. Stillmann, A.R. Calawa and G. Metze: Appl. Phys.
       Lett. 44, 240 (1984)
9   R. Chow and Y.G. Chai: J. Vac. Sci. Technol. A 1, 49 (1983)
10   A.R. Calawa: Appl. Phys. Lett. 33, 1020 (1979)
11   Y.G. Chai and R. Chow: Appl. Phys. Lett. 38, 796 (1981)
12   E. Tokumitsu, Y. Kudou, M. Konagai and K. Takahashi: J. Appl. Phys. 55,
       3163 (1984)
13   E. Tokumitsu, Y. Kudou, M. Konagai and K. Takahashi: Jap. J. Appl. Phys.
       24, 1189 (1985)
14   Y. Kawaguchi, H. Asahi and H. Nagai: Jap. J. Appl. Phys. 23, L737 (1984)
15   W.T. Tsang: Appl. Phys. Lett. 45, 1234 (1984)
16   W.T. Tsang: Appl. Phys. Lett. 46, 1086 (1985)
17   W.T. Tsang: Appl. Phys. Lett. 46, 742 (1985)
18   N. Vodjdani, A. Lemarchand and H. Paradon: J. Phys. Colloq. C5, 43, 339
       (1982)
19   N. Pütz, E. Veuhoff, H. Heinecke, M. Heyen, H. Lüth and P. Balk: J. Vac.
       Sci. Technol. B 3, 671 (1985)
20   N. Pütz, H. Heinecke, M. Heyen, P. Balk, M. Weyers and H. Lüth: J. Cryst.
       Growth, in press
21   M. Weyers, N. Pütz, H. Heinecke, M. Heyen, H. Lüth and P. Balk: J. Elec-
       tronic Mat., in press

22 J.D. Wiley,in: Semiconductors and semimetals, Vol. 10, edited by R.K.
        Willardson and A.C. Beer, Academic Press, New York (1975), p. 154
23 H. Heinecke, M. Weyers, K. Werner, H. Lüth and P. Balk: to be published
24 H. Poth, H. Bruch, M. Heyen and P. Balk: J. Appl. Phys. $\underline{49}$, 285 (1978)
25 G.H. Döhler, H. Künzel, D. Olego, K. Ploog, P. Ruden, H.J. Stolz and
        G. Abstreiter: Phys. Rev. Lett. $\underline{47}$, 864 (1981)
26 P. Balk, H. Heinecke, N. Pütz, A. Brauers, C. Plass, M. Weyers and
        H. Lüth: J. Cryst. Growth, to be published
27 R. Kaplan: J. Vac. Sci. Technol. A $\underline{1}$, 551 (1983)
28 N. Pütz, H. Heinecke, E. Veuhoff, G. Arens, M. Heyen, H. Lüth and P. Balk:
        J. Cryst. Growth $\underline{68}$, 194 (1984)

Recent Developments in MBE Growth and Properties of $Hg_{1-x}Cd_xTe/CdTe$ Superlattices

J.P. Faurie, K.C. Woo, and S. Rafol

University of Illinois at Chicago, Department of Physics, P.O. Box 4348, Chicago, IL 60680, USA

$Hg_{1-x}Cd_xTe-CdTe$ superlattices of both Type I and Type III have been grown for the first time using the molecular beam epitaxy technique. The superlattices were grown at 190°C. They have been characterized by electron and X-ray diffraction, infrared transmission and magneto-transport measurements. The presence of satellite peaks in the X-ray spectra shows the superlattices to be of high quality. Infrared transmission spectra show that HgCdTe-CdTe superlattices have narrower bandgaps than equivalent HgCdTe alloys. These superlattices are p-type. Shubnikov-de Haas oscillations are observed in the high field magneto-transport measurements. The temperature dependence of such oscillations indicate the heavy holes are the dominating carriers at low temperatures. The Hall characterizations over a large range of alloy concentration in HgCdTe-CdTe superlattices seem to indicate that the high hole mobilities observed in p-type HgTe-CdTe superlattices are due to some type of relationship between the heavy hole gas and the interface state existing in Type III superlattice. Quantized Hall Effect is observed in these p-type superlattices.

1 Introduction

HgTe-CdTe superlattices have received a great deal of attention since they were first proposed as a new infrared material in 1979.[1] They were grown for the first time by molecular beam epitaxy (MBE) in 1982.[2] Since then this superlattice (SL) system has been grown by several other groups[3-7] and much work, both theoretical and experimental, has been done.

Most of the studies have focused primarily on the determination of the superlattice bandgap as a function of layer thicknesses and as a function of temperature. Also, the description of the electronic and optical properties at energies close to the fundamental gap has received much attention.[8-10] Some differences have been observed in the past between theoretical predictions and experimental determinations of the SL bandgaps. Since then, the theory has been refined, the control of the layer thicknesses has been improved and the understanding of the interpretation of the experimental data used to determine the bandgap has deepened.

Interest in this superlattice system arises for two main reasons, (i) its potential as a material for far infrared detectors, (ii) its creation of a new type of superlattice structure containing a semiconductor, CdTe, and a semimetal, HgTe. The Γ_8 light hole band in CdTe becomes the conduction band in HgTe. As a consequence of the matching-up of bulk states belonging to the conduction band in HgTe with the light-hole valence band in CdTe there

exists a quasi-interface state.[11]   Because of the peculiar character of
this superlattice it constitutes a new type of superlattice system called a
Type III superlattice.[12]

In spite of the large amount of work that has been done on HgTe-CdTe superlattices, there are still several unanswered questions. One such question
is the existence of high hole mobilities in p-type SLs. These high hole
mobilities were previously reported,but they are not predicted   or understood.[13]   Hole mobilities are all above 10^3 cm^2 V^{-1}s^{-1} and some values as
high as 3×10^4 cm^{2}V^{-1}s^{-1}[14] have been reported. From the band structure
calculated for this SL,[15,16] the heavy holes are expected to dominate the
transport properties in p-type SLs. For heavy holes we would expect much
lower mobilities.

In order to investigate this interesting problem we have recently grown
for the first time a related superlattice system, Hg$_{1-x}$Cd$_x$Te-CdTe.[17]   When
x is less than 0.14 Hg$_{1-x}$Cd$_x$Te is a semimetal at 77K.   Thus, this new
system is a Type III SL at T = 77K similar to HgTe-CdTe.   But when x is
larger than 0.14 it is a Type I SL, similar to GaAs-AlGaAs, since at 77K
HgCdTe is then a semiconductor. Even though the semiconductor to semimetal
transition has been intensively studied, the transition of Type III to Type
I superlattice has never before been investigated. In fact,it has not even
been proposed before now.

In addition to serving as a tool to investigate HgTe-CdTe SLs, HgCdTe-CdTe
are interesting in themselves. Not only will the bandgap be controlled by
the layer thicknesses,but the value of x will serve as an additional parameter
to adjust the band structure of the superlattice.

2   Growth and Characterization

In our laboratory at the University of Illinois at Chicago, the growth experiments are currently carried out in a RIBER 2300 MBE machine that was
specially designed to handle mercury. To grow HgTe-CdTe SLs, we used three
different effusion cells containing CdTe for the growth of CdTe, Te and Hg
for the growth of HgTe. In order to grow Hg$_{1-x}$Cd$_x$Te-CdTe SLs, we used four
different cells. One containing CdTe was used for the growth of the CdTe
layers. The Hg$_{1-x}$Cd$_x$Te layers were grown using three cells containing Hg,
Te and CdTe. The growth of HgTe-CdTe SLs have been extensively reported
before.[14]

Hg$_{1-x}$Cd$_x$Te-CdTe superlattices were all grown on GaAs(100) substrates.
The substrates were purchased prepolished. They were chemically etched in
a 4:1:1 solution of H$_2$SO$_4$:H$_2$O$_2$:H$_2$O. They were then placed in concentrated
HCl to remove any oxide on the surface. Finally the substrates were rinsed
in methanol, dried under nitrogen and quickly loaded into the vacuum chamber.
The GaAs substrate was preheated to 580°C prior to the growth at 290°C of
the CdTe buffer layer. This treatment assured that the CdTe grew in the
(111) orientation. In order to minimize any adverse effects of the large
lattice mismatch between GaAs and CdTe (14.6%), buffer layers of 2-2.5 μm
were grown.

The HgCdTe-CdTe superlattices were grown at 190°C. Growth rates of 5-6Å
s^{-1} for HgCdTe and 1Å s^{-1} for CdTe were used. The superlattices exhibited
a shiny, mirror-like surface.

In situ electron diffraction investigations were carried out during the
growth. They exhibit short thin streaks and well-defined Kikuchi lines and

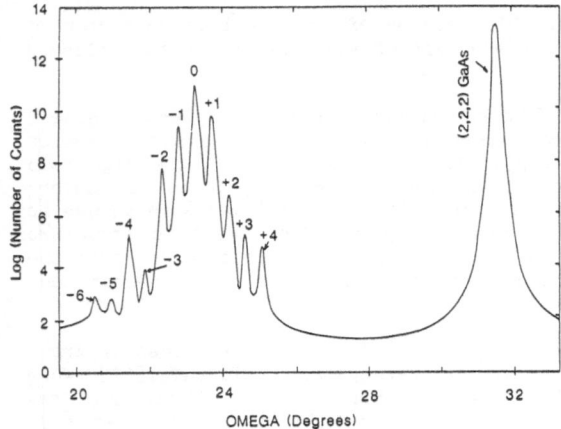

Fig. 1 Room temperature X-ray diffraction profile about the (2,2,2) reflection of $Hg_{0.92}Cd_{0.08}Te$-CdTe superlattice with a period of 102Å

bands. This confirms the epitaxial growth in the (111) orientation, along with the high crystalline quality of the superlattices.

X-ray diffraction experiments carried out on these superlattices also show their high crystal quality.' Fig. 1 displays a room temperature X-ray diffraction profile about the (222) reflection of sample 20842. The presence of up to six satellite peaks in these experiments confirms that these alternate microstructures are indeed high quality superlattices. Using the relation $Sin\theta_{\pm L} = Sin\theta_B \frac{L\lambda}{2d}$, where θ_L and θ_B are the position of the Lth satellite and Bragg peaks respectively, we can calculate the period, $d = d_1 + d_2$, of the superlattices. The HgCdTe layer thickness (d_1) and CdTe layer thickness (d_2) are determined from energy-dispersive X-ray analysis (EDAX) measurements of the superlattice.

For all samples infrared (IR) transmission spectra were measured at 300K between 500 and 4000 cm^{-1}. The cutoff wavelength (λ_c) ranged from 4 to 10 μm depending on the layer thicknesses and x value. To determine λ_c we have calculated the absorption coefficient (α) vs. energy and defined the energy bandgap to be the energy where α is equal to 1000 cm^{-1}. The absorption coefficient was obtained by taking the negative of the natural logarithm of the transmission spectrum and then dividing by the thickness of the SL. From our experience on HgTe-CdTe superlattices we know that this is only a crude determination of the bandgap, especially for thin superlattices.[18] Nevertheless these investigations confirm our expectation that HgCdTe-CdTe superlattices have narrower bandgaps than the equivalent alloy. For example consider sample 20943 which has a bandgap of about 0.177 eV. The equivalent alloy would have x = 0.465. From the empirical expression for $E_g(x,T)$ presented by Hansen, Schmit and Casselman[19] the room temperature bandgap of this alloy would be 0.515 eV. This narrowing of the bandgap in the SL has already been seen in HgTe-CdTe superlattices.[20] Our investigations also show that the bandgap of the HgCdTe-CdTe SL is larger than that of a HgTe-CdTe SL with the same layer thicknesses. This can be seen by comparing samples 20943 and 18124.

3 Transport Properties

We have grown and characterized these SLs by Hall measurements. At 300K, all of these SLs have negative Hall coefficients at low magnetic fields.

Table I: Hall characterization of HgTe-CdTe SL's at 10K with B = 0.3T. R_H is equal to zero for the transition temperature observed in Hall characterization for p-type samples where R_H is the Hall coefficient.

$$d_1 = \text{HgTe layer thickness}$$
$$d_2 = \text{CdTe layer thickness}$$

Sample	Substrate	$d_1[A]$	$d_2[A]$	T[K] for $R_H = 0$	μ_H [cm^{3}V^{-1}s^{-1}]
17821	GaAs	200	50	< 5	
1074	CdTe	180	44	20	3.0x10^4
4213	GaAs	97	60	80	1.5x10^3
844	CdZnTe	70	30	100	6.0x10^3
18124	GaAs	70	45	30	2.5x10^3
18326	GaAs	47	30	200	1.1x10^3
18225	GaAs	58	35	220	1.0x10^3
314	CdTe	45	17	150	1.1x10^4
28	CdZnTe	40	20	150	1.1x10^4

As the temperature is lowered, their sign changes to positive. At liquid helium temperature, all of our SLs exhibit positive Hall coefficient at sufficiently high magnetic fields. Such n to p-type transition is not unexpected, since the valence band offset Λ is positive.[15] Otherwise charge transfer would occur from CdTe to HgTe, always leading to a n-type structure. It is not clear today why p-type SLs have only been grown in our laboratory. This question, which is currently under investigation, has been previously discussed.[14]

Table I presents some p-type SLs grown on different substrates along with the transition temperature (T_T) observed in Hall effect and the hole mobilities observed at 10K. It can be seen that T_T tends to increase when the HgTe layer thickness (d_1) decreases or when the CdTe layer thickness (d_2) increases. In other words, this means that the p-type character of the superlattice is more easily revealed by Hall characterization when the amount of CdTe is increased in the SL. This is not surprising, since the CdTe material that we are growing by MBE is currently p-type.

Also it can be seen that the SLs grown on GaAs systematically exhibit lower p-type mobilities than those grown on CdTe or CdZnTe substrates.

It is not clear today why a factor of five or more is observed at low temperature between the hole mobilities in these two groups of p-type SLs. This could be due to the large mismatch between GaAs and HgTe-CdTe SLs inducing a temperature-dependent residual strain which could affect the hole mobility when the temperature is lowered.

The third observation, which has been reported previously[13] on a few samples, concerns the values of the hole mobilities. They are all above 10^3 cm^{2}v^{-1}s^{-1}.

According to the band structure calculated for this SL system,[15,16] the heavy holes should dominate the transport properties in p-type SLs. For heavy holes we should expect much lower mobilities.

Several hypotheses have been formulated to explain these high mobilities; (i) incorrect interpretation of the Hall measurements for a p-type SL, (ii) band-mixing effect, caused by strain in the HgTe layers, which allows the participation of the light holes in the transport properties, (iii) existence of a (2D) heavy hole gas at the interface, due to a positive value for $\Lambda = \Gamma_{8HgTe} - \Gamma_{8CdTe}$, which causes a mobility enhancement, (iv) presence of an interface state in the plane parallel to the layers, which could significantly contribute to the transport properties. Up to now no convincing theoretical explanation has been given.

To study these hypotheses we have made magneto-transport measurements on several of these HgTe-CdTe and $Hg_{1-x}Cd_xTe$-CdTe superlattices in magnetic fields up to 23 tesla and temperatures as low as 0.5K with a 12 tesla superconducing magnet and the bitter magnets at the Francis Bitter National Magnet Laboratory at Cambridge, MA, U.S.A. by pumping on liquid He^4 or He^3. Samples were cut into Hall bar and Van de Pauw geometries, and ohmic contacts were put onto the samples. Standard d.c. technique was used to measure exx and exy. To avoid ohmic heating of the samples, current less than 1 µA was applied to the sample.

Shubnikov-de Haas oscillations are observed at magnetic fields above 4 tesla for sample 20842 at 4.2K. From the $\omega_C \tau = 1$ condition (or $\mu H = 10^4$, where μ is in $cm^2/V.sec$ and H is in tesla) for observing quantum oscillations, our sample should have mobility about 2500 $cm^2/V.sec$. We have also made magnetotransport measurements on some of our best p-type epitaxy layer and failed to observe any Shubnikov-de Haas oscillations. These are in excellent agreement with our low magnetic fields Hall meaurements and rule out the possibility of incorrect interpretation of Hall measurements.

At low temperatures, the temperature dependence of the amplitude of the Shubnikov-de Haas oscillations is related to the effective mass of the carrier by

$$A \; \alpha \; T/\sinh \; (\frac{2\pi^2 kT}{k\omega_C})$$

$$\omega_C = \hbar eH/m^*c$$

(1)

where A is the amplitude of the oscillation, T is the temperature, ω_C is the cyclotron frequency, H is the magnetic field, m^* is the effective mass, $\hbar$ is the Planck's constant, e is the electronic charge and c is the velocity

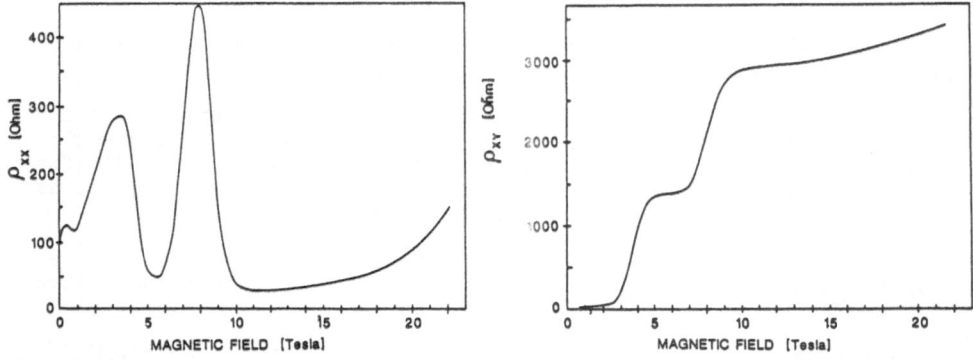

Fig. 2   Temperature dependence of the Shubnikov-de Haas oscillations between 1.7K and 4.2K.

Table II: Characteristics of $Hg_{1-x}Cd_xTe$-CdTe superlattices grown at 190°
on CdTe(111)/GaAs(100) substrates. The energy bandgaps were determined
from room temperature infrared transmission spectra. The Hall mobilities
were measured at 30K except for sample #18124 which was measured at 10K.
R_H is equal to zero for the transition temperature observed in Hall characterization of p-type samples where R_H is the Hall coefficient.

d_1 = $Hg_{1-x}Cd_xTe$ layer thickness
d_2 = CdTe layer thickness
n = numbers of periods
x = Cadmium composition in $Hg_{1-x}Cd_xTe$ layers

Sample	x	$d_1[A]$	$d_2[A]$	n	$E_g[meV]$	$\mu_H[cm^2V^{-1}s^{-1}]$
18124	0	70	45	70	138	2.5×10^3
20539	0.01	82	34	120	124	1.8×10^3
20842	0.08	70	32	100	155	2.5×10^3
20943	0.16	70	40	100	177	3.5×10^2
18929	0.23	48	22	90	310	1.3×10^2
18728	0.27	69	22	100	248	5.0×10^1

of light. Fig. (2) shows the temperature dependence of Shubnikov-de Haas
oscillation of sample 20842, a $Hg_{0.92}Cd_{0.08}Te$/CdTe superlattice. Using the
(1), we obtained the effective mass of the carrier to be 0.36 m_e. For
HgTe/CdTe superlattices, an effective mass of 0.30 m_e is obtained for sample
28. Depending on the experimental method, the effective mass of the heavy
hole of HgTe has been reported to be between 0.3 to 0.7 m_e.[21] The mass
of light hole in HgTe is about the same as the electron, which is 0.03 m_e.
Our results indicate conclusively, the carrier that shows the Shubnikov-de
Haas oscillations is the heavy hole and not the light hole in the HgTe or
$Hg_{1-x}Cd_xTe$ layer in the superlattices.

Although the carriers in these superlattices are two-dimensional above
0.3 tesla,[15] there is no strong reason to believe that a transition from
three to two-dimensional should result in a mobility enhancement as high as
a factor of thirty.

This leaves us with the effect of interface states in these Type III superlattice In order to investigate this interesting problem, Hall measurements
have been done on $Hg_{1-x}Cd_xTe$-CdTe SLs. The Hall mobilities that have been
obtained at 30K are reported in Table II. There are several points that
should be noted from these values. First, for Type III SLs, (those samples
with x < 0.14), the mobilities are in the 10^3-10^4 $cm^2V^{-1}s^{-1}$ range. This is
the same range as HgTe-CdTe superlattices grown on GaAs (See Table I). It
can also be seen from the information presented in Table II that the mobility
for Type I SLs, where x > 0.14, is from one to two orders of magnitude
lower than for Type III SLs. In other words, the Type I superlattices show
a much lower mobility than the Type III superlattices. In fact the values
obtained for the Type I superlattices are about what one would expect for
heavy hole mobilities.

Thus, the question is: What is different between $Hg_{1-x}Cd_xTe$-CdTe Type
III SLs (x = 0-.14) and $Hg_{1-y}Cd_yTe$-CdTe Type I SLs (y = .14-1)?

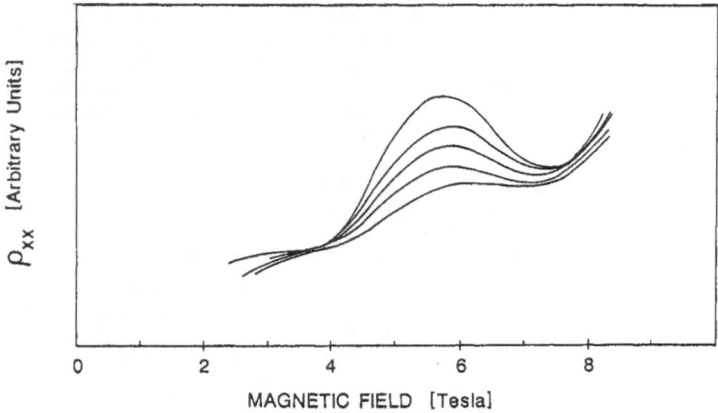

Fig. 3   Quantized Hall Effect in p-type $Hg_{0.92}Cd_{0.08}Te$-CdTe superlattice

(i)   Hall measurements are done in the same way. Therefore, such a drastic change cannot be attributed to the measurement but must be due to the sample. This indirectly proves that the high hole mobility in the Type I case is real.

(ii)   The strain is about the same in these two types of SLs. This experimentally rules out this hypothesis. This agrees with recent theoretical calculations investigating the strain effect on the SL's band structure.[16]

(iii)   The valence band offset Λ is also about the same in both cases, i.e., small and positive. Thus, the 2D hole gas should be present at the HgTe interface in both cases.

(iv)   It has been shown that the existence of the interface state is due to the inverted band structure of HgTe.[11]   Hence the interface state present in Type III SLs disappears in Type I SLs.

Even though it is not clear why these heavy hole mobilities are so high compared to those in HgTe or HgCdTe,the fact that we observed a drastic change between Type III SLs and Type I SLs implies that this phenomenon is likely to be somehow connected with the existence of the interface state.

Fig (3) shows the observation of the quantized Hall Effect in these p-type semiconductor superlattices. Since the observation of the Quantized Hall Effect in silicon inversion layer in 1980, most of the other systems that are found to show the Quantized Hall Effect are in the n-type III-V semiconductor heterojunctions. There is only one report of p-type Quantized Hall Effect. Detailed analysis of this data will be published elsewhere.

4   Conclusion

In conclusion,we have reported here on the growth and characterization of $Hg_{1-x}Cd_xTe$-CdTe superlattices. These superlattices have been grown to study the Type III to Type I superlattice transition for the first time. Electron diffraction done during growth and X-ray diffraction done after growth confirm that these superlattices are of good crystal quality. Infrared transmission spectra confirm that these novel superlattices have energy bandgaps narrower than the equivalent HgCdTe alloy, but larger than the corresponding HgTe-CdTe SL with the same layer thicknesses. By changing x we have been able to grow both Type III and Type I $Hg_{1-x}Cd_xTe$-CdTe super-lattices. The drastic change of up to two orders of magnitude for the hole mobilities between Type III and Type I, along with magnetotransport experiments, seem to indicate that the high hole mobility in HgTe-CdTe and Type III $Hg_{1-x}Cd_xTe$-CdTe superlattices comes from a heavy hole gas, and that it is somehow connected with the existence of an interface state in the Type III superlattices.

Even though the research into $Hg_{1-x}Cd_xTe$-CdTe superlattices is just beginning, they have shown themselves to be a valuable tool to investigate the differences between Type I and Type III superlattices. Also they are of great interest as a material for infrared detectors, because x serves as an additional parameter to tailor the band structure.

Acknowledgements

This work is supported in part by Defense Advanced Research Projects Agency under contract No. MDA-903-85K-0030. Part of of work was performed at the National Magnet Laboratory at Cambridge, Massachusetts.

References

1   J. N. Schulman and T. C. McGill. Appl. Phys. Lett. 34, 663 (1979).
2   J. P. Faurie, A. Million and J. Piaguet, Appl. Phys. Lett. 41, 713 (1982).
3   J. T. Cheung, J. Bajaj and M. Khoshnevisan, Proceedings of Infrared Information Symposia, Detector Specialty, Boulder (1983).
4   P. P. Chow and D. Johnson, J. Vac. Sci. Technol. A3, 67 (1985).
5   K. A. Harris, S. Hwang, D. K. Blanks, J. W. Cook Jr., and J. F. Schetzina, 1985 U. S. Workshop on the Physics and Chemistry of Mercury Cadmium Telluride - San Diego.
6   D. J. Leopold, M. L. Wroge, J. M. Ballingall, B. J. Morris, D. J. Peterman and J. G. Broerman, 1985 U. S. Workshop on the Physics and Chemistry of Mercury Cadmium Telluride - San Diego.
7   R. Koestner, Private communication (1985).
8   G. Bastard, Phys. Rev. B25, 7584 (1982).
9   D. L. Smith, T. C. McGill and J. N. Schulman, Appl. Phys. Lett. 43, 180 (1983).
10 Y. Guldner, G. Bastard and M. Voos, J. Appl. Phys. 57, 1403 (1985).
11 Y. C. Chang, J. N. Schulman, G. Bastard, Y. Guildner and M. Voos, Phys. Rev. B, 31, 2557 (1985).
12 L. Esaski, Proceedings of the 17th International Conference on the Physics of Semiconductors, Edited by J. D. Chadi and W. A. Harrison, Springer-Verlag, Inc., New York (1985) p. 473.
13 J. P. Faurie, M. Boukerche, S. Sivananthan, J. Reno and C. Hsu, Super-lattices and Microstructures, 1, 237 (1985).
14 J. P. Faurie, IEEE Journal of Quantum Electronics (in press).
15 Y. Guldner, G. Bastard, J. P. Vieren, M. Voos, J. P. Faurie and A. Million, Phys. Rev. Lett. 51, 907 (1983).
16 J. N. Schulman and Y. C. Chang, Phys. Rev. B (in press).

17 J. Reno, I. K. Sou, P. S. Wijewarnasuriya and J. P. Faurie, Appl. Phys. Lett. (Submitted).

18 C. E. Jones, T. N. Casselman, J. P. Faurie, S. Perkowitz and J. N. Schulman, Appl. Phys. Lett. $\underline{47}$, 140 (1985).

19 G. L. Hansen, J. L. Schmit and T. N. Casselman, J. Appl. Phys. $\underline{53}$, 7099 (1982).

20 J. P. Faurie, J. Reno and M. Boukerche, J. Cry. Growth, $\underline{72}$, 111 (1985).

21 R. Dornhaus and G. Nimtz, Narrow-Gap Semiconductors, Springer Tracts in Modern Physics Vol. 98.

Transport Properties of Two-Dimensional Electron and Hole Gases in GaAs/AlGaAs Heterostructures

G. Weimann and W. Schlapp

Forschungsinstitut der Deutschen Bundespost, Am Kavalleriesand 3, D-6100 Darmstadt, Fed. Rep. of Germany

Two-dimensional electron and hole gases confined at the interface of selectively doped semiconductor heterostructures show enhanced mobilities due to the spatial separation of the free carriers from their parent dopant atoms. These high mobilities are of interest for device applications and for the observation of physical phenomena in two dimensions. The molecular beam epitaxial growth and low field transport properties of single interface heterojunctions are described.

1. Introduction

Heterojunctions between two dissimilar semiconductor materials have found application in numerous electronic and optoelectronic devices. The properties of abrupt heterojunctions are governed by the lineups of the bands at the interface, depending on bandgaps and electron affinities of the two materials in contact. Few interface defects result only if the semiconductors are lattice matched, the most widely studied system AlGaAs/GaAs being almost ideal in this respect. Only this semiconductor system will be treated in this paper. It has a "straddled" lineup, i.e. the wide band gap semiconductor AlGaAs has the smaller electron affinity, so that the conduction band of the narrow band gap material GaAs lies lower in energy that that of the wide gap semiconductor. The valence band, on the other hand, lies higher in energy than in the wide gap material.

Selective doping of the wide gap material only - usually termed modulation doping (MD) - results in the diffusion of free carriers into the narrow gap material with higher carrier affinity. The band structure of a selectively doped N^+-AlGaAs/GaAs heterostructure is shown in Fig. 1b. The single interface MD heterostructure consists of a thin doped AlGaAs layer, (the usual donor in MBE-grown structures is Si) in contact with an undoped GaAs layer (Fig. 1a). The alignment of the Fermi levels leads to the transfer of conduction electrons into the GaAs, leading to a positively charged depletion layer in the AlGaAs near the interface. This space charge is the origin of a strong electric field causing band bending in the GaAs and the formation of a nearly triangular potential well confining the two-dimensional electron gas (2DEG). This 2DEG has two degrees of free motion parallel to the interface, whereas quantization occurs in the perpendicular direction. The band diagram shows that the doped AlGaAs layer is not depleted by this transfer of electrons into the GaAs only, but also by surface band bending, so that the doped AlGaAs layer is, ideally, free of mobile carriers and conduction takes place in the 2DEG channel only.

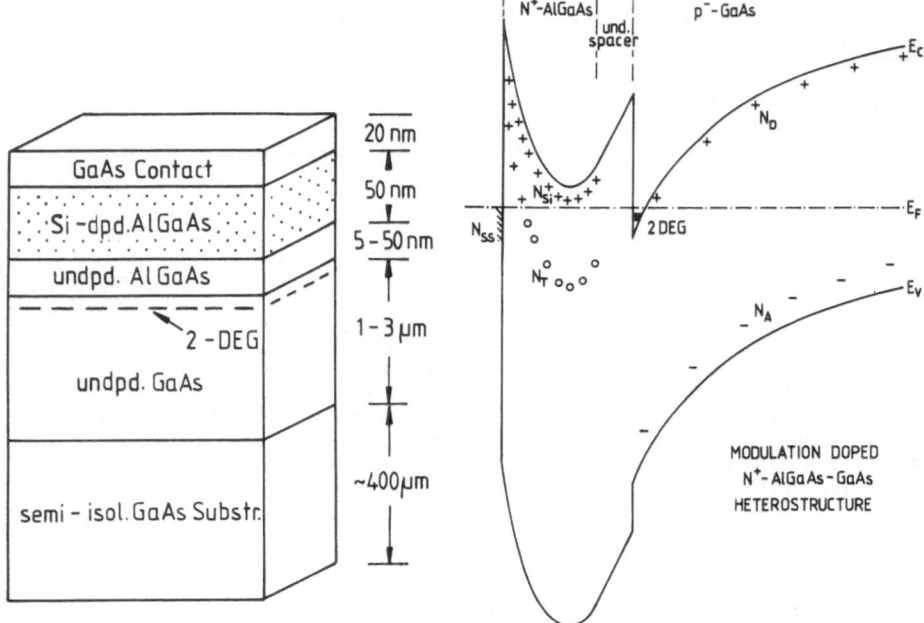

Fig. 1a: Selectively doped
N+-AlGaAs/GaAs heterostructure
with "normal" interface, i.e.
with the doped ternary grown
on top of the undoped GaAs.

Fig. 1b: Band diagram of heterostructure, showing depletion of
doped AlGaAs by transfer of electrons into the 2DEG and into
surface states.

The electrons in this channel are spatially separated from their parent Si donors and so experience only little scattering from ionized impurities, this leading to the mobility enhancement, which makes these
2DEG structures so interesting. An additional undoped AlGaAs spacer
layer is usually introduced between the doped N+-AlGaAs layer and the
undoped GaAs, reducing the coulombic interaction between electrons and
donor ions further, thus giving an additional increase in mobility. Modulation doped N+-AlGaAs/GaAs heterojunctions grown in our laboratory
have shown electron mobilities exceeding $2 \cdot 10^6$ cm^2/Vs at 4.2K. Doping
the AlGaAs with an acceptor results - since the band lineup is
"straddled" - in the formation of a two-dimensional hole gas (2DHG) at
the interface. Our 2DHG structures exhibited low temperature hole mobilities exceeding 200 000 cm^2/Vs.

MD heterostructures require interfaces of high quality, i.e.
abrupt interfaces with the band structures changing within one or
two atomic layers. Low defect densities at the interface are a further
prerequisite. This interface quality is achieved by modern growth
techniques, such as molecular beam epitaxy (MBE) or metal organic vapour
phase epitaxy (MOCVD). The increase in the 2DEG and 2DHG mobilities in
the past three years is due to improvement of the MBE growth processes
i.e. mainly the reduction of background impurities. Optimized growth
conditions and structural parameters will be given in Section 2 for
single interface N+-AlGaAs/GaAs heterojunctions.

2. Single Interface N⁺-AlGaAs/GaAs Heterostructures

Figure 1a shows the typical MD heterojunction with "normal" interface, i.e. with the doped ternary grown on top of the undoped GaAs in which the 2DEG is confined. The structure consists of an undoped GaAs layer between 1 and 5 µm thick grown on the semi-insulating (Cr-doped) GaAs substrate, followed by the undoped spacer layer of 2 to 40 nm thickness, the Si-doped AlGaAs and the final thin GaAs contact layer, which is usually around 20nm thick. The MBE growth procedure has been described in detail elsewhere [1] . Highest mobilities have been obtained with structures grown at substrate temperatures between 600 and 640°C, i.e. at growth temperatures resulting in optimized GaAs properties. This is not the optimal growth temperature for AlGaAs, which was found to be around 700°C. The doping level of the AlGaAs (N^+ = 1 to $1.4 \cdot 10^{18}$ cm^{-3}) was verified by growing thick AlGaAs layers under identical conditions. GaAs grown with unchanged Si-flux had electron densities of n = $3.5 \cdot 10^{18}$ cm^{-3}, revealing one of the major problems of highly N-doped AlGaAs. Whereas Si - and other n-dopants - from shallow donors in GaAs, one finds a marked increase of the donor ionization energy with increasing Al content x in AlGaAs, or to be more precise, two donor levels are observed. One donor level is shallow and the deep lying level, associated with the L conduction band (this level behaves as the DX centre) becomes significant for x > 0.2 [2] .

The basic difference in the low field transport of electrons in the 2DEG channel of a MD heterojunction and electrons in bulk GaAs is the reduction or elimination of coulombic scattering by ionized impurities. Ionized impurity scattering is the dominant process limiting the mobility at low temperatures, so that the 2DEG mobility enhancement is most

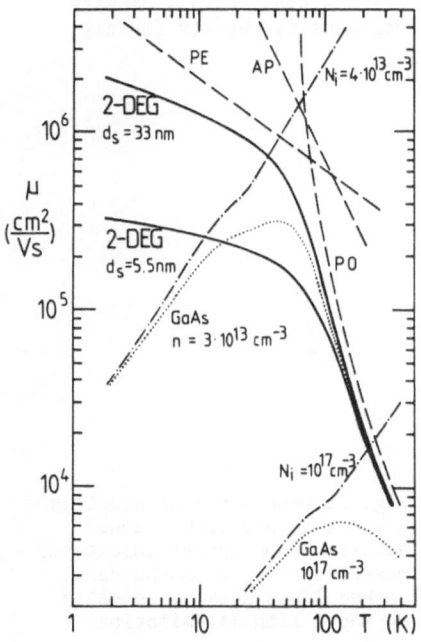

Fig. 2: Comparison of electron mobilities in bulk GaAs (with n = $3 \cdot 10^{13}$ cm^{-3} and 10^{17} cm^{-3}) and in MD heterostructures (with spacer widths of 33 and 5.5 nm and electron densities of $4 \cdot 10^{11}$ cm^{-2} under illumination and $6.3 \cdot 10^{11}$ cm^{-2} in dark, respectively). Ionized impurity scattering is eliminated.

pronounced at low temperatures. Figure 2 shows the temperature dependence of electron mobility in two 2DEG heterojunctions and in bulk n-type GaAs. The three-dimensional or bulk samples shown are very pure GaAs (n = $3 \cdot 10^{13}$ cm^{-3}, total ionized impurity concentration = $4 \cdot 10^{13}$ cm^{-3}) and n-type GaAs with an electron concentration of 10^{17} cm^{-3}, which is the concentration range of interest for field effect transistors. These bulk samples are compared with two MD heterostructures with different spacer thicknesses. The ionized impurity scattering is almost totally eliminated, the mobility of the samples with the thick spacer of 33 nm and sheet density of $2.3 \cdot 10^{11}$ cm^{-2} in the 2DEG channel reaching $1.2 \cdot 10^{6}$ cm^2/Vs. The sample with a spacer thickness of 5.5 nm (N_S = $6.9 \cdot 10^{11}$ at 4.2K) is representative for structures suited for device applications and has a low temperature mobility of $3 \cdot 10^5$ cm^2/Vs. However, even at room temperature one measures an enhanced mobility of around 8000 cm^2/Vs, typical of very pure GaAs, which is approximately two times higher than the mobility in conventional MESFET structures with channel widths of 0.1 μm and doping levels of 10^{17} cm^{-3}. This demonstrates the potential of 2DEG channels for field effect device applications.

The comparison of the MD heterojunctions of Fig. 2 demonstrates that increasing the spacer thickness effectively reduces scattering by the ionized Si donors, at the expense, however, of charge transfer from the doped AlGaAs into the 2DEG channel.

Figure 3 shows this dependence of mobility μ and sheet concentration N_S of the thickness of the spacer layer in detail. Measured values of N_S and μ are given for a number of MD heterostructures, these values were obtained by v.d. Pauw measurements in the dark. Illumination of the samples increases the electron concentration by the photoexcitation of deep levels in the AlGaAs, the increased electron density results in a further enhancement of the low temperature mobility, shown by the dashed lines in Fig. 3. Two mechanisms are responsible for this density dependence of mobility. An increased electron density results firstly

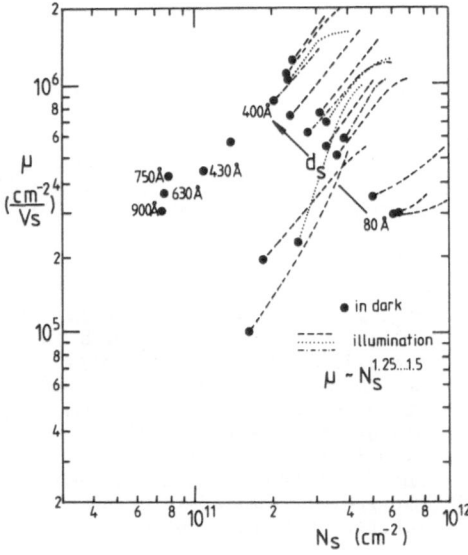

Fig. 3: Dependence of electron mobility μ and 2DEG channel density N_S on spacer thickness, measured at 4.2K in the dark. Dashed lines show increase in N_S and μ with illumination.

in enhanced screening of the ionized impurity potential and, secondly, leads to an increased Fermi wave vector k_F so reducing the effectiveness of scattering processes with given wave vector [3]. Figure 3 shows that the density dependence of the electron mobility is of the form $\mu \sim N^{\gamma}$, with observed γ values between 1.25 and 1.5. This is in reasonable agreement with theoretical calculation; ANDO [4] gave values of $\gamma = 1$ for structures without undoped spacers and $\gamma = 1.5$ for $d_S = 20$ nm. It should be noted that the samples investigated here had no current "bypass" in the AlGaAs, i.e. the doped AlGaAs layers were totally depleted by electron transfer into the 2DEG and into surface states. This has been verified by Shubnikov-de Haas measurements. In the high mobility samples shown in Fig. 3, with carrier concentrations of less than $5 \cdot 10^{11}$ cm^{-2}, only the first electronic subband is populated. With higher electron densities the population of higher subbands sets in with intersubband scattering and decreasing mobilities [4].

Controversial reports on the influence of the spacer layer thickness on the observed mobilities have been given [5, 6], with optimum spacer thicknesses between 10 and 15 nm. Figure 3, however, shows that a decrease in mobility can only be observed for spacer thicknesses exceeding 30 to 40 nm, or, to be more correct, for N_S values below 10^{11} cm^{-2}. The existence of an optimum spacer thickness is the result of the competition between mobility enhancement due to reduced coulombic scattering and the mobility reduction due to decreasing electron density with increasing spacer thickness. A comparison of electron mobilities valid for a normalized electron density (e.g. $N_S = 3 \cdot 10^{11}$ cm^{-2}) can be obtained from Fig. 3 by using the μ–N_S dependence observed under illumination. Figure 4 gives such a dependence of measured mobility on spacer thickness d_S for two different series of MD heterostructures, the only difference between the two sample series being the thickness of the undoped GaAs buffer layer. The heterojunctions grown on thick buffer layers

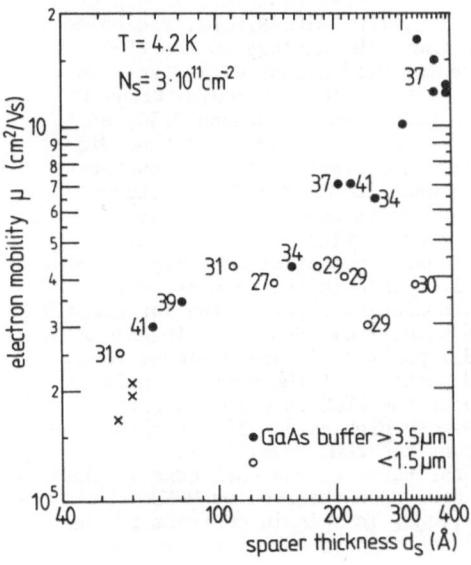

Fig. 4: Mobility dependence on spacer thickness for normalized electron density of $N_S = 3 \cdot 10^{11}$ cm^{-2}. Numbers give Al-content in percent. Thick buffer layers result in extremely high mobilities. All samples are grown at 600–640°C, with exception of those indicated by crosses, which were grown at 700°C.

37

show the expected increase in mobility with increasing spacer thickness
at least upto reasonable thicknesses between 30 and 40 nm. The corresponding mobilities average to approximately $1.5 \cdot 10^6$ cm²/Vs. The second
series of heterostructures, grown with buffer layers around 1 µm thick,
shows an increase in mobility upto $0.5 \cdot 10^6$ cm²/Vs, corresponding to
spacer thicknesses of around 15 nm. We can conclude from these results,
that the samples grown with thick buffer layers are purer at the heterointerface than those with thin buffer layers. It can be assumed that impurities are introduced on the surface of the GaAs substrate prior to
growth. Thick buffer layers serve to incorporate these impurities
well away from the interface; with thin buffers a higher impurity concentration results near the 2DEG channel, limiting mobility. Increasing
the spacer thickness in these samples gives no further increase in mobility. Although it cannot be unambiguously decided whether these impuri-
purities are located in the undoped AlGaAs spacer or in the GaAs channel,
we believe that residual impurities in the latter are presently limiting
the mobilities in our samples. Recent calculations by WALUKIEWICZ et
al. [7] , who calculated the electron mobility in MD heterostructures
taking all major scattering processes into account, showed that a total
ionized impurity concentration in the GaAs of $7 \cdot 10^{14}$ cm⁻³ and 10^{14} cm⁻³
result in optimum spacer thicknesses of 15 nm and 35 nm and mobilities
(for $N_S = 3 \cdot 10^{11}$ cm⁻²) of $0.5 \cdot 10^6$ cm²/Vs and $1.5 \cdot 10^6$ cm²/Vs, respectively. This is in good agreement with our results. These impurity
contents are compatible with v.d. Pauw measurements of undoped GaAs
layers of several microns thickness grown under identical conditions,
they show p-type conductivity ($p < 10^{14}$ cm⁻³, $\mu_{77} = 8000$ cm²/Vs).
The major impurity, revealed by photoluminescence, is carbon. Thick
spacer layers are thus necessary for 2DEG structures with extremely
high mobilities, at least under our present growth conditions.

3. Two-dimensional Hole Gas Structures

Two-dimensioanl hole gases (2DHG) are formed at the interface of
selectively doped P⁺-AlGaAs/GaAs heterostructures. Since the lineup of
the bands is "straddled" in the material system AlGaAs/GaAs, the holes
diffuse from the doped AlGaAs into the GaAs, the analogy to the 2DEG
structures is straightforward. The acceptor used almost exclusively in
the MBE growth is Be. One modification usual in 2DHG heterojunctions is
the increase in the Al content in the AlGaAs layer to around 0.50, as
the valence band discontinuity is smaller than the conduction band discontinuity [8] . There is, additionally, a difference in the confinement of the two-dimensional carriers between 2DEG and 2DHG structures.
The background doping of pure MBE grown GaAs is usually of p-type in
the low 10^{14} cm⁻³ range. Electrons are thus confined in an inversion
layer, i.e. the width of the quasi-triangular potential well typical of a
single interface heterojunction is smaller than in the case of 2DHG
structures, where the holes form an accumulation layer at the interface.
Figure 5 shows the band diagram of a selectively doped P⁺-AlGaAs/GaA
heterostructure. Scattering by ionized impurities is again almost totally suppressed, as the temperature dependence of the measured hole
mobility in Fig. 6 shows. The mobility of the 2DHG is compared with
that of very pure bulk p-type GaAs grown by high vacuum liquid phase
epitaxy [9] . This comparison shows that the 2DHG mobility is
approximately 25 to 30% lower than in the three-dimensional case in the
regime of phonon scattering. This has also been observed by MENDEZ and
WANG in high quality 2DHG structures [10] . This is in contrast to the
behaviour of 2D electrons. We measured the highest mobilities in the

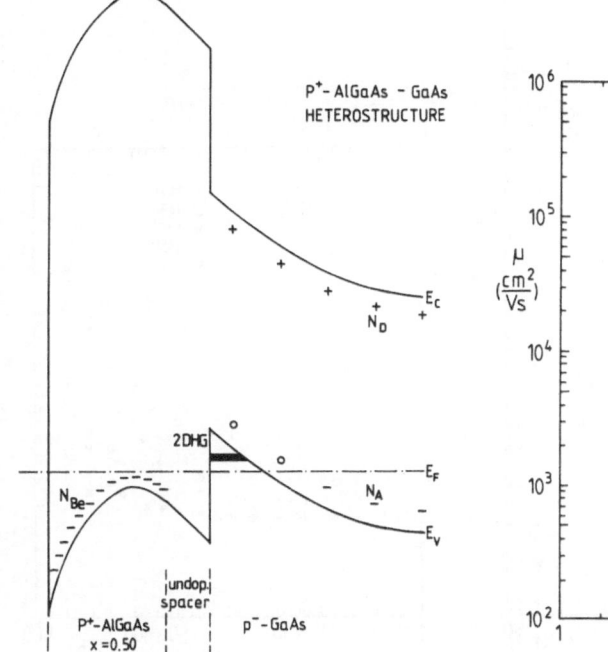

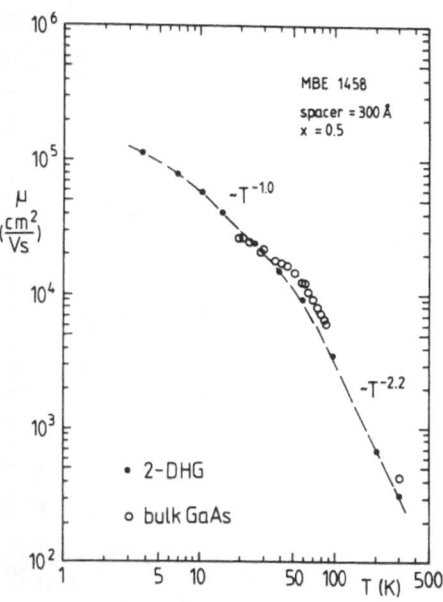

Fig. 5: Band structure of selectively
doped p-type heterostructure. 2DHG
is confined in an accumulation layer
at interface.

Fig. 6: High mobility as
function of temperature
of 2DHG, compared with
extremely low doped p-
type bulk GaAs.

case of 2DEG heterojunctions to be 8500 cm^2/Vs at 300K and 205 000 cm^2/
Vs at 77K for electron densities of $4.6 \cdot 10^{11}$ cm^{-2} and $2.6 \cdot 10^{11}$ cm^{-2},
respectively. These values coincide with the corresponding mobilities
measured on very pure n-type GaAs and are clearly higher than the theoretical 2DEG mobilities calculated by VINTER [11] for the given
channel densities. This could be an indication that screening of electron-
phonon interaction might be important, which has been neglected in [11].
The thickness of the undoped spacer layer can be used to control the
hole density at the interface. Figures 7a and 7b give the temperature
dependence of mobility and hole density for several structures with
different spacer thicknesses. A general result is that the lower the
hole density the higher the mobility turns out to be. The highest 4.2K
mobilities were found to be 113 000 cm^2/Vs and 94 000 cm^2/Vs for two
samples with spacers of 30 nm (sample 1458) and 48 nm (sample 1394),
respectively. The summary of low temperature values of P_S and μ given
in Fig. 8 as functions of spacer thickness show that the scatter of
the measured data is more pronounced than with 2DEG structures. It is,
however, possible to control P_S in the range from 10^{12} cm^{-2} to the
low 10^{11} cm^{-2} range. A striking result is, however, that the mobility
increase is still marked if the temperatures are lowered below 4.2K,
with the two best samples mentioned above reaching values of 220 000
cm^2/Vs, and 180 000 cm^2/Vs, respectively, at 50 mK.

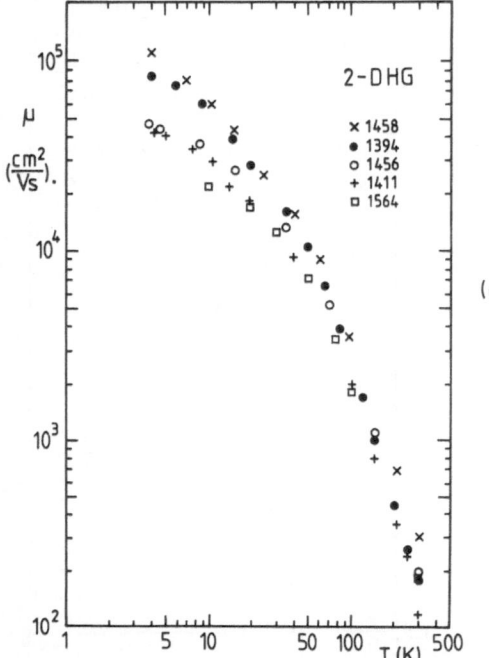

Fig. 7a: Hole mobilities as function of temperature for modulation doped 2DHG structures with different spacer thicknesses (sample 1394: d_S = 48 nm, 1564: d_S = 19 nm)

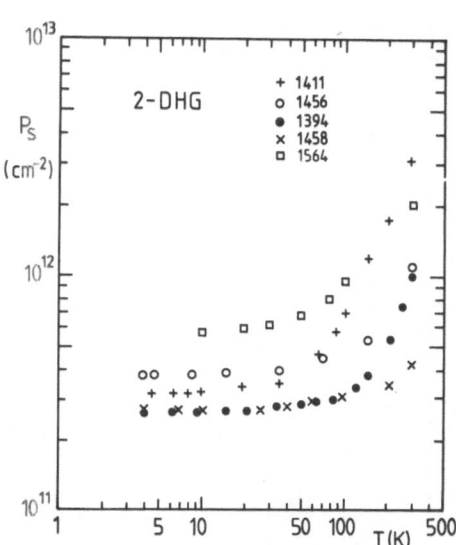

Fig. 7b: Corresponding hole sheet concentration as function of temperature for samples shown in Fig. 7a.

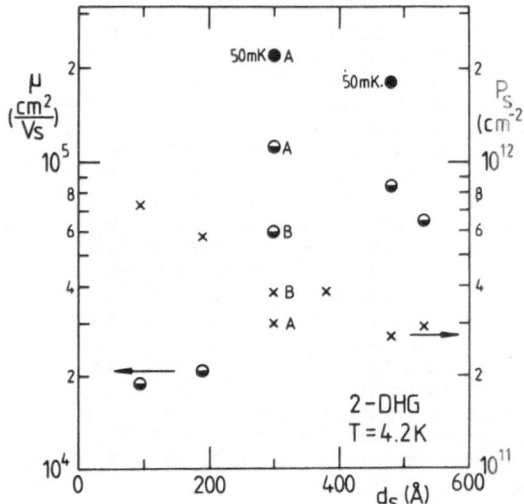

Fig. 8: Low temperature values of μ and P_S as function of spacer thickness. Measurements at 4.2K and 50 mK show marked increase in mobility with decrease in temperature.

These extremely high mobilities show the excellent quality of these samples, which is very much the same as in the 2DEG structures, since practically equal scattering times $\tau = \frac{m^*}{e} \mu$ are measured for 2DEG and 2DHG structures. (The effective carrier masses are taken to be $0.067 \cdot m_0$ for electrons and $0.4 \cdot m_0$ for holes). We can thus conclude that both electron and hole mobilities at low temperatures are presently limited by ionized impurity scattering in the GaAs channel. At temperatures of 77K or higher, ionized impurity scattering is negligible and phonon scattering is dominant. The electrons in the 2DEG exhibit mobilities as high as in the three-dimensional case, making screening of the electron-phonon interaction probable. This does not seem to apply for 2D holes.

4. Summary and Conclusions

The low field transport properties of MD heterostructures have been described, showing that the transport in pure bulk GaAs and MD structures is comparable at high temperatures. At low temperatures mobility enhancement is observed in both 2DEG and 2DHG structures, with Hall mobilities exceeding 10^6 cm^2/Vs and 10^5 cm^2/Vs, respectively. These values reflect the improvement in the growth of the semiconductor heterostructures in the past five years, opening possibilities not only on device applications but also for novel physics in two dimensions. The transport under high electric fields has not been examined in this paper, it is again comparable to that of pure bulk GaAs. The peak electron velocity of $3 \cdot 10^7$ cm s^{-1} at 77K should make MD field effect transistors superior to conventional MESFETs [12] .

References

[1] G. Weimann and W. Schlapp, Appl. Phys. Lett. 46, 411 (1985)

[2] M. Tachikawa, M. Mizuta, H. Kukimoto and S. Minomura, Jap. Journ. Appl. Phys. 24, L821 (1985)

[3] H.L. Störmer, Surf. Sci. 142 130 (1984)

[4] T. Ando, Journ. Phys. Soc. Jap. 51, 3900 (1982)

[5] P. Delecluse, M. Laviron, J. Chaplart, D. Delagebeaudeuf and N.T. Linh, Electron. Lett. 17 344 (1981)

[6] T.J. Drummond, H. Morkoç and A.Y. Cho, J. Appl. Phys. 52, 1380 (1981)

[7] W. Walukiewicz, H.E. Ruda, J. Lagowski and H.C. Gatos Phys. Rev. B30, 4571 (1984)

[8] There is still some controversy about the correct band offsets in AlGaAs/GaAs. New data since 1984 seem to favour values of $0.6 \cdot \Delta E_g$ and $0.4 \cdot \Delta E_g$ for the conduction band and the valence band offset, where ΔE_g is the band gap difference. See e.g. J. Batey and S.L. Wright, Journ. Appl. Phys. 59, 200 (1986)

[9] K.H. Zschauer, Proc. 4th Int. Symp. on GaAs and Related Compounds, Boulder 1972, Int. Phys. Conf. Series, London 17, 3 (1973)

[10] E.E. Mendez and W.I. Wang, Appl. Phys. Lett. 46, 1159 (1985)

[11] B. Vinter, Appl. Phys. Lett. 45, 581 (1984)

[12] H. Morkoç in: The Technology and Physics of Molecular Beam Epitaxy, ed. by E.H.C. Parker (Plenum Press, New York and London 1985), pp. 185-232

In Situ Study of MBE Growth Mechanisms Using RHEED Techniques – Some Consequences of Multiple Scattering

B.A. Joyce, P.J. Dobson, J.H. Neave, and J. Zhang*

Philips Research Laboratories, Redhill, Surrey RH1 5HA, England

Abstract

Very few methods are available for direct in-situ investigation of crystal growth mechanisms at the atomic level, but the forward scattering geometry of reflection high energy electron diffraction (RHEED) is particularly well suited to studies of molecular beam epitaxy (MBE) growth dynamics. After describing the basic concepts of the RHEED intensity oscillation technique, the influence of multiple scattering effects on the form of the oscillations is demonstrated. It is then shown how the relative contributions of the various diffraction processes involved can be used to derive models of growth dynamics, treating GaAs (001) films as an example.

1. Introduction

Reflection high-energy electron diffraction (RHEED) is widely used for monitoring the surface structure of films grown by molecular beam epitaxy (MBE). The work was initially aimed at establishing the surface symmetry of the large number of reconstructions which occur on MBE-grown semiconductor surfaces [1,2], and subsequently to show the presence of one- and two-dimensional surface disorder effects [3]. More recently, RHEED has been used to study thin film growth dynamics by means of temporal changes in diffracted intensity (intensity oscillations) which occur during growth [4,5].

Very few methods are available for the direct, real-time investigation of crystal growth mechanisms at the atomic level, but the forward scattering geometry of RHEED makes it particularly well suited to in-situ studies of MBE growth. In this article we will therefore be concerned exclusively with the intensity oscillation technique and its application to the study of the growth dynamics of GaAs films on (001) oriented GaAs substrates. After describing the basic occurrence and phenomenon of temporal intensity oscillations in RHEED during MBE growth, we will comment on the changing form of the oscillations as diffraction conditions are changed and discuss how this results from the dynamical (multiple scattering) nature of the diffraction process. Although this introduces some difficulties in data analysis, we will show how it is possible to make use of the effects to extend the presently accepted two-dimensional (2-D) growth model [4,6,7], which has been derived from kinematic (single scattering) considerations alone.

---

* Permanent address: Department of Physics, Imperial College of Science and Technology, Prince Consort Road, South Kensington, London SW7 2BZ, England.

2. RHEED Intensity Oscillations – the Basic Effects

Observations of the oscillation of the intensity of all RHEED features immediately following the initiation of growth by MBE are now well established [e.g. 4,5,8,9]. The effect has been observed for GaAs, (Al,Ga)As, (InGa)As, Ge and Si but it is unlikely to be limited to these materials. The period of the oscillation usually corresponds to the growth of precisely one monolayer, i.e. for GaAs a complete molecular layer of Ga and As atoms or $a_o/2$ in the <001> direction, but apparently more complex behaviour has been observed with Si [9]. Similar oscillatory effects during thin film growth have been observed in Auger line shapes [10], LEED patterns and spot shapes [11], electrodeposition overpotentials [12] and the intensities of beams of diffracted He atoms [13]. In each case the oscillatory behaviour was associated with 2-D layer-by-layer growth.

In Fig. 1 we show a set of oscillations obtained from the growth of GaAs films under fixed growth conditions (substrate temperature, $T_s \sim 580°C$, arsenic flux, $J_{As_2} \sim 2 \times 10^{14}$ molecules $cm^{-2}s^{-1}$, gallium flux, $J_{Ga} \sim 1 \times 10^{14}$ atoms $cm^{-2}s^{-1}$) which gave a GaAs (001)-2x4 surface reconstruction. Measurement of the intensity of the specular spot on the 00 rod were made with the beam in the [110] azimuth for a range of

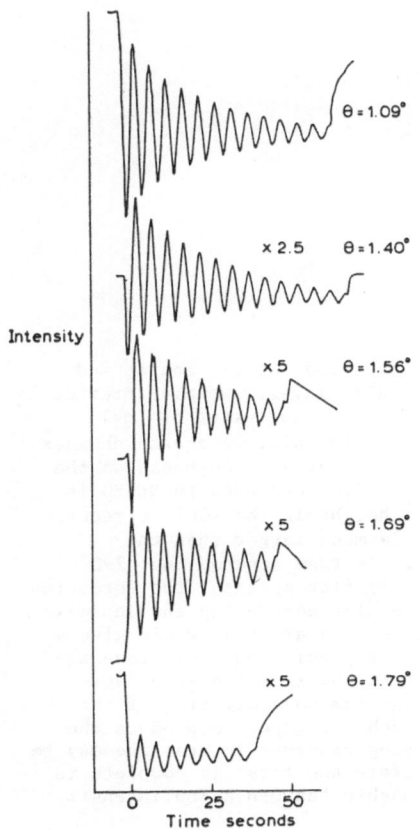

$\theta = 1.09°$

x 2.5 $\theta = 1.40°$

Intensity

x 5 $\theta = 1.56°$

x 5 $\theta = 1.69°$

x 5 $\theta = 1.79°$

0   25   50
Time seconds

Fig. 1 RHEED oscillations of the specular beam taken at different primary beam incident angles from a GaAs (001)-2x4 reconstructed surface; [110] azimuth; 12.5 keV electron beam energy

43

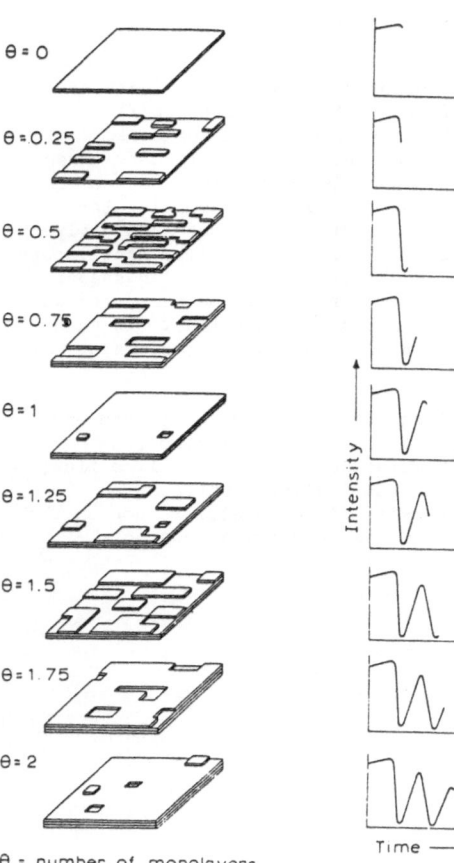

θ = 0

θ = 0.25

θ = 0.5

θ = 0.75

θ = 1

θ = 1.25

θ = 1.5

θ = 1.75

θ = 2

Intensity ⟶

Time ⟶

θ = number of monolayers
deposited

Fig. 2 First-order growth
model (two monolayers) in
relation to the RHEED intensity behaviour

incidence angles (θ). The same period is observed for all angles and
corresponds to the growth of a single molecular layer. We have previously
presented a very simple model, based on a single scattering optical
analogue, to explain these oscillations (4). In this, we equate changes
in the specular beam intensity with changes in surface roughness on the
atomic scale. The de Broglie wavelength of electrons used in RHEED is
typically < 0.1Å, while the monolayer step height in the <001> direction
in GaAs is 2.83Å ($a_0/2$), so the scatterer is much larger than the
wavelength of the probing radiation. If it is then assumed that 2-D
layer-by-layer growth starts on an atomically flat surface, the formation
of monolayer-high steps will reduce the specular scattering and increase
diffuse scattering. Provided that growth is regular, this would give a
minimum in specular scattering at approximately half-monolayer coverage
and a corresponding maximum in diffuse scattering from the step edges.
The specular intensity would return to a maximum on completion of the
layer, and the process would repeat with each new layer to produce the
oscillatory intensity behaviour. The damping observed in practice may be
due to the initiation of a second layer before the first is complete in
the area sampled by the beam. The relationship between RHEED intensity

and the growing layer is illustrated schematically in Fig. 2. This simple
model can apparently account for the form of the oscillations at θ = 1.09°
in Fig. 1, but not at other angles, especially where there is an initial
intensity increase.

Lent and Cohen [6] have extended this kinematic treatment, and suggest
that the specular intensity results from Bragg-type interference of beams
scattered by terraces which are at two levels, a monolayer apart. They
refer to a "Bragg" condition when the beams reflected from the two levels
are in phase and an "off-Bragg" condition when the beams are out of
phase. In phase conditions will therefore occur when $n\lambda = 2d_s \sin \theta$,
where n is an integer, d_s the step height (= $a_0/2$) and λ is the electron
wavelength. For the 12.5 keV beam energy we used, such conditions would
occur at θ ~ 1.1°, 2.2°, 3.3° and 4.4°, and the out-of-phase conditions
would be between these angles, i.e. at θ ~ 1.6°, 2.8° and 3.9°. The
results shown in Fig. 1 do not support this model, however. At θ = 1.09°
we would expect an increase in intensity as growth is initiated, since
more terraces will be formed, thus increasing the extent of constructive
interference. Conversely, at 1.69°, close to the out-of-phase condition,
the model would predict an initial decrease, again contrary to
observation.

3. Multiple Scattering Considerations

To interpret the complete data set shown in Fig. 1 it is in fact necessary
to examine the diffraction process itself in much more detail, and since
intensities are involved we must consider dynamical (or multiple
scattering) events. Maksym and Beeby [14] have provided a theoretical
framework, but the most direct experimental method is to record a rocking
curve, i.e. to set the electron beam along a particular azimuthal
direction and then measure the specular scattered intensity as the angle
of incidence is varied. If only single scattering events are involved,
the intensity will show maxima at the corresponding Bragg positions, but
in this case we are referring to true Bragg conditions for beams inside
the crystal, and refraction effects arising from the inner potential have
to be taken into account. The angles at which these maxima occur will
therefore differ from any maxima produced by the step interference effect
discussed by Lent and Cohen [6], since electrons involved in that process
will not experience the same "bulk" inner potential.

In Fig. 3 we show rocking curves of the specular intensity along the 00
rod in the [110] azimuth for three differently reconstructed GaAs (001)
surfaces, C(4x4), (2x4) and (3x1). All measurements of diffracted
intensities were made at substrate temperatures (T_s) and As_2 fluxes
(J_{As_2}) where these structures are stable: for C(4x4), T_s = 475°C,
J_{As_2} = 5x10^14 molecules cm^-2 s^-1; for (2x4), T_s = 565°C, J_{As_2} = 2x10^14
molecules cm^-2 s^-1; for 3x1, T_s = 630°C, J_{As_2} = 2x10^14 molecules
cm^-2 s^-1. No gallium flux was supplied, i.e. there was no growth. These
curves could be accurately reproduced in two separate systems using quite
different measurement techniques [15]. To complement the rocking curve
information, we show in Fig. 4 diffraction patterns taken at various
critical points along the rocking curve from the 2x4 reconstructed
surface.

The dynamical nature of the diffraction is immediately apparent from
both sets of data. Dealing first with the rocking curves, the most
obvious point is the large amount of structure within them, whereas there

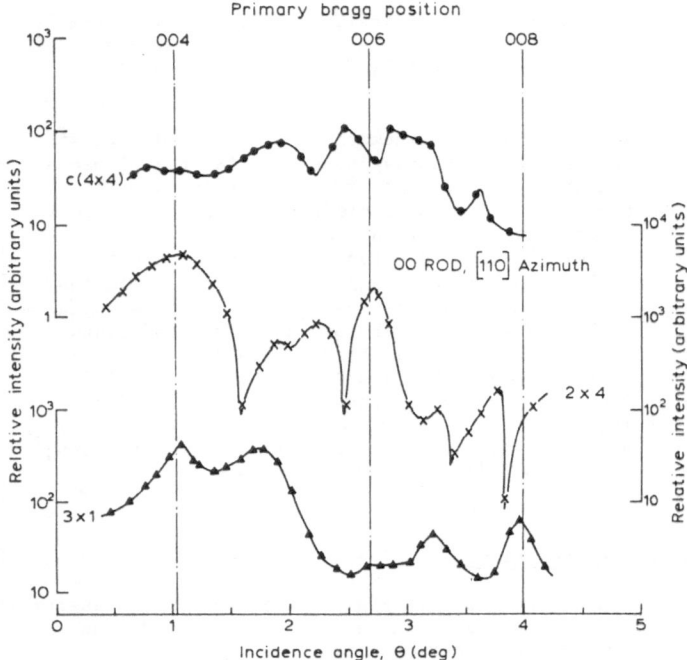

Fig. 3 Rocking curves for the 00 rod in the [110] azimuth for three differently reconstructed GaAs (001) surfaces. The calculated positions of the primary Bragg peaks, with a 14.5 eV inner potential, are indicated. Primary beam energy = 12.5 keV

would have been peaks only at the indicated primary Bragg positions with a single scattering process. It can also be seen that the different surface reconstructions give substantially different results, indicating that there is a strong interaction between the outgoing specular beam and the fractional order surface-related beams. Finally, although we do not illustrate it here, much of the structure is very sensitive to the incident beam azimuth, which is a further indication of the importance of dynamical effects. A detailed interpretation of these and other rocking curves is presented elsewhere [15].

Considering now the diffraction patterns, we have previously described the detailed shape of the streaks in relation to surface structure [3], but here we will draw attention only to those aspects which are related to dynamical effects. The most significant is the changing distribution of intensity between the specular spot (on the 00 rod) and the various diffracted beams. We can see this very clearly between patterns B2 and B3, where the incidence angle changed from 1.05° to 1.65°. At the lower angle the specular intensity is very high, but on increasing the angle most of the intensity is transferred to the 01 and 0$\bar{1}$ rods as these beams just emerge from the crystal. This coincides with a sharp minimum in the rocking curve at 1.65°. As the incidence angle is further increased to ⩾ 2.80°, a second multiple scattering effect becomes apparent, since

46

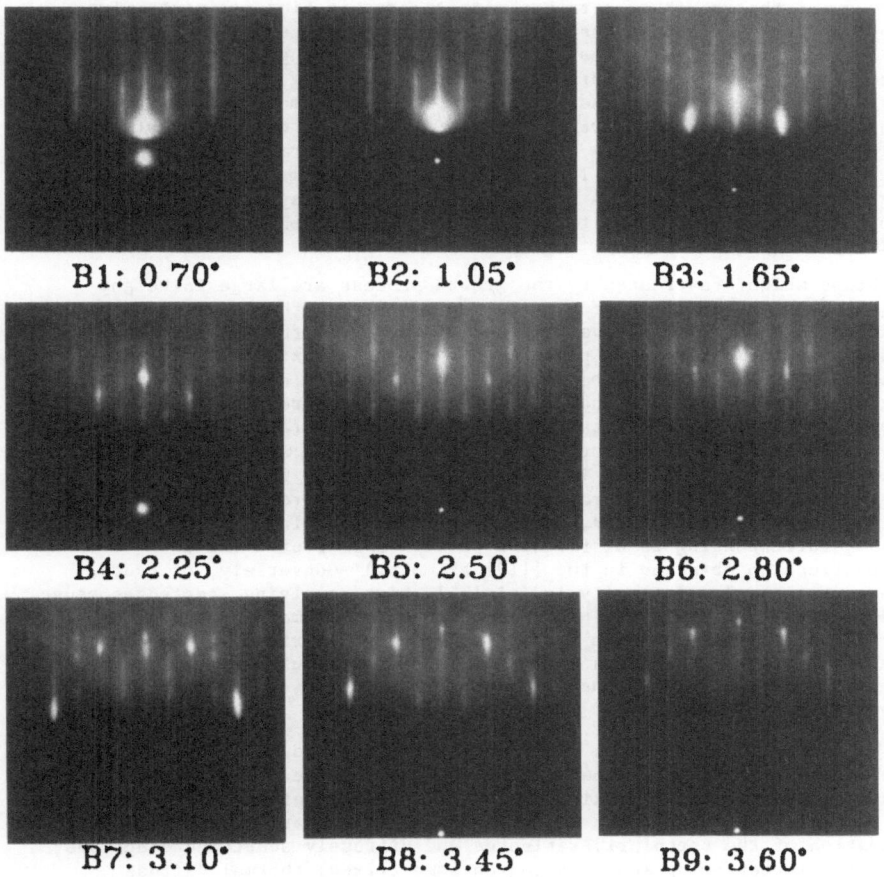

Fig. 4   RHEED patterns from a GaAs (001)-2x4 reconstructed surface at
different angles of incidence; 12.5 keV electron beam energy

several bulk beams, (006, 224 and 222) are simultaneously excited.   Then
at θ = 3.10°, the 02 and 0$\bar{2}$ beams emerge and there is a corresponding
reduction in the specular beam intensity.

4.   Effects of Multiple Scattering on Intensity Oscillations and the Development of Growth Models

Having established the strong dynamical nature of diffraction we are in a
position to investigate the influence it has on the intensity oscillations
which occur during growth and from this use the changing oscillation
waveform with angle of incidence (Fig. 1) to establish a more detailed
growth mechanism.

We will consider first the initiation of growth of, say, GaAs, which is
brought about by allowing a Ga flux to impinge on the substrate surface

47

already at the growth temperature with an arsenic flux incident. Two important changes can be postulated to occur at the surface: firstly, the Ga surface population increases, which tends to change the reconstruction from a more As-rich to a more Ga-rich structure, e.g. from (2x4) to (3x1) at least transiently; secondly, the comparatively smooth "equilibrium", non-growing surface is converted to one with a high density of steps, which may be of height $a_0/4$ if the terrace is Ga terminated or $a_0/2$ if it is terminated by arsenic. These two changes would manifest themselves together in the diffraction patterns in a complex way, but by careful choice of diffraction conditions it should be possible to separate their effects.

Steps which are created by the growth process are large in height relative to the electron de Broglie wavelength, whether they are atomic or molecular, so they will have a marked effect on diffuse scattering. Any feature in the diffraction pattern which owes its existence to diffuse scattering will therefore be enhanced. This can be seen to occur for angles of incidence corresponding to minima in the rocking curves, i.e. where the specular intensity is so weak that most of the electron flux is due to thermal diffuse scattering. The presence of additional scattering from step edges would then increase the intensity. This condition is clearly seen in Fig. 1 where the oscillations were recorded for an incidence angle of $\theta = 1.69$, very close to the rocking curve minimum at 1.65°, corresponding to diffraction pattern B3. (For a (2x4) reconstruction observed in the [110] azimuth.) Conversely, of course, where the specular intensity is initially high, evolving step edges cause a decrease in intensity. This occurs close to maxima in the rocking curves and is the condition we have previously described in relation to the simple model [4]. This may be thought of as equivalent to a phase change of 180° in the intensity change, between the two sets of conditions.

We summarise the effects of diffuse scattering in relation to intensity oscillations in Fig. 5. The main point to emphasise is that diffuse scattering will be at a maximum when the specular reflectivity from terraces is a minimum. The measured intensity variations result from the summation of the specularly reflected and diffusely scattered electrons, with the latter being superimposed on some (fixed) thermal diffuse background. In addition, it can be seen in Fig. 5 that the presence of a "harmonic content" is also predicted by this summation, depending on the level of the thermal diffuse background.

By measuring intensity oscillations at various points along a RHEED streak at a fixed angle of incidence we are able to examine further the effect of (thermal) diffuse scattering. A set of oscillations taken at various positions along the 00 rod relative to the specular spot at an incidence angle of $\theta = 0.7°$ in the [010] azimuth is shown in Fig. 6. By choosing this lower angle of incidence we have changed the relative contributions of the various scattering processes and we can now see a strong "harmonic content", as predicted in Fig. 5. Although the apparent periodicity is no longer equivalent to the growth of a single molecular layer, while for other diffraction conditions (see Fig. 1) it is, it has been accounted for in our model purely on the basis of diffraction, as opposed to growth effects. We should point out, however, that by use of digital data acquisition combined with Fourier analysis of the waveform we have shown that all oscillations appear to have a measurable harmonic content [16], and this may require some additional explanation, possibly related to transitory reconstruction changes and a growth mechanism we will now consider.

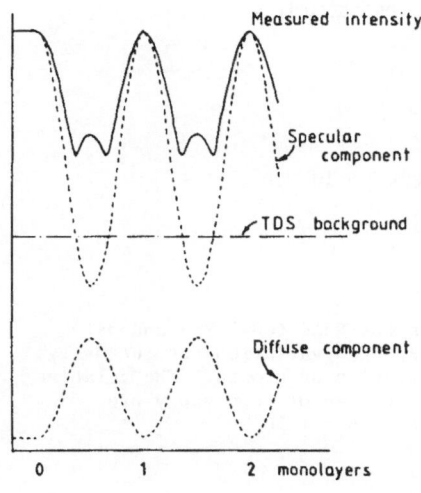

Fig. 5 Schematic diagram showing how measured intensity variations are the result of summation of the specularly reflected electrons and those diffusely scattered by topographic features. The latter are superimposed on a background due to thermal diffuse scattering (TDS). The appearance of a "harmonic content" depends on the TDS level

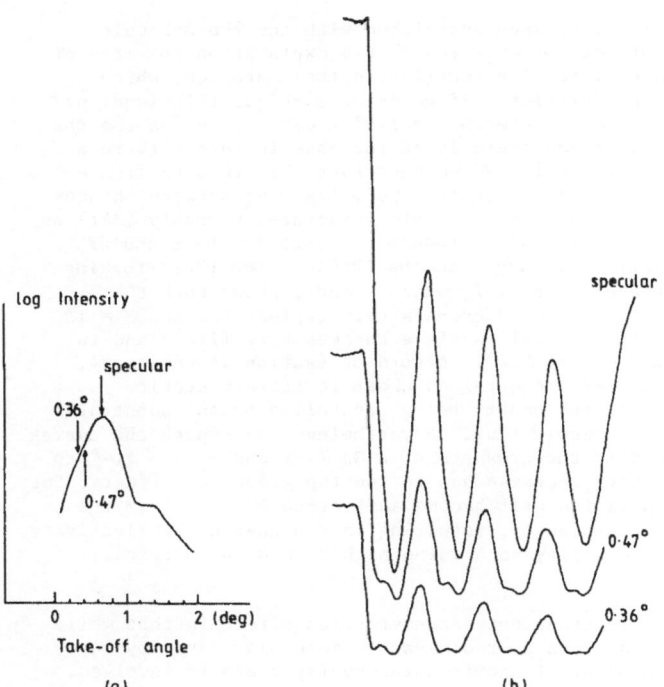

Fig. 6 (a) The intensity distribution along the 00 streak for a GaAs (001)–2x4 reconstructed surface; [010] azimuth; 0.7° incidence angle; 12.5 keV electron beam energy

Fig. 6 (b) Intensity oscillations which occur at the take-off angles indicated when growth is initiated. Note the appearance of "harmonics" in the oscillations

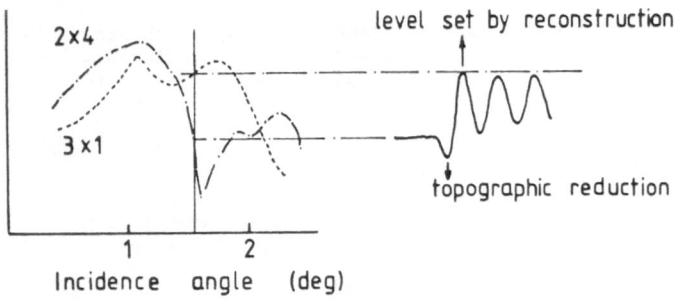

Incidence angle (deg)

Fig. 7 Superposition of the rocking curves for GaAs (001)-2x4 and 3x1 reconstructed surfaces to illustrate the expected variations of intensity with changing reconstruction following .initiation of growth. The relative intensity changes set by reconstruction and change of topography are indicated for an assumed angle of incidence of θ = 1.56°

Whilst the changes in topography associated with the 2-D molecular layer-by-layer growth process offer a reasonable explanation for many of the features in the RHEED intensity oscillation, there are some which require more careful consideration. If we examine the oscillations in Fig. 1 for angles of incidence between θ = 1.4° and θ = 1.56° we see that after an initial rapid decrease there is an increase in intensity to a level above that of the original. We suggest that this results from a transient change in the form of reconstruction, i.e. the surface changes from the As-stable (2x4) to a more Ga-stable structure, probably (3x1) as the Ga flux is initiated. This will produce a change in the specular reflectivity as we effectively move from the (2x4) to the (3x1) rocking curve. This is illustrated in Fig. 7, which clearly shows that the reflectivity of the (2x4) surface decreases very rapidly between θ = 1° and 1.7°, whereas that of the (3x1) surface changes very little and in fact reaches a maximum close to 1.7°. A word of caution is necessary, however. The rocking curves are obtained under relatively static conditions, the surface stoichiometry being controlled by the substrate temperature and incident arsenic flux. Nevertheless, we expect the curves to follow the same trend in the presence of a Ga flux and at θ = 1.4° to 1.56° there will be a small decrease due to the topographical effects, but then the Ga surface population will become sufficient to transform the surface reconstruction transiently, resulting in a higher net reflectivity and an increase in the intensity to a value higher than the original, as observed.

The implication of this transient reconstruction effect is that while the RHEED oscillations indicate a predominantly molecular layer-by-layer process, there is a component of atomic layer-by-layer growth involved. This is in keeping with the presently accepted surface reaction mechanisms [7,17,18], in that the sticking coefficient of As_2 or As_4 is zero in the absence of a Ga surface population. It could also account for some part of the harmonic content of the RHEED oscillation signal, since there would be a doubling of the frequency for an atomic cf. a molecular layer process. The continued presence of the "harmonic content" beyond the first oscillation (Fig. 6b) indicates that this process continues with each new layer grown.

5. Conclusions

We have shown how the RHEED intensity oscillations observed during MBE growth are critically dependent on diffraction conditions as a result of the multiple scattering processes involved. We have used the form of the oscillations to derive models for the growth dynamics of GaAs films.

References

1. A.Y. Cho: J. Appl. Phys 47, 2841 (1976)
2. J.H. Neave and B.A. Joyce: J. Crystal Growth 44, 387 (1978)
3. B.A. Joyce, J.H. Neave, P.J. Dobson and P.K. Larsen: Phys. Rev. B29, 814 (1984)
4. J.H. Neave, B.A. Joyce, P.J. Dobson and N. Norton: Appl. Phys. A31, 1 (1983)
5. J.M. Van Hove, C.S. Lent, P.R. Pukite and P.I. Cohen: J. Vacuum Sci. Technol. B1, 741 (1983)
6. C.S. Lent and P.I. Cohen: Surface Sci. 139, 121 (1984)
7. S.V. Glaisas and A. Madhukar: J. Vac. Sci. Technol. B3, 540 (1985)
8. B.F. Lewis, T.C. Lee, F.J. Grunthaner, A. Madhukar, R. Fernandez and J. Maserjian: J. Vac. Sci. Technol. B2, 419 (1984)
9. T. Sakamoto, N.J. Kawai, T. Nakagawa, K. Ohta and T. Kojima: Appl. Phys. Lett. 47, 617 (1985)
10. Y. Namba, R.W. Vook and S.S. Chao: Surface Sci. 109, 320 (1981)
11. K.D. Grönwald and M. Henzler: Surface Sci. 117, 180 (1982)
12. V. Bostanov, R. Roussinova and E. Buderski: J. Electrochem. Soc. 119, 1346 (1972)
13. L.J. Gómez, S. Bourgeal, J.Ibáñez and M. Salmeron: Phys. Rev. B31, 2551 (1985)
14. P.A. Maksym and J.L. Beeby: Surface Sci. 110, 423 (1981)
15. P.K. Larsen, P.J. Dobson, J.H. Neave, B.A. Joyce, B. Bölger and J. Zhang: Surface Sci., in the press
16. B.A. Joyce, P.J. Dobson, J.H. Neave, K. Woodbridge, J. Zhang, P.K. Larsen and B. Bölger: Surface Sci., in the press
17. C.T. Foxon and B.A. Joyce: Surface Sci. 50, 435 (1975)
18. C.T. Foxon and B.A. Joyce: Surface Sci. 64, 298 (1977)

Growth Mode and Interface Structure of MBE Grown SiGe Structures

E. Kasper, H.-J. Herzog, H. Dämbkes, and Th. Ricker

AEG Research Center Ulm, Sedanstr. 10,
D-7900 Ulm, Fed. Rep. of Germany

Abstract

SiGe heterostructures and supperlattices were grown on Si substrates by molecular beam epitaxy (MBE). The observed growth mode and interface structure exhibited strong deviations from equilibrium theory predictions. A concept of strain adjustment in Si/SiGe superlattices is proposed. The first succesful realized n-channel SiGe-MODFET is given as example of device application of this concept.

1. Motivation for Growth of SiGe Superlattices

Besides the general interest in superlattice systems there are specific technological and scientific reasons for investigating SiGe superlattices on Si substrates.

(i) The possibility of monolithic integration of SiGe superlattices with conventional integrated circuits on Si substrates. An exciting example of this type of integration is an optoelectronic receiver given by S. LURYI /1/ in this series.

(ii) Theoretical predictions of a band gap conversion from indirect type to direct type by Brillouin zone folding of the superlattice /2,3/. One expects minizones in the momentum space, not observed in the host crystal (Fig. 1).

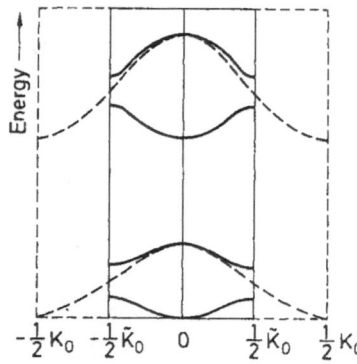

Fig. 1: Band structure of the superlattice cystal (solid line) and the band structure (indirect gap) of the host crystal (dashed line) after GNUTZMANN /2/

The probability of direct optical transitions is strongly enhanced in this minizone scheme: Simple estimates give a ratio Q of the transition probability of optical direct and indirect transitions, which exceeds one hundred for photon frequencies near the band gap energy /2,3/. Zone folding effects were demonstrated for the phonon spectrum by G. ABSTREITER's /4/ Raman scattering experiments. An experimental proof for the predicted direct optical transition in the SiGe superlattice system is lacking at the moment.

(iii) Adjustment of the band line up by the strained layer SiGe superlattice. The conduction band line up of a SiGe superlattice directly grown on a Si substrate is roughly zero, as was demonstrated by J. BEAN's group /5/. The conduction band line up increases considerably toward a staggered band gap by applying our concept of strain symmetrization /6,7/ as was experimentally demonstrated by H. JORKE /8,9/. The technique of strain adjustment will also be explained in more detail in following sections.

(iv) Si/SiGe is a model system for studies of misfit dislocation generation, strain relaxation and surface morphology. In the following chapters we will mainly concentrate on these topics.

2. Molecular beam epitaxy (MBE)

Molecular beam epitaxy is now a well-known technology for growth of advanced material structures, for a review see e.g. E. PARKER's edition /10/ of the technology and physics of molecular beam epitaxy. The principal arrangement of our MBE apparatus which was used for growth of the SiGe superlattices is given in Fig. 2.

The following subsystem are shown; (i) material sources (Si, Ge, Sb), (ii) substrate oven, (iii) secondary implantation equipment (ionization ring, substrate voltage), (iv) in situ monitoring equipment (quartz microbalance, mass spectrometer). These subsystems are installed inside an ultrahigh vacuum chamber.

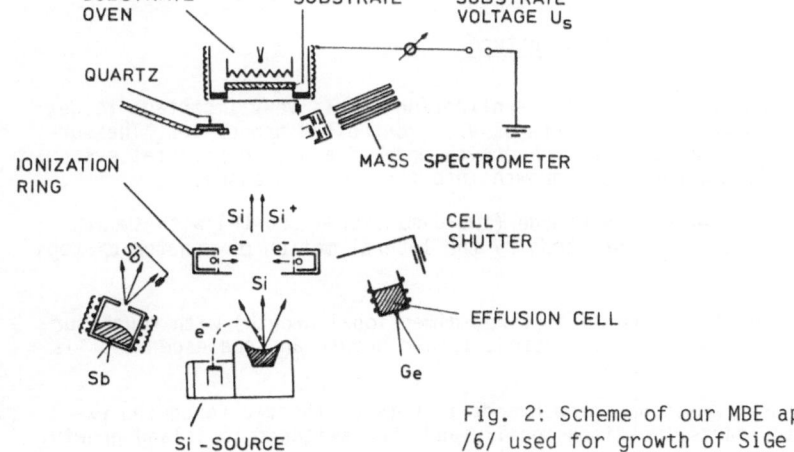

Fig. 2: Scheme of our MBE apparatus /6/ used for growth of SiGe superlattices

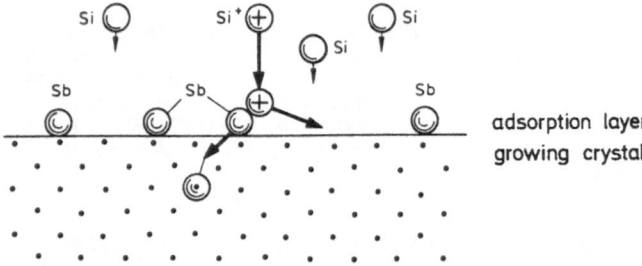

adsorption layer
growing crystal

Fig. 3: Doping by secondary implantation (DSI) /13/. The adsorbed Sb atoms
are implanted by recoil momentum from Si ions impinging on the growing
crystal. Incorporation depth is only some atomic distances.

Three distinct features are typical for silicon molecular beam epitaxy
(Si MBE).

(i) The industrial requirements for throughput and wafer size /11/. We
used 3" wafers for growth of the SiGe structures.

(ii) The existence of strong electromagnetic radiation (thermal, X-rays),
electron fluxes, and ion fluxes inside the growth chamber. This radiation and
charged particle fluxes which influence strongly surface physics are caused
by the operation of the electron gun evaporator for silicon /12/.

(iii) The usage of ionized matrix or dopant atoms to enhance the incorporation of dopant materials. We used the method of secondary implantation
/13/ for n-type doping with antimony (Fig. 3).

With this method Si ions generated by the electron gun evaporator and/or
by an ionizer ring (see Fig. 2) are accelerated toward the substrate by
an applied potential. At the substrate surface these Si ions implant the
adsorbed Sb atoms (Fig. 3).

3. Growth Mode and Interface Structure

From the view point of industrial applications it is very important to get
smooth growth surfaces and interfaces with controlled properties. The surface smoothness is governed by the growth mode of a heteroepitaxial system.
A rough classification divides growth into three mode classes.

(i) Frank-v.d. Merwe growth mode (two-dimensional growth) with smooth
surfaces. Vertical growth proceeds by the lateral motion of monatomic steps
/14/.

(ii) Volmer-Weber growth mode (three-dimensional growth) with rough surfaces. Growth proceeds via nucleation, island growth and coalescence of islands.

(iii) Stranski-Krastanov growth mode is between the two foregoing ex-
tremes. Growth starts two-dimensionally and then switches to island growth.

The mismatch between layer and substrate material governs the interface structure. Mismatch accommodation can either be performed by elastic forces or by introduction of misfit dislocations /15/. For a thick substrate the following simple relation is valid /16/

$$\eta_0 + \varepsilon = c/p \quad , \tag{1}$$

with mismatch η_0, layer strain ε parallel to the interface, misfit dislocation distance p, effective Burger's vector c. For an ideal misfit dislocation in silicon this effective vector c amounts to 0.384 nm. For other configurations only a smaller component of the Burger's vector is effective which actually accommodates mismatch.

Theories about growth mode and interface structure usually consider thermodynamic equilibrium conditions. The model system SiGe on Si proves that there are severe deviations from thermodynamic equilibrium at least at typical MBE growth temperatures achieved.

3.1 Equilibrium Theories

Homoepitaxial growth is governed by the Frank - V.d. Merwe growth mode. With heteroepitaxy two forces can drive the growth into other modes, namely mismatch and different chemical binding energies. The pure influence of the mismatch was investigated by L. STOOP /17/ for a film with cap-shaped islands. He found a continuous increase of the contact angle θ with increasing mismatch, e.g. the contact angle θ is roughly 10° for $\eta_0 = 10^{-2}$ and roughly 20° for $\eta_0 = 4 \times 10^{-2}$. The exact value of the contact angle is size dependant. The influence of the chemistry on the contact angle is given by /18/

$$\sigma_s = \sigma_i + \sigma_f \cos \theta \quad , \tag{2}$$

where σ_s, σ_f and σ_i are the surface energies of substrate, film and interface, θ is the contact angle of the film islands deposited on the substrate.

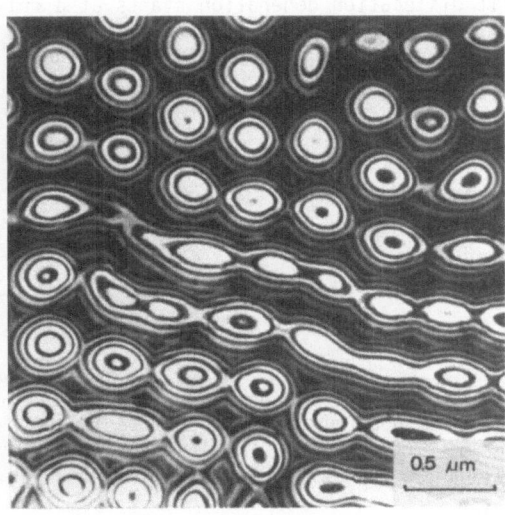

Fig. 4: TEM image of a SiGe superlattice grown at 750° C by the Volmer-Weber mechanism. Contrast is enhanced by preferential etching. The distance of the pseudoregularly spaced three-dimensional nuclei is about 400 nm

0.5 μm

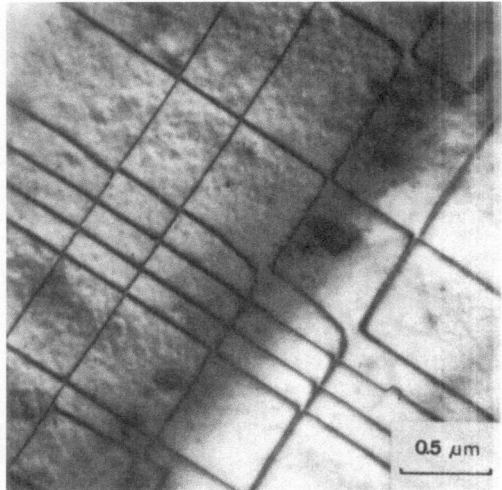

Fig. 5: TEM micrograph of the misfit dislocation network lying in the interface between the (100) Si-substrate and the SiGe layer. The straight dislocation segments are along < 110> directions.

Two-dimensional growth is predicted for $\sigma_s - \sigma_i \geqslant \sigma_f$, whereas three-dimensional island growth is predicted for $\sigma_s - \sigma_i < \sigma_f$. The combined influence of lattice mismatch and chemistry was shown for a model substance with Lennard-Jones type interactions by T. HALICIOGLU /19/.

Depending on the energy and distance parameters of the Lennard-Jones potential Frank-V.d. Merwe, Volmer-Weber and Stranski-Krastanov growth regimes were defined. But in all cases Volmer-Weber growth dominates with increasing lattice mismatch. This was also confirmed by growth experiments at 750 °C. $Si_{1-x}Ge_x$ layers on Si substrate were grown island-like for x > 0.2 (Fig. 4).

The lattice mismatch is accommodated by strain for thin layers and by misfit dislocations for thick layers. Misfit dislocation generation starts at a critical thickness t_c of the layer. Applying V.d. MERWE's /20/ theory to the case of SiGe /16,21/ yields the following relation between mismatch n_0 and critical thickness t_c (measured in nm)

$$n_0 t_c = 1.175 \times 10^{-2} nm \ (2.19 + \ln t_c) \ . \qquad (3)$$

This equilibrium theory gives rather low values of critical thickness, e.g. t_c = 1 nm for $Si_{0.4}Ge_{0.6}$ films on Si (n_0= 0.026). The misfit dislocation network is mainly lying in the interface (Fig. 5).

3.2 Deviation from equilibrium at low growth temperatures

Under the equilibrium conditions described in the foregoing section, high quality growth with smooth surfaces and without misfit dislocations is only allowed in a very limited range of composition and layer thickness. The experimental investigations have shown rather strong quantitative deviations from the equilibrium theories.

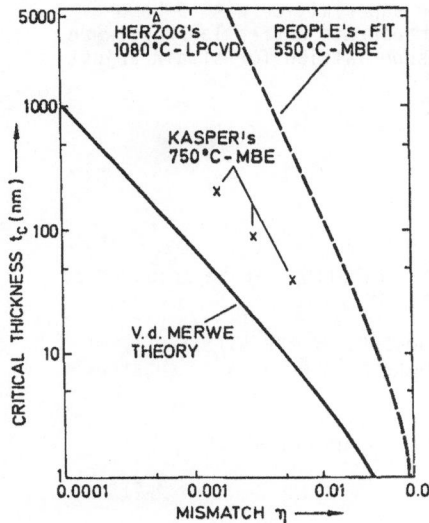

Fig. 6: Critical thickness t_C as function of mismatch η_0 /6/. Equilibrium theory (V.d. Merwe) and experimental results compared are for different growth temperatures.

(i) The value of the critical thickness is higher than predicted by V.d. MERWE's theory (Fig. 6) already at high growth temperatures, as demonstrated by H.-J. HERZOG's /22/ LPCVD experiments at 1080 °C.

(ii) The critical thickness increases with decreasing growth temperature (Fig. 6). E.g., at a mismatch $\eta_0 = 0.005$ equilibrium theory predicts a critical thickness of 10 nm, 750 °C-MBE yields 50 nm and 550 °C - MBE /23/ yields 600 nm critical thickness.

(iii) The growth mode switches from Volmer-Weber type to Frank-V.d. Merwe type with smooth surfaces by decreasing the growth temperature from 750 °C to 550 °C as was demonstrated by J.C. BEAN /24/.

The deviations from equilibrium prediction fortunately increase the range of high quality growth of SiGe layers on Si substrate. There is no clear explanation of these deviations at the moment. To our opinion temperature dependent nuclei distances /6/ and kinetic barriers against dislocation generation /21/ play a key role. But further investigations about this model system are needed.

4. Strained layer superlattices

Strained layer SiGe superlattices on Si substrates were successfully grown by our group /7,9,15,21/ and J.C. BEAN's group /24/.

4.1 Adjustment of strain distribution within the superlattice

The SiGe superlattice directly grown on Si substrate is asymmetrically strained with unstrained Si layers and compressed SiGe layers. The minimum strain energy of a superlattice is achieved with symmetrical strain distri-

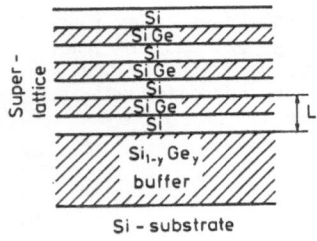

Fig. 7: Si/SiGe superlattice with a homogeneous, incommensurate buffer layer between substrate and superlattice for strain adjustment /21/

bution /6/. It was shown that the strain distribution can be adjusted by using a buffer layer of uniform composition /21/.

According to this concept the structure consists of the $Si/Si_{1-x}Ge_x$ superlattice on an incommensurate $Si_{1-y}Ge_y$ buffer layer on a Si substrate (Fig. 7).

The buffer layer provides an in-plane lattice constant a" of

$$a" = a_0 (1 + 0.042 y + \varepsilon_B),\tag{4}$$

with Si lattice constant $a_0 = 0.543$ nm, Ge content y of the buffer, and residual strain ε_B of the buffer layer. The residual strain ε_B is dependant on Ge content, layer thickness and technological parameters like growth temperature. It has to be determined experimentally because of deviations from equilibrium theory mentioned above. The influence of strain distribution on superlattice properties can be investigated systematically using this concept of strain adjustment. As a first success, room temperature mobility enhancement in symmetrically strained SiGe superlattices /7,9/ was achieved.

4.2 Symmetrically strained n-channel MODFET

The first successful operation of an n-channel SiGe MODFET demonstrates the capability of device tailoring by strain adjustment. The structure and characteristics of this MODFET are shown in Figs. 8,9.

The good device characteristics and the high mobility of the electrons collected in the wide band gap Si-channel are assumed to be caused by the chosen strain distribution.

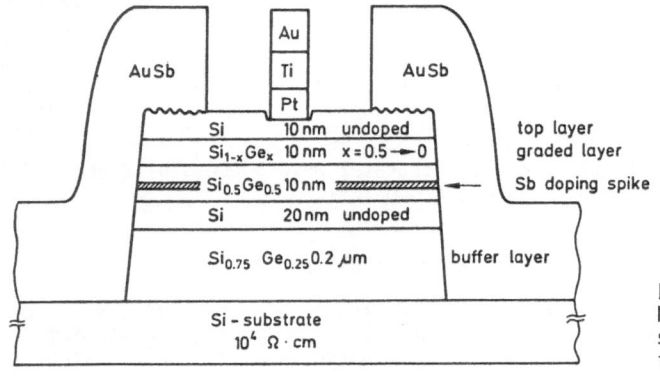

Fig. 8: n-channel SiGe-MODFET structure with symmetrical strain distribution /26/

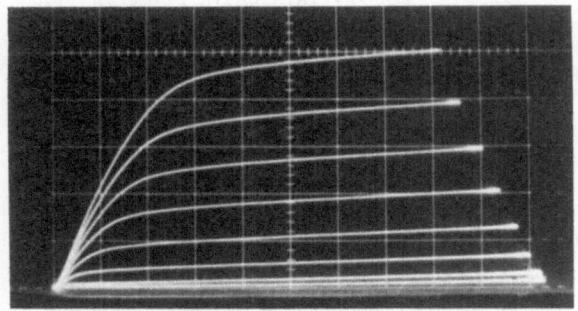

Fig. 9: Room temperature characteristics of an n-channel SiGe-MODFET with 1.6 μm gate length. Current (500 μmA/div) versus voltage (500 mV/div) for different gate voltages (200 mV/step).

5. Conclusion

SiGe superlattices on Si substrates offer unique possibilities for optoelectronic and electronic devices monolithically integrated with conventional Si-IC's. The rather severe limitations for the range of high quality growth which are predicted by equilibrium theories can be overcome by low temperature - MBE. The material properties of the strained layer SiGe superlattices can be tailored by applying our concept of strain adjustment. This was successfully tested by an n-channel SiGe-MODFET.

References

1. S. Luryi and F. Capasso: Springer series in solid state sciences, this volume
2. U. Gnutzmann and K. Clausecker: Appl. Phys. 3, 9 (1974)
3. S.A. Jackson and R. People: Proc. MRS, Boston, Dec. 85, to be published

4. G. Abstreiter, H. Brugger, T. Wolf, R. Zachai, and Ch. Zeller: this volume
5. R. People, J.C. Bean, and D.V. Lang: Proc. 1st Int. Symp. Si-MBE, p. 360, Proc. vol. 85-7, Ed.: J.C. Bean, The Electrochem. Soc., Pennington (USA), 1985
6. E. Kasper: Proc..2nd Int. Conf. Mod. Sem. Struct. (MSS-II), Kyoto, Sept. 85, to be published in Surf. Sci.
7. Th. Ricker and E. Kasper: J. Physique MRS-Europe, p. 192 (1986)
8. H. Jorke and H.-J. Herzog: Same proceedings as [5], p. 352
9. G. Abstreiter, H. Brugger, T. Wolf, H. Jorke and H.-J. Herzog: Phys. Rev. Lett. 54, 2441 (1985)
10. The Technology and Physics of Molecular Beam Epitaxy, Ed.: E. Parker, Plenum, New York, London, 1985
11. E. Kasper and K. Wörner: J. Electrochem. Soc. 132, 2481 (1985)
12. E. Kasper and K. Wörner: Proc. 2nd Int. Symp. VLSI Science Technol., p. 429, Proc. vol. 84-7, Ed.: K.E. Bean and G.A. Rozgonyi: Electrochem. Soc., Pennington (USA), 1984
13. H. Jorke, H.-J. Herzog, and H. Kibbel: Appl. Phys. Lett. 47, 511 (1985)
14. E. Kasper: Appl. Phys. A28, 129 (1982) or E. Kasper: Wiss. Ber. AEG-Telefunken 53, 170 (1980)
15. E. Kasper, H.-J. Herzog, and H. Kibbel: Appl. Phys. 8, 199 (1975) or E. Kasper and H.-J. Herzog: Wiss. Ber. AEG-Telefunken 49, 213 (1976)
16. E. Kasper and H.-J. Herzog: Thin Solid Films 44, 357 (1977)
17. L.C.A. Stoop: Thin Solid Films 24, 243 (1974)

18. J.W. Matthews, D.C. Jackson, and A. Chambers: Thin Solid Films 26, 129
    (1975)
19. T. Halicioglu: J. Crystal Growth 29, 40 (1975)
20. V.d. Merwe, Surf. Sci. 31, 198 (1972)
21. E. Kasper, H.-J. Herzog, H. Dämbkes, and G. Abstreiter: Proc. MRS,
    Boston, Dec. 85
22. H.-J. Herzog, L. Csepregi, and H. Seidel: J. Electrochem. Soc. 131,
    2969 (1984)
23. R. People and J.C. Bean: Appl. Phys. Lett. 47, 322 (1985)
24. J.C. Bean, T.T. Sheng, L.C. Feldman, A.T. Fiory, and R.T. Lynch:
    Appl. Phys. Lett. 44, 102 (1984)
25. J.C. Bean, L.C. Feldman, A.T. Fiory, S. Nakahara, and I.K. Robinson:
    J. Vac. Sci. Technol. A2, 436 (1984)
26. H. Dämbkes, H.-J. Herzog, H. Jorke, H. Kibbel and E. Kasper: IEDM 85,
    Techn. Dig., p. 768 (1985) and IEEE-ED, April 86, to be published

Part II

Band Discontinuities

Elementary Tight-Binding Theory
of Schottky-Barrier and Heterojunction Band Line-Ups

W.A. Harrison

Department of Applied Physics, Stanford University,
Stanford, CA 94305, USA

1 Tight-Binding Theory

The electronic structure of a solid is ordinarily represented in terms of one-electron states, each represented by a wavenumber associated with the electron's propagation through the lattice. The energy bands give the energy of such states as a function of their wavenumber. Each of these states may be approximated as a combination of local atom-like states, with the wavenumber determining the coefficient of each such state in the propagating electronic state. If real atomic states are used, this is called the method of *Linear Combinations of Atomic Orbitals*. This idea is as old as Bloch's original paper [1]; it has generally only been considered of qualitative significance, though very often the results of more complete and accurate calculations are described in terms of such an atomic basis. Then the parameters which determine the bands are not obtained from real atomic states, but are adjusted to accord with the more accurately calculated bands; such an approach is called *Tight-Binding Theory*.

Recently it has been realized that the systematics of such parameters for sp-bonded systems is so great that the parameters can readily be estimated for any system without the use of a full calculation [2,3]. These then allow an approximate first-principles calculation of the electronic structure and thus the properties of a wide range of materials. They are almost guaranteed to be close to the results of the more complete and accurate computer-intensive predictions, but because of their relative simplicity they can be much more quickly and generally applied and the predictions are usually much more directly and deeply understandable.

The parameters which enter tight-binding calculations are the energy to be associated with each atomic state and the magnitude of the energy of coupling between it and neighboring atomic states. For the sp-bonded materials it has proven adequate for most purposes to include only the corresponding s- and p-states on each atom, and to use the free-atom term value for their energies.

Table 1. Magnitudes of the free-atom term values from MANN [4]. The first value is $-\epsilon_s$, the second is $-\epsilon_p$. [eV]

Li	Be	B	C	N	O	F
5.34	8.42	13.46	19.38	26.22	34.02	42.79
-	5.81	8.43	11.07	13.84	16.77	19.87
Na	Mg	Al	Si	P	S	Cl
4.96	6.89	10.71	14.79	19.22	24.02	29.20
-	3.79	5.71	7.59	9.54	11.60	13.78
Cu	Zn	Ga	Ge	As	Se	Br
6.49	7.96	11.55	15.16	18.92	22.86	27.01
3.31	3.98	5.67	7.33	8.98	10.68	12.44
Ag	Cd	In	Sn	Sb	Te	I
5.99	7.21	10.14	13.04	16.03	19.12	22.34
3.29	3.89	5.37	6.76	8.14	9.54	10.97

For our *universal tight-binding parameters* [2] we have chosen to use Hartree-Fock term values given by MANN [4]. These are approximately equal to the energy required to remove the electron from the free atom. Table 1 provides such values for most elements of interest here. More complete sets appear in Ref. 2 (p. 534), Ref. 3, Ref. 4, and Ref. 5. These are all that are required to make almost all of the predictions described here.

It has also proven adequate to include only the interactions between nearest-neighbor atoms. The magnitude of these interactions could be obtained by noting that the true energy bands of semiconductors, which are reasonably well describable in tight-binding theory, are also quite close to the bands obtained by assuming that the electrons were completely free. [The corresponding free-electron parabola, $E = \hbar^2 k^2 / 2m$, must be folded back into the Brillouin Zone to make the comparison.] To the extent that this is true, the nearest-neighbor couplings must be of the form

$$V_{ll'm} = \eta_{ll'm} \, \hbar^2/md^2 \qquad (1)$$

where d is the internuclear distance, and $\eta_{ll'm}$ is a geometrical constant depending on the total angular momentum quantum numbers l and l' of the two states and the quantum number m for the angular momentum around the internuclear vector. For $l = l' = m = 0$, for example, we find $\eta_{ss\sigma} = -9\pi^2/64$ A better description of the bands is in fact obtained if these coefficients are adjusted slightly. Fitting the known bands for germanium (and in fact adding an excited atomic s-state which is not utilized in the subsequent analysis), we obtained [3] $\eta_{ss\sigma} = -1.32$, $\eta_{sp\sigma} = 1.42$, $\eta_{pp\sigma} = 2.22$, and $\eta_{pp\pi} = -0.63$, and these values are used here.

These universal parameters have given good descriptions of a wide range of properties of bulk semiconductors. Even in polar semiconductors, such as GaAs, where there is considerable charge transfer between the gallium and arsenic atoms, the free-atom term values remain appropriate.   However, in some circumstances, such as at heterojunction interfaces, it becomes necessary to include a shift in the free-atom term value due to redistribution of the charge. A procedure for doing this has recently been codified in Ref. 5. Such a procedure is needed if we are to calculate the redistribution of charges among atoms in detail, but we shall find that it will be sufficient here to describe the principal coulomb effects using a dielectric constant.

2 Natural Band Line-Ups

A band calculation based upon tight-binding parameters yields all energies relative to the same scale upon which the atomic term values were given, the energy relative to an electron carried an infinite distance from the atom. [It was seen in   Ref. 5   that the energy to remove an electron from a bulk semiconductor is less than this by a few electron volts due to dielectric relaxation, but that does not affect the present discussion.] The valence-band maximum, in particular, is given by

$$\epsilon_v = (\epsilon_p^+ + \epsilon_p^-)/2 - \{[(\epsilon_p^+ - \epsilon_p^-)/2\,]^2 + (1.28\hbar^2/md^2)^2\}^{1/2} \tag{2}$$

where the super $+$ and $-$ refer to the metallic and nonmetallic atoms in a compound semiconductor, respectively.   Thus the universal tight-binding parameters give the valence-band maxima of two semiconductors on the same scale and by subtracting we obtain the difference in the *natural line-ups*, which will correspond to the valence-band discontinuity at a heterojunction between those two semiconductors, assuming there are no dipole layers due to charge redistribution at the interface, shifting the bands relative to each other.

Table 2. Values of $\epsilon_v(B) - \epsilon_v(A)$ [eV] from natural band line-ups, by shifting the average hybrid energies on the two sides to equal each other, and from experiment, mostly [4].

A/B	Natural	Matched Hybrids	Experiment
AlAs/GaAs	0.03	0.12	0.50
InAs/GaSb	0.81	0.42	0.46
GaAs/InAs	0.16	-0.13	0.17
Si/Ge	0.38	0.29	0.20
ZnSe/GaAs	1.42	1.35	0.96
ZnSe/Ge	2.09	2.01	1.52
Ge/GaAs	-0.67	-0.66	-0.56
CdS/InP	1.86	1.37	1.63
HgTe/CdTe	0.05	-0.48	-

Such a set of natural band line-ups was made some years ago [6]. In Table 2 we give such values for the semiconductors for which band line-ups have been collected by KRAUT [7]. We see that it gives a reasonably good account of the experimentally observed line-ups. However, we shall also see that in some cases very large dipoles arise at the interface and change these predictions considerably and, in fact, improve the agreement between the predictions and experiment.

3 Interface Dipoles

HEINE [8] some time ago noted that at a metal-semiconductor interface the metallic states with energies in the semiconductor band gap will have exponentially decaying tails extending into the semiconductor which produce dipoles, similar to the way real surface states in the semiconductor gap would; they have come to be called *metal-induced gap states*. LOUIE and COHEN [9] calculated the electronic structure of an interface between a number of semiconductors and jellium, the latter having parameters chosen to represent aluminum. They found sizable dipole shifts which they interpreted in terms of these metal-induced gap states. Similar effects could be expected at interfaces between two semiconductors.

TERSOFF [10] noted, as had Heine, that the states tailing into the semiconductor could be expanded mathematically in a basis of valence-band and conduction-band states. The tails of occupied states near in energy to the conduction-band edge are predominantly conduction-band like and produce a dipole layer lowering the energy on the metallic side. Tails of empty states near in energy to the valence-band edge produce a dipole of the opposite sign. Thus these dipoles tend to shift the metallic Fermi energy away from the band edges. TERSOFF[10] suggested that the shift is large and that the point toward which the Fermi energy is shifted is a branch-point energy, E_B, which he calculated from the energy bands. He suggested that the Fermi energy in the metal was effectively pinned at that point, fixing the Schottky-barrier height for p-type materials at E_B minus the valence-band maximum.

We have argued [11] that these effects are not large. We noted that two rather different contributions could be distinguished, one arising from the exponential tails in the semiconductor, as envisaged by Heine, and the other from the transfer of charge between two atomic layers which form the interface. We analyzed both rather carefully, and found that while the latter were larger, they were too small to account even qualitatively for the formation of Schottky barriers. In a subsequent more careful and complete analysis of the dielectric response,we have found our earlier conclusion to be in error. Our replacement of the bulk of the semiconductor by a dielectric continuum, rather than by a collection of polarizable bonds, had led to a considerable error.

4 The Simplest Heterojunction

There is one case for which one can see rigorously the strength of the dielectric screening. That is a heterojunction between two semiconductors which, in terms of tight-binding theory, differ only in a uniform shift Δ between the free-atom energies on the two sides. This will shift the band edges relative to each other on the two sides of the heterojunction by Δ. This shift may be modified by interface dipoles, but if Δ is small the *net* shift will be proportional to it. We write the net shift in valence bands on the two sides as $\delta \epsilon_v = \lambda \Delta$. Now we can immediately see that λ is equal to the reciprocal of the dielectric constant for the semiconductor. We do this by applying the same tiny shift between every adjacent pair of atom planes, which in tight-binding theory is exactly equivalent to applying a uniform electric field. That field will be reduced by a factor of the dielectric constant ϵ and thus each $\delta \epsilon_v$ will be reduced by that factor, corresponding to $\lambda = 1/\epsilon$.

Of course for this case the *natural* line-ups will correspond to a valence-band discontinuity of Δ and thus are seriously in error. TERSOFF [12] had argued that in a heterojunction the E_B levels in the two semiconductors would line up, just as the metallic Fermi energy is pinned to the E_B level. This means that Tersoff was essentially correct for this case; his E_B will differ also by Δ and dipoles will arise to bring the two values, and the two valence (or conduction) bands, very nearly into coincidence. It remains to consider a more general pair of semiconductors in tight-binding theory and to find the counterpart of his E_B in that context. We do that now.

5 Real Heterostructures

We shall begin with the simplest case discussed above and then add features to the electronic structure on the two sides. The rigid shift of the term values by Δ is to be screened by the dielectric constant which we take to be infinite, reducing the discontinuity to zero. Then, holding the average term value on each side fixed (that average is $(\epsilon_S + 3\epsilon_p)/4$ which is equal to the hybrid energy, or the average of the two hybrid energies $\langle\epsilon_h\rangle$ if it is a compound semiconductor), we may change the sp-splitting on one side. It is not difficult to see [13] that this change in $\epsilon_S - \epsilon_p$ does not in itself produce a dipole at the surface; it does modify the λ which relates $\delta\epsilon_v$ to Δ, but that simply means we are to use a slightly different dielectric constant than before, but we are taking it infinite in any case so the relative position of $\langle\epsilon_h\rangle$ on the two sides is unaffected. The modification of $\epsilon_S - \epsilon_p$ is a change in the metallic energy, $V_1 = (\epsilon_S - \epsilon_p)/4$ in the language of tight-binding theory [2]. Changing it modifies the band gap and introduces discontinuities in the band edges, but no dipole, so these discontinuities are calculated after matching the average hybrid energies on the two sides.

Similarly we might change the internuclear distance on one side with respect to the other, a modification of the covalent energy $V_2 = -3.22\, \hbar^2/md^2$, which again does not in itself introduce dipoles, though it modifies the discontinuities in the band edges, as did the change in V_1, and in principle it also modifies the dielectric constant. Finally we may introduce a polar energy $V_3 = (\epsilon_h^+ - \epsilon_h^-)/2$ on one or both sides, with the same result, modified band discontinuities but no additional dipole. These three additional modifications of the tight-binding parameters on the two sides allow us complete generality in the specification of the semiconductors on the two sides, with the result that dipole shifts will arise to reduce the difference in the average hybrid energies on the two sides very nearly to zero.

In tight-binding theory the average hybrid energy $<\epsilon_h>$ for a compound (or elemental) semiconductor thus plays exactly the role of Tersoff's E_B. In the second column of Table 2 we have given the valence-band line-ups predicted by matching the average hybrid energies on the two sides. For example, from Table 1 we obtain hybrid energies ($(\epsilon_s + 3\epsilon_p)/4$) for Ge. Ga and As of -9.29, -7.14, and -11.47 eV, respectively. The average hybrid energy in GaAs is then -9.30 eV. This is almost equal to the germanium value, but suggests dipoles arise to raise the GaAs bands 0.01 eV in comparison to germanium. The values for ϵ_v obtained from Eq. 1 (with d = 2.44 A) for Ge and GaAs are -8.99 and -9.66 eV, respectively, giving the natural band line-up discontinuity in the valence bands of -0.67 eV, but reduced by the dipole shift to -0.66 eV.

We note that in many cases, as in the example given, the dipole shifts are quite small, so the two predictions are not so different and both theories work rather well. In most cases, however, the agreement with experiment is improved by including the dipole shifts. The particular case of HgTe/CdTe is interesting. We did not include the Hartree-Fock term values for mercury in Table 1. Mann's values [4] for Hg are in fact quite close to those for Cd giving almost no valence-band discontinuity from natural band energies, and in fact almost no dipole shift, with both predictions giving no discontinuity. However, the reason values for the Hg row were not given in Table 1 is that relativistic effects, not included by Mann, are large for this row, the principal effect being a lowering in the s-state energy of Hg in comparison to that of Cd. Since only the p-state term values enter Eq. 1 for the valence-band maximum, this does not affect the natural band line-ups. However, it lowers the average hybrid energy in the HgTe considerably, and the dipoles required to raise it to the CdTe value bring the valence band maximum in HgTe above that in CdTe by a half an electron volt. This is in accord with an earlier prediction by TERSOFF [14] who has examined the experimental indications, finding them not inconsistent with this large discontinuity.

The analysis of this case also pointed up the uncertainty in the predictions based upon universal-parameter tight-binding theory, as well perhaps as by other techniques. Use of different sets of relativistic term values, of seemingly equal validity, gave discontinuities differing by a few tenths of volts. The differences were simply in the manner that exchange and correlation were incorporated in the calculation of the term values. Such differences in the treatment of these many-electron effects would presumably show up also in full quantitative self-consistent calculations of the heterojunction electronic structures if different treatments were used.

6 Schottky Barrier Heights

As TERSOFF [10] used the difference between his E_B values and the valence-band maximum as a prediction of Schottky barrier heights ϕ_{bp} (with p-type semiconductors), we may use the average hybrid energy minus the valence band maximum, $\langle\epsilon_h\rangle - \epsilon_V$, as such a prediction. This does not follow as neatly from tight-binding theory as did the heterojunction line-ups, because of the difficulties of representing the metal in these same terms. However, to the extent that the metal is represented by partially occupied atomic levels, and these are to be identified with the Fermi energy, the prediction follows. This is essentially Tersoff's argument. Predictions are immediate using Table 1 and Eq. 1. Values for the systems listed by SZE [15] are given in Table 3. There are discrepancies of a few tenths of an electron volt but there is some general accord between theory and experiment. We may say the general features of Schottky-barrier formation are in reasonable accord with the simple tight-binding theory. However, there are many cases for which the Schottky-barrier heights depend upon the exact preparation technique. In such a case the differences must come from nonideal features of the interface which are not present in the simple theory, nor in fact in the more quantitative first-principles theories.

Table 3. Tight-binding predictions of Schottky-barrier heights, compared with experiment [15] [eV]

Semiconductor	$\langle\epsilon_h\rangle - \epsilon_V$	ϕ_{bp} [Sze]
Si	-0.03	0.32
Ge	-0.32	0.07
GaP	0.66	0.94
InP	0.77	0.77
AlAs	0.46	0.96
GaAs	0.34	0.52
InAs	0.47	0.47
AlSb	0.23	0.55
GaSb	0.05	0.07
InSb	0.27	0.00

LANGER and HEINRICH [16] have noted that a particular deep level, such as an iron donor (0,+) level in a III-V semiconductor, may be taken as a reference level, matched in two semiconductors for which the band edges are known relative to that level, to predict the band line-ups in a heterojunction. They in fact found that such a scheme gave predictions in rather good accord with experiment.

Such deep levels can also be treated in the context of tight-binding theory, but it requires going somewhat beyond what we gave above. We have done this in an unpublished joint effort with J. Tersoff.   A transition-metal atom, substituted for Ga in GaAs, will have three valence d-states, of symmetry t_2 coupled to the surrounding four As hybrids. Two other d-states, of symmetry e, will be uncoupled. In contrast to states in simple atoms, the d-state energies are sensitive to any charge transferred to them [2], shifting by an energy $U\delta z_d$ if the d-state occupation changes by δz_d. Values of U and ϵ_d for each transition-metal atom ($U = 5.1\ eV$ for iron) have been given by FROYEN [17]. Coupling energies between these d-states and neighboring s- and p-states are given in Ref. 2.   They can be combined to give the coupling between a t_2-symmetry d-state and the corresponding combination of four hybrids as $V_{dh_4} = -4.78\ \hbar^2 r_d^{3/2}/md^{7/2}$, with r_d values given for each element in Ref. 2. For iron $r_d = 0.80\ A$, and $d = 2.45\ A$ for GaAs, giving $V_{dh_4} = -1.13\ eV$. It is not difficult to solve this two-level system self-consistently, evaluating the occupation z_d of the d-states in terms of the ϵ_d, ϵ_h, and V_{dh_4} and evaluating ϵ_d in terms of the z_d. Obtaining z_d depends upon the fact that in Fe^0, for example, there are four electrons in e-symmetry d-states, six in hd-bonds, and one in an hd-antibond (and two in a nonbonding hybrid state). It is found that the d-state tends to follow the hybrid-state if the energy of the latter is shifted; shifting ϵ_h downward drains electrons from the d-state so ϵ_d also drops. The difference $\epsilon_d - \epsilon_h$ due to such a shift is found for this case to be reduced by a factor $1/[1 + 5U\alpha_c^3/4V_{dh_4}]$, with $\alpha_c = |V_{dh_4}|/[V_{dh_4}^2 + (\epsilon_d - \epsilon_h)^2/4]$, a covalency for this bond. For the case of Fe^0 in GaAs, with α_c near one, this is a reduction by a factor of 1/7.   This means that the d-state energy, *and therefore the deep level energy*, follows closely the hybrid energy.

The only difficulty is that we have used the arsenic hybrid energy, not the average. However, if we include the coupling $V_1(As)$ between this hybrid and the hybrids in the neighboring bonds and antibonds, we find that the corresponding *metallized* arsenic hybrid does lie much closer to the average hybrid (ten times closer in the case of GaAs) than to the As hybrid (which is also why α_C is near one). Thus the deep level *does* track the average hybrid and becomes, according to tight-binding theory, a reasonable empirical predictor of the band line-ups, exactly as proposed by LANGER and HEINRICH [16].

In summary, the tight-binding theory of ideal heterojunction and metal-semiconductor interfaces predicts behavior very close to that observed for laboratory junctions. Though interface defects can play a role, they should be regarded corrections to a seemingly well-understood ideal behavior.

8 Acknowledgement

This work was supported by the Office of Naval Research under Contract N00014-85-0167. Part of the research was carried out while the author was a guest of the T. J. Watson Research Laboratory of IBM.

References

1. F. Bloch, Z. Physik 52, 555 (1928)
2. W. A. Harrison: Electronic Structure and the Properties of Solids (Freeman, New York (1980))
3. W. A. Harrison, Phys. Rev. B24, 5835 (1981)
4. J. B. Mann, Atomic Structure Calculations, 1: Hartee-Fock Energy Results for Elements Hydrogen to Lawrencium. Distributed by Clearinghouse for Technical Information, Springfield, VA 22151, USA (1967)
5. W. A. Harrison, Phys. Rev. B31, 2121 (1985)
6. W. A. Harrison, J. Vac. Sci. and Technol. 14, 1016 (1977)
7. E. A. Kraut, J. Vac. Sci. and Technol. B2, 486 (1984)
8. V. Heine, Phys. Rev. A138, 1689 (1965)
9. S. G. Louie and M. L. Cohen, Phys. Rev. B13, 2461 (1976)
10. J. Tersoff, Phys. Rev. Lett. 52, 465 (1984)
11. W. A. Harrison, J. Vac. Sci. and Technol. B3, 1231 (1985)
12. J. Tersoff, Phys. Rev. B30, 4874 (1984)
13. W. A. Harrison and J. Tersoff, to be published in Proc. PCSI, Pasadena, January, 1986, in J. Vac. Sci. and Technol.
14. J. Tersoff, to be published in Proc. PCSI, Pasadena, January, 1986, in J. Vac. Sci. and Technol.
15. S. M. Sze, Physics of Semiconductor Devices, (Wiley, New York, (1969))
16. J. M. Langer and H. Heinrich, to be published in Proc. Int. Conf. Hot Electrons in Semiconductors, Innsbruck, July, 1985, in Physica
17. S. Froyen, Phys. Rev. B22, 3119 (1980)

Electrical Measurements of Band Discontinuities at Heterostructure Interfaces

T.W. Hickmott

IBM T.J. Watson Research Center, Yorktown Heights, NY 10598, USA

Experimental methods of measuring band discontinuities at heterostructure interfaces are discussed using the $GaAs/Al_xGa_{1-x}As$ system as an example. Electrical measurements of current-voltage and capacitance-voltage characteristics of single barrier systems are emphasized.

1. INTRODUCTION

An important factor determining the properties of devices based on semiconductor heterojunctions is the way in which the difference in band gap energies of the semiconductors divides between the conduction band discontinuity ΔE_C and the valence-band discontinuity ΔE_V. For any heterojunction in which two semiconductors, characterized by band gap energies E_{G1} and E_{G2}, are in contact, $E_{G2} - E_{G1} = \Delta E_C + \Delta E_V$. Determination of ΔE_C and ΔE_V for any pair of semiconductors is primarily an experimental problem; present theoretical models are not able to predict values of ΔE_C and ΔE_V to the accuracy needed in understanding devices [1,2].

The most extensively studied heterojunction system has been $GaAs/Al_xGa_{1-x}As$. (x is the mole fraction of AlAs in the alloy.) It is a nearly ideal system; the lattice constants of GaAs and $Al_xGa_{1-x}As$ match within 0.1 % over the whole composition range. In spite of the ideality of the $GaAs/Al_xGa_{1-x}As$ system experimental determination of the band discontinuities has been a controversial subject. Examination of the experimental problems in determining band offsets in this system shows the pitfalls that may arise in studying other heterojunction systems.

Broadly, the experimental methods used for determining band discontinuities can be divided into two categories, optical and electrical. DUGGAN [3] has recently reviewed optical methods for determining ΔE_C. This review emphasizes electrical methods, particularly current-voltage (I-V) and capacitance-voltage (C-V) measurements. The initial experimental results of DINGLE *et al.* [4] on infrared absorption of $GaAs/Al_{0.2}Ga_{0.8}As$ multiple quantum-well structures gave

$\Delta E_C{:}\Delta E_V$ = 85:15, a value that seemed firmly established until 1984–1985 when a series of papers was published that gave $\Delta E_C{:}\Delta E_V \sim$ 60:40 [5–15]. · Nearly all recent experimental measurements, both electrical and optical, of band discontinuities have given this result for $0.1{<}x{<}0.4$. The choice of a particular experimental method often depends on the kind of structure available. Two kinds of structures are considered in detail here; semiconductor-insulator-semiconductor (SIS) capacitors that contain single undoped $Al_xGa_{1-x}As$ dielectric layers between two GaAs layers [9–11,15], and n-N or p-P isotype heterojunctions [5,12,13]. Either structure can be used for independent measurements of ΔE_C and ΔE_V. The limitations and experimental problems are emphasized since the same problems will occur when these electrical methods are applied to other heterojunction systems.

2. THERMIONIC EMISSION IN UNDOPED $Al_xGa_{1-x}As$ CAPACITORS

The SIS capacitor using undoped $Al_xGa_{1-x}As$ as the dielectric is analogous in its structure and in the kinds of information one obtains to the metal-oxide-semiconductor capacitor with polysilicon gate. A typical GaAs-gate capacitor grown by molecular-beam epitaxy is shown schematically in Fig. 1a. An n^- GaAs substrate layer, ~1 μm thick, is grown on a <100> oriented n^+ GaAs wafer. The

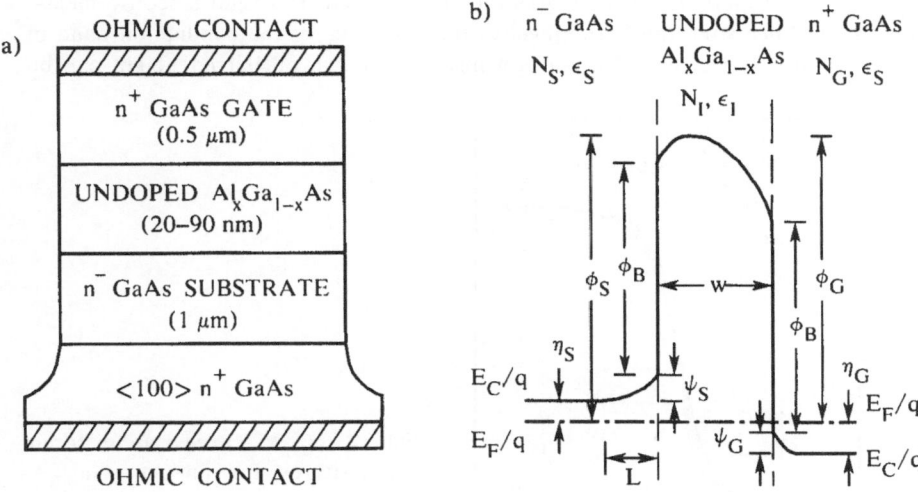

Fig. 1. (a) Schematic diagram of layer structure of GaAs-gate capacitor used to measure conduction-band discontinuities. (b) Schematic potential diagram of an n^- GaAs–undoped $Al_xGa_{1-x}As$–n^+ GaAs capacitor when the bands in the insulator are bent due to negative charge, $N_I(cm^{-3})$. (After Ref. [9].)

$Al_xGa_{1-x}As$ layer of desired composition and thickness is grown, and then an n^+ GaAs layer, ~0.5 μm thick is grown to serve as gate electrode. After the layers are grown, ohmic contacts are made to the top and bottom n^+ layers and capacitors of known area are formed by etching mesas. The schematic potential diagram for such a GaAs-gate capacitor is shown in Fig. 1b [9]. The conduction band discontinuity in Fig. 1b is $\Delta E_C = q\phi_B$ where q is the electron charge. The magnitude of the potential differences between the Fermi level, E_F, and the conduction-band edge, E_C, are η_S and η_G for substrate and gate, respectively. They depend on substrate doping, N_S, and gate doping, N_G, as well as on temperature. The undoped $Al_xGa_{1-x}As$ of thickness w is assumed to have a uniform density of negative charge, N_I, which bends the conduction band of the $Al_xGa_{1-x}As$ and also produces band bending in substrate and gate, ψ_S and ψ_G, even when the applied gate bias $V_G = 0$ as in Fig. 1b. Such charge is observed experimentally.

Thermionic emission of electrons over the barrier in the insulator is the principal conduction mechanism in SIS capacitors at low voltages and high temperatures. The current density at temperature T due to thermionic emission, $J(A/cm^2)$, is given by

$$J = A^* T^2 \exp(-\phi_E/kT) \tag{1}$$

where A^* is the effective Richardson constant and ϕ_E is an effective barrier height at the negatively biased electrode. For $V_G > 0$ V, $\phi_E = \phi_S$ in Fig. 1b while $\phi_E = \phi_G$ for $V_G < 0$ V. N_I increases the effective barrier height since ϕ_E measures the distance from the Fermi level to the maximum of the conduction band in the insulator. The effect of a uniform insulator charge on barrier height can be

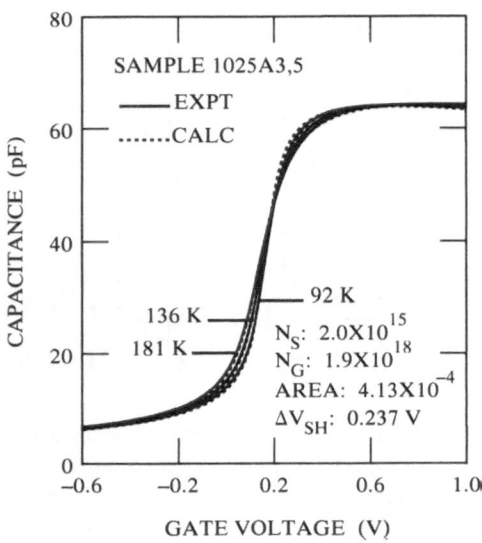

Fig. 2. C–V curves at different temperatures for an $Al_{0.4}Ga_{0.6}As$ capacitor. Solid curves are measured, dotted curves are calculated using model that includes a uniform distribution of negative charge in $Al_{0.4}Ga_{0.6}As$ dielectric. [9].

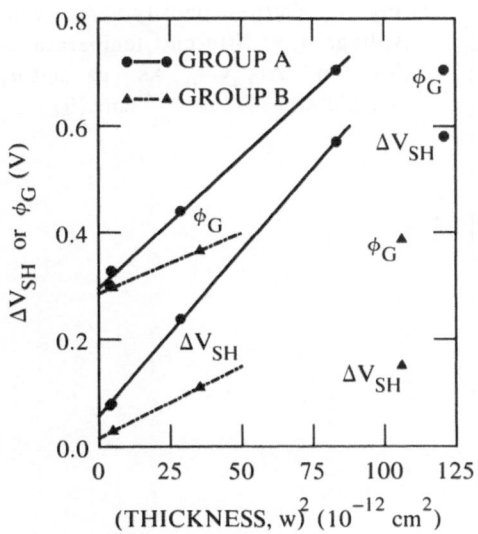

Fig. 3. Dependence of ΔV_{SH} and ϕ_G on square of $Al_{0.4}Ga_{0.6}As$ thickness w. ΔV_{SH} derived by comparing experimental and calculated C–V curves. ϕ_G obtained from I–V curves. w calculated from C–V curves [9].

eliminated by measuring samples of different thickness, w, plotting ϕ_E versus w^2, and extrapolating to zero thickness. For $V_G < 0$ V the resulting barrier height is

$$\phi_G(0) = \phi_B + \psi_G - \eta_G. \tag{2}$$

η_G and ψ_G must be known to extract $\phi_B = \Delta E_C/q$ from barrier heights measured by I–V curves. These quantities are obtained by measuring C–V curves on the same SIS capacitors for which barrier heights are measured by I–V curves.

Capacitance-voltage curves for GaAs gate capacitors are nearly ideal. Figure 2 shows typical C–V curves at three temperatures for a capacitor with an $Al_{0.4}Ga_{0.6}As$ dielectric thickness of ~50 nm [9]. The solid lines are experimental; the dotted lines are calculated from classical SIS theory using Fermi-Dirac statistics. Substrate doping, N_S, at a distance L from the interface, is obtained from the measured capacitance in the depletion region [16]. N_G, w, ψ_S, ψ_G and C_I, the insulator capacitance, are derived by comparing calculated and experimental C–V curves. An advantage of using SIS capacitors is that η_S and η_G can be calculated if N_S, N_G and T are known. The flat-band voltage for ideal C–V curves, V_{FB}, equals $\eta_G - \eta_S$ and should be negative for C–V curves of samples such as those in Fig. 1. In general, C–V curves of SIS capacitors are shifted to positive voltages by an amount ΔV_{SH} due to negative charge in the undoped $Al_xGa_{1-x}As$ layer. In Fig. 2, $\Delta V_{SH} = 0.237$ V is required to obtain agreement between experimental and calculated C–V curves. In Fig. 3, ΔV_{SH} for two groups of samples is plotted as a function of w^2 [9]. Each group of samples was prepared with as nearly identical layers as possible but with different $Al_xGa_{1-x}As$ thicknesses. The proportionality between ΔV_{SH} and w^2 for thicknesses below ~90 nm shows that negative charge

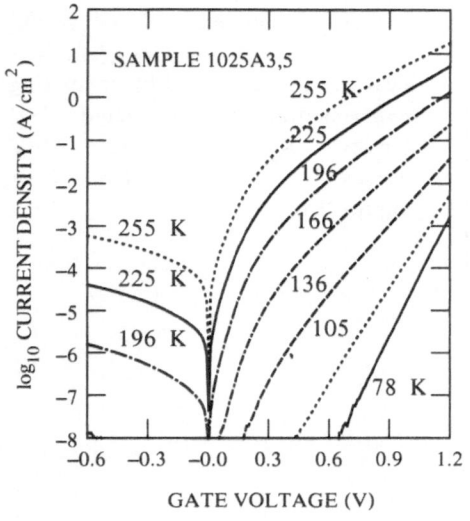

Fig. 4. Current density versus gate voltage for different temperatures for an $Al_{0.4}Ga_{0.6}As$ capacitor. Sample area, 4.13×10^{-4} cm^2 [9].

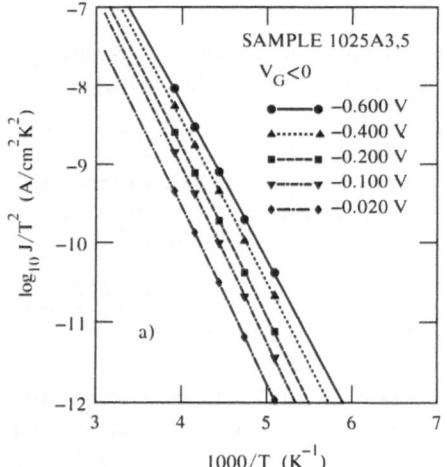

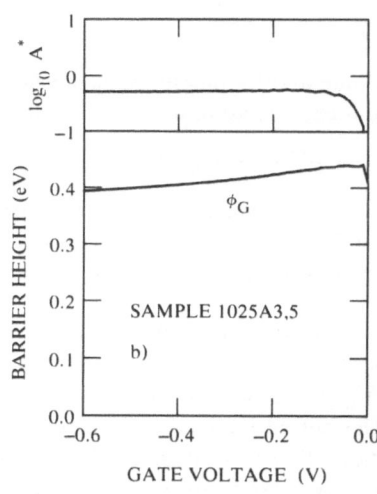

Fig. 5. (a) Thermionic emission analysis of I–V data. Plot of $\log_{10}(J/T^2)$ versus $1/T$ at different voltages for $V_G < 0$ V. Lines are linear least–squares fit of data points. (b) Thermionic emission analysis of I–V data for $V_G < 0$ V showing barrier height versus gate voltage and $\log_{10} A^*$ versus gate voltage [9].

is uniformly distributed through the $Al_{0.4}Ga_{0.6}As$. The difference in slopes of the two lines shows that the value of N_I is process dependent.

I-V characteristics at different temperatures for the sample of Fig. 2 are shown in Fig. 4. Three features are characteristic of I-V curves; rectification, thermal activation at low voltages and tunneling at low temperatures and high positive gate voltages. Rectification occurs because of the difference in doping of n^- substrate and n^+ gate. Figure 5a shows the analysis according to (1) of the data of Fig. 4 for $V_G < 0$ V. The linearity of the plots of log (J/T^2) versus $1/T$ is good for all voltages. The voltage dependence of the barrier height ϕ_G, derived from the slopes of the plots of log (J/T^2) versus $1/T$, is shown in Fig. 5b. The effective barrier height for V_G extrapolated to 0 V is $\phi_G = 0.44$ V. In Fig. 3 values of ϕ_G for samples of different thicknesses are plotted as a function of w^2. The plot is linear, extrapolating to $\phi_G(0) \sim 0.30$ V. It is apparent that negative charge in the $Al_xGa_{1-x}As$ layer can strongly affect values of ϕ_G measured from I-V curves. Results for two groups of samples are shown in Fig. 3. Using values of η_G and ψ_G derived from C-V curves, a value of $\Delta E_C / \Delta E_G = 0.63$ is obtained [9], in excellent agreement with determinations by other methods.

Two kinds of structures have been made that eliminate the problem of the effect of negative charge in undoped $Al_xGa_{1-x}As$ on measured barrier heights. N_I appears to be due to a deep acceptor in $Al_xGa_{1-x}As$. One approach is to place a small number of donors in the middle of the $Al_xGa_{1-x}As$ layer during growth [10]. This neutralizes acceptors in undoped $Al_xGa_{1-x}As$. The second approach is to use p-type GaAs instead of n-type GaAs on both sides of the $Al_xGa_{1-x}As$ barrier layer. Using p-type GaAs as substrate and gate layers places the Fermi level of the system below the acceptor level and eliminates shifts in C-V curves [11,15].

BATEY and WRIGHT [15] have used symmetrical p-type GaAs-gate capacitors to measure the valence band discontinuity, ΔE_V, as a function of the AlAs mole fraction x. They find that ΔE_V is proportional to x;

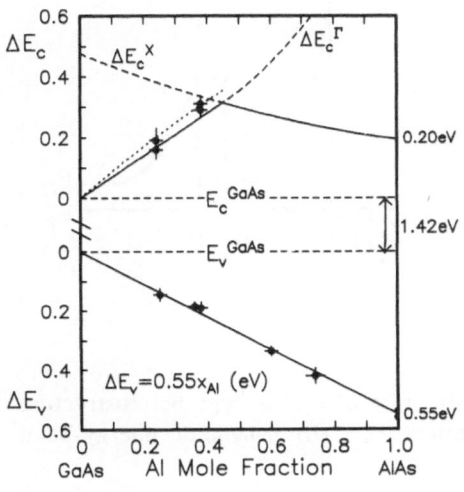

Fig. 6. Energy band alignment for GaAs/$Al_xGa_{1-x}As$ deduced using (3) and room-temperature band-gap data of [17] [Ref. 15].

$$\Delta E_V \sim 0.55 \, x_{Al} \quad (eV) \qquad (0 < x_{Al} \leq 1) \tag{3}$$

BATEY and WRIGHT used these data to extract the full energy band picture from known band-gap data, as shown in Fig. 6. Figure 6 shows the measured variation of ΔE_C and ΔE_V as a function of x. Values of $\Delta E_C{}^\Gamma$ and $\Delta E_C{}^X$ are obtained by subtracting (3) from the band-gap data of CASEY and PANISH [17]. $Al_xGa_{1-x}As$ changes from a direct to an indirect gap semiconductor at $x \sim 0.43$. Below this value of x, ΔE_C increases linearly with x, above it ΔE_C decreases. The values of ΔE_C plotted in Fig. 6 are those of BATEY et al. [10]. Separately measured values of ΔE_C and ΔE_V at a constant value of x add up to the band-gap difference at that composition. Thus combining C-V and I-V measurements on SIS capacitors leads to values of ΔE_C and ΔE_V by relatively simple data analysis. The method is sensitive to charge in the insulator; however, it can measure the amount of insulator charge and correct for it.

3. C-V PROFILING OF ISOTYPE HETEROJUNCTION BARRIERS

The isotype heterojunction barrier consists of two semiconductor layers of similar type with a Schottky barrier contact to one of them to provide a depletion layer which can be moved through the heterojunction boundary; it is the simplest structure for studying band discontinuities. KROEMER and co-workers [5,18,19] first used C-V profiling through a heterojunction to measure ΔE_C. OKUMURA et al. [12] measured ΔE_C for several compositions of $Al_xGa_{1-x}As$, and WATANABE

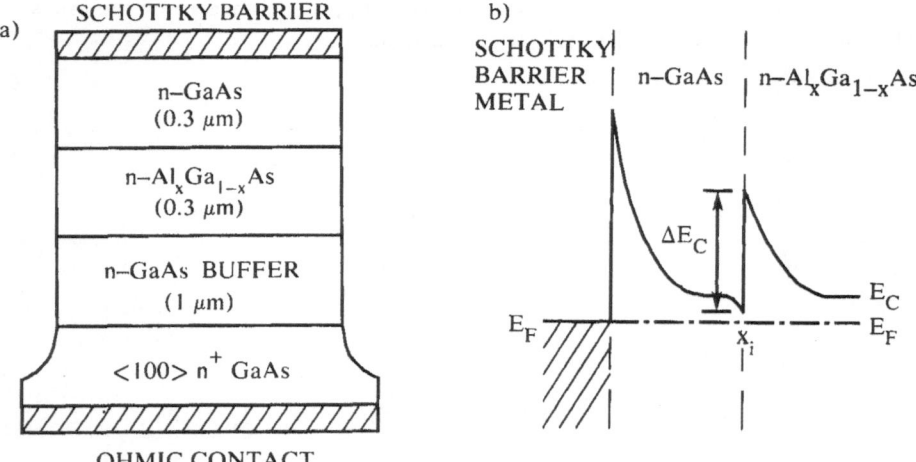

Fig. 7. (a) Schematic diagram of layer structure of n-N isotype heterostructure used to measure conduction-band discontinuities. (b) Schematic energy-band diagram for n-N isotype heterostructure.

et al. [13] have systematically investigated both *p-P* and *n-N* heterostructures to determine the effect of composition on band discontinuity.

A typical *n-N* sample structure is shown in Fig. 7a. (*N* refers to the doping of the wider band-gap material.) The GaAs and $Al_xGa_{1-x}As$ layers are uniformly doped, have $N_D \sim 10^{16}$ to 10^{17} /cm^3 and, ideally, have nearly equal values of N_D. The band diagram for the heterojunction is shown schematically in Fig. 7b. *C-V* curves are measured for the sample as increasing bias on the Schottky barrier moves the depletion layer of the semiconductor through the heterostructure. Carrier concentration as a function of depth is determined from *C-V* curves [16]. Figure 8, taken from WATANABE *et al.* [13], shows a typical curve of carrier concentration versus depth. The effect of the band discontinuity is to produce an apparent increase in the carrier concentration on the low band-gap side of the heterojunction and a decrease on the wide band-gap side. The peak of the distribution is on the low band-gap side and is very close to the position of the heterojunction, x_j.

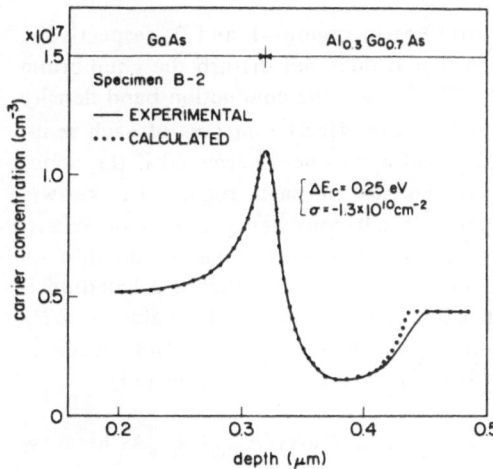

Fig. 8. Experimental and simulated carrier profile for a GaAs/Al$_{0.3}$Ga$_{0.7}$As *n-N* isotype heterostructure. The solid and dotted curves show measured and calculated results, respectively [13].

KROEMER and CHIEN [5,18] have shown that although the apparent carrier profile is distorted by Debye length effects the total number of carriers and the first moment of the total number of carriers are conserved. If $n(x)$ is the true local carrier concentration and $\hat{n}(x)$ is the apparent carrier concentration derived from *C-V* profiling through the heterojunction, they showed that

$$\int_0^\infty n(x)dx = \int_0^\infty \hat{n}(x)dx \qquad (4a)$$

and

$$\int_0^\infty n(x)xdx = \int_0^\infty \hat{n}(x)xdx . \qquad (4b)$$

Electrical neutrality requires that the integral (4a) be equal to the integral of the net donor concentration plus the density σ_i of any fixed interface charges that may be present. This neutrality condition can be expressed as

$$\sigma_i = - \int_0^\infty [N_D(x) - \hat{n}(x)]dx. \tag{5}$$

If $N_D(x)$ is known independently, the interface charge can be extracted from (5).

The electrostatic potential $\Phi(\infty) - \Phi(0) = \Delta\Phi$ between the two sides of the heterojunction can be obtained from (4b);

$$\Delta\Phi = \frac{q}{\varepsilon_S} \int_0^\infty [N_D(x) - \hat{n}(x)](x-x_i)dx. \tag{6}$$

ε_S is the dielectric constant of the semiconductor. $\Delta\Phi$ can also be expressed as

$$\Delta\Phi = (\frac{kT}{q}) \ \ln \ [(\frac{n(\infty)}{n(0)}) \cdot (\frac{N_C(0)}{N_C(\infty)})] \ + \ \frac{\Delta E_C}{q} \tag{7}$$

where $n(0)$ and $n(\infty)$ are carrier concentrations in regions 1 and 2, respectively, far enough away from the heterojunction that it does not disturb the equilibrium bulk carrier concentration, and $N_C(0)$ and $N_C(\infty)$ are the conduction band density of states in regions 1 and 2 which depend on the effective masses of each semiconductor. Thus, by measuring $\hat{n}(x)$, ΔE_C and σ_i can be determined if the donor doping profile $N_D(x)$ and the effective masses in each region are known. KROEMER has emphasized the importance of self-consistency checks of experimental data. It is essential to calculate what experimental C-V curves should look like from the band offset data and compare to experiment. Such a calculation is shown in Fig. 8 and agrees well with experimental data. The value of ΔE_C derived from C-V profiling depends sensitively on the choice of x_i. Deep levels in either semiconductor can change N_D, particularly at low temperatures [13].

KROEMER et al. [5] first used C-V profiling on GaAs/Al$_{0.2}$Ga$_{0.8}$As heterostructures and found $\Delta E_C/\Delta E_G \sim 0.66$. However, this did not agree with the accepted value of $\Delta E_C/\Delta E_G$, 0.85; they attributed the discrepancy to compositional grading during sample growth. KROEMER [19] has since shown that compositional grading would not affect the experimental value of ΔE_C. WATANABE et al. [13] have measured both n-N and p-P heterojunctions ($0.15 \leq x \leq 0.3$) by C-V profiling. They find $\Delta E_C/\Delta E_G = 0.62$ for this composition range, in remarkably close agreement with [9], [10] and [15]. PEOPLE et al. [20] find $\Delta E_C/\Delta E_G = 0.71$ for In$_{0.52}$Al$_{0.48}$As/In$_{0.53}$Ga$_{0.47}$As heterojunctions. It is not clear how reliable this value is since they checked the accuracy of their method by measuring GaAs/Al$_x$Ga$_{1-x}$As heterojunctions and found $\Delta E_C/\Delta E_G = 0.88$ for the latter system. This agrees well with the accepted value at the time but not with recent results. FORREST et al. [21] find $\Delta E_C/\Delta E_G \sim 0.4$ for a number of lattice-matched compositions in the InP/InGaAsP heterojunction system.

4. OTHER METHODS

Two other structures have been used for electrical measurements of $\Delta E_C/\Delta E_G$ for the GaAs/Al$_x$Ga$_{1-x}$As system. WU and YANG [6] obtained $\Delta E_C/\Delta E_G$=0.64 by measuring temperature dependence of I-V curves of N-p heterojunctions. Their value did not agree with the then accepted value, $\Delta E_C/\Delta E_G$=0.85, so they assumed that the composition of the Al$_x$Ga$_{1-x}$As in their structures was not the expected value. WANG et al. [8,22] used modulation-doped p-Al$_x$Ga$_{1-x}$As/GaAs heterojunctions to study the valence-band discontinuity. They measured hole carrier density in a conducting channel at the GaAs/Al$_x$Ga$_{1-x}$As interface as a function of the undoped Al$_x$Ga$_{1-x}$As spacer thickness between the heavily doped Al$_x$Ga$_{1-x}$As layer and the GaAs substrate. For x=0.5 they find ΔE_V=0.21±0.03 eV; for x=1, they find ΔE_V=0.45±0.05 eV, in good agreement with BATEY and WRIGHT [15].

5. CONCLUSION

Electrical measurements on a variety of heterostructures can provide consistent measurements of $\Delta E_C/\Delta E_G$ and $\Delta E_V/\Delta E_G$ for GaAs/Al$_x$Ga$_{1-x}$As heterostructures. Of the four methods presented, the combination of C-V and I-V measurements on SIS capacitors as a function of temperature appears to be most accurate over the whole Al composition range. Charged defects in the system, which can give spurious results if they are not taken into account, can be determined by the method. Confidence in the value of the band discontinuity for any heterostructure depends on obtaining consistent values by several different methods.

6. REFERENCES

1. H. Kroemer, Surface Science **132**, 543 (1983).
2. H. Kroemer, Surface Science, in press (1986).
3. G. Duggan, J. Vac. Sci. Technol. B **3**, 1224 (1985).
4. R. Dingle, W. Wiegmann and C. H. Henry, Phys. Rev. Lett. **33**, 827 (1974). R.Dingle, *Festkörperprobleme/Advances in Solid State Physics*, edited by H. J. Queisser (Vieweg, Braunschweig, 1975), Vol 15, p. 21.
5. H. Kroemer, W. Y. Chien, J. S. Harris, Jr. and D. D. Edwall, Appl. Phys. Lett. **36**, 295 (1980).
6. C. M. Wu and E. S. Yang, J. Appl. Phys. **51**, 2261 (1980).
7. R. C. Miller, D. A. Kleinman and A. C. Gossard, Phys. Rev. B **29**, 7085 (1984).
8. W. I. Wang, E. E. Mendez and F. Stern, Appl. Phys. Lett. **45**, 639 (1984).
9. T. W. Hickmott, P. M. Solomon, R. Fischer and H. Morkoç, J. Appl. Phys. **57**, 2853 (1985).
10. J. Batey, S. L. Wright, and D. J. DiMaria, J. Appl. Phys. **57**, 484 (1985).

11. D. Arnold, A. Ketterson, T. Henderson, J. Klem and H. Morkoç, J. Appl. Phys. **57**, 2880 (1985).

12. H. Okumura, S. Misawa, S. Yoshida, and S. Gonda, Appl. Phys. Lett. **46**, 377 (1985).

13. M. O. Watanabe, J. Yoshida, M. Mashita, T. Nakanisi, and A. Hojo, J. Appl. Phys. **57**, 5340 (1985).

14. G. Duggan, H. I. Ralph and K. J. Moore, Phys. Rev. B **32**, 8395 (1985).

15. J. Batey and S. L. Wright, J. Appl. Phys. **59**, 200 (1986).

16. E. H. Nicollian and J. R. Brews, *MOS Physics and Technology*, (J. Wiley, New York, 1982), p. 385.

17. H. C. Casey and M. B. Panish, *Heterostructure Lasers: Part A; Fundamental Principles*, (Academic Press, New York, 1978).

18. H. Kroemer and W. Y. Chien, Solid-State Electron. **24**, 655 (1981).

19. H. Kroemer, Appl. Phys. Lett. **46**, 504 (1985).

20. R. People, K. W. Wecht, K. Alavi, and A. Y. Cho, Appl. Phys. Lett. **43**, 118 (1983).

21. S. R. Forrest, P. H. Schmidt, R. B. Wilson, and M. L. Kaplan, Appl. Phys. Lett. **45**, 1199 (1984).

22. W. I. Wang and F. Stern, J. Vac. Sci. Technol. B3, 1280 (1985).

Heuristic Approach to Band-Edge Discontinuities in Heterostructures

H. Heinrich[1] *and J.M. Langer*[2]

[1]Institut für Experimentalphysik, Johannes Kepler Universität Linz,
 A-4040 Linz, Austria
[2]Institute of Physics, Polish Academy of Sciences, Al. Lotnikôw 32/46,
 PL-02-668 Warsaw, Poland

After a brief presentation of the various experimental techniques of band-
-offset determination in heterojunctions, we present our recent heuristic
procedure for band-offset prediction. According to it, the valence-band-
-edge discontinuity in heterojunctions is given by the difference in the
energy level positions of a transition metal deep impurity in the two compounds forming the heterojunction.

1. Introduction

Device engineering and the physics of heterojunctions (HJ's) requires a
fairly accurate knowledge of the most important parameter in these structures,
namely the band-edge offset (BEO). The interface barriers in HJ's resulting
from the BEO's are vital for most devices employing HJ's, since all the
electrical transport in HJ's is governed by them. There exist at least two
extreme approaches to this problem. One, very empirical, tests the possibility of attaining a reasonable accuracy of predictions of the BEO's in HJ's
on a purely theoretical basis [1]. The other, favoured by us, opts for a
search for a proper (and, if possible, not very complicated) approach yielding the values of the HJ barrier heights with enough accuracy. The main
argument for a purely empirical approach (make a HJ and measure the barrier
heights) is the possible dependence of HJ properties on the way it was made
(crystallographic orientation, grading, strains, etc.). The large scatter
of the data on the HJ barrier heights may favour such an approach. There is,
however, a great deal of evidence that the scatter and even irreproducibility
of the data is a simple consequence of the still insufficiently mature enough
technology of preparation of a given HJ, and also of improper experimental
techniques used to measure the barrier height. The GaAs/Ga$_{1-x}$Al$_x$As HJ story
may serve as the best illustration of this statement. For example, there is
a report [2] on a dependence of the BEO on the growth sequence, while, very
recently, quite opposite behaviour was found [3]. It is quite likely that
the earlier results are a consequence of the antiphase disorder at the interface [4]. Although the experimental consensus on the value of the BEO in
some technologically well-developed HJ's does not seem to be far away, there
is still an urgent need for an accurate procedure of predicting the "canonical" BEO's in HJ's. All the "first principles" calculations cannot serve
this purpose, due to their inherent 0.1 - 0.2 eV accuracy limit, typical for
current band-structure calculations. The only alternative is thus a heuristic
approach to this problem. The earliest approach of this kind is the Anderson-
Shockley Electron-Affinity Rule [5]. This states that the band offset is
just the difference of the electron affinities χ of the two constituents of
a HJ. Even if this rule were the correct one, the largely unknown surface

corrections to the measured values of χ have made it impractical (see e.g. the illuminating discussion of this approach by J. VAN VECHTEN [6]). Much more justified is the KATNANI-MARGARITONDO [7] approach. Measuring the differential external photothresholds of HJ's formed by Si or Ge overlayers on various semiconductors, they have obtained the relative BEO's for the various semiconductors from which, in principle, any HJ BEO could be obtained. Even if the HJ's grown by them were of device quality, the measuring technique suffers from its accuracy limit of, on average, 0.2 eV. It may be quite useful, however, in predicting the chemical trends in BEO's.

Very recently we [8], and independently ZUNGER [9], have proposed a novel empirical approach, which utilizes transition metal (TM) deep impurity energy levels as a reference in the band alignment in HJ's.

Before proceeding to a description of this procedure and its predictions, a brief summary of the methods of BEO determination is appropriate. The GaAlAs/GaAs lesson (when after many sophisticated measurements in the last two years, the old widely accepted DINGLE [10] value of the BEO ratio E_{cb}/E_{vb} = 85:15 has effectively been replaced by a more balanced 3:2) has taught us that one must be very cautious in taking any other measured discontinuity as a well-established, or even the correct one. For a value of a BEO to qualify as belonging to even the secondary-test class, several restrictive conditions must be fulfilled. One concerns the growth technology. As is well established now, the device quality is of primary importance. Moreover, junction abruptness and lack of grading (including impurity distribution) are prerequisites for most experimental techniques of band-offset determination. The above conditions favour all the lattice matched combinations. A non-lattice matched HJ may also be used, but then a way should be found of estimating the contribution from the inherent strain in these structures to the relative band shifts.

The reliability of the band-offset determination depends very much on the choice of the measurement technique. Again, as learned from GaAlAs/GaAs, among the techniques more immune to interpretation pitfalls are the Kroemer-type [11] C(V) measurements on isotype HJ's and internal photoemission [12] in single HJ's. There are several ways of estimating the BEO's by employing single or multiple quantum wells (QW). Most of them, especially absorption, rely very much on identification of the observed structure, as well as on the values of the effective masses used in the fitting procedure. The spatial potential profile of the QW also influences the accuracy of the band-offset determination [13]. Triangular and parabolic QW were shown [14] to serve this purpose much better than rectangular QW's used e.g. by DINGLE et al. [10]. (The problem of the influence of the measurement technique on the accuracy of band-offset determination has been the subject of several recent reviews [15-17] and we would like to refer the reader to them for further details and a critical analysis.)

2. Transition-Metal Impurity Levels in Semiconductors

Transition metals (TM) are known to form localized deep impurity states in semiconductors. Their properties are markedly different from those of the defects not possessing unfilled d (or f) electronic shells. They are usually magnetically active and may possess several stable charge states in semiconductors. Many of their properties (internal absorption, magnetism, spin-scattering) indicate the still d-like character of their wave function [18],

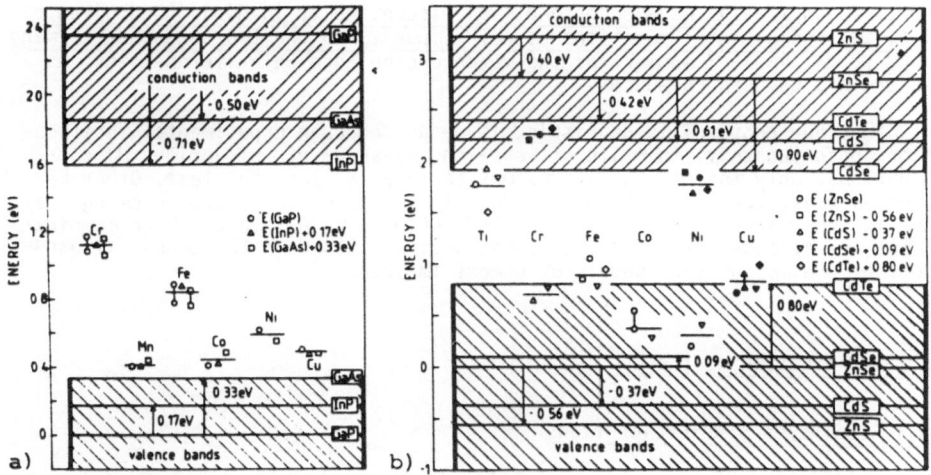

Fig. 1.(a) Average energy levels of TM acceptors (-/0) in GaP, InP and GaAs relative to the top of the valence band of GaP. They were determined by the appropriate vertical shift of the experimental TM energy pattern in all three compounds so as to minimize the overall mean square deviations. (b) The average energy levels of TM donors (0/+) (open symbols) and acceptors (-/0) (filled symbols) in II-VI compounds obtained in the same way as in (a). All band-edge shifts are relative to the ZnSe band edges [8].

even in compounds as covalent as Si or III-V compounds. This, in turn, leads to the conclusion that to construct a proper impurity wave function even several of the nearby host bands may not be enough. Therefore, the ground energy levels of TM's cannot, in principle, follow the nearby band edges, but instead, some other more global characteristic reference level would have to be used. This remark is consistent with the observed preservation of the relative ordering of the ionization energies for a whole series of TM's when the host is changed, provided they are isovalent (Fig. 1).

The observed alignment suggests the existence of a common bulk reference level for them in isovalent semiconductor compounds [8]. This observation has led us to postulate that the same reference level can be used for band alignment in a HJ made of a pair of isovalent compounds (i.e. III-V on III-V

Table 1. Relative energies of the valence-band edges of II-VI and III-V compounds as determined from TM ionization energies [19].

III-V	II-VI
GaAs + 0.33 eV	CdTe + 0.80 eV
InP + 0.17 eV	CdSe + 0.09 eV
GaP 0	HgSe 0-0.2 eV
AlAs - 0.12 eV	ZnSe 0
	CdS - 0.37 eV
	ZnS - 0.56 eV

or II-VI on II-VI). A valence band (vb) discontinuity in HJ is then given by just the energy level positions of a TM impurity in the two compounds forming a HJ [8,9]. Table I summarizes the valence-band edge relative energies, as determined by our procedure.

Since our proposal is heuristic, it needs thorough experimental verification. Unfortunately, in spite of year-long measurements of various HJ combinations, only the GaAlAs/GaAs HJ may serve as a clear-cut test. Other HJ combinations, for which the device-quality HJ structures start to emerge, are the lattice-matched InGaAsP/InP and CdHgTe/CdTe HJ's. All these three combinations are already used in device construction, therefore the canonical band-offset prediction for them is of utmost importance.

3. Verification of the Model

3.1 GaAlAs/GaAs Heterojunctions

Until 1984, the 85:15 - $\Delta E_{cb}:\Delta E_{vb}$ band-edge-discontinuity ratio was considered to be well established and was used in device modeling [20] as well as in testing various band-offset theories [15]. A flood of new results, published during just the last two years, has definitely proved that the BEO ratio $\Delta E_{cb}:\Delta E_{vb}$ is close to 3:2. Unfortunately, there are no published data on the TM energy levels in AlAs. There exist, however, very precise DLTS [21] and photocapacitance [22] data on the Fe^{2+} acceptor level in $Ga_{1-x}Al_xAs$ bulk crystals. The composition dependence of this level is given by

$$E_{Fe} - E_{vb} = 0.516 \text{ eV} + (0.453 \pm 0.011)x \text{ eV}.$$

According to our model, the x-dependent part of the energy level position is just the valence-band offset between GaAs and $Ga_{1-x}Al_xAs$. Since for the x < 0.4 region the energy gap change is: $\Delta Eg = 1.247x$ eV, the predicted band-edge discontinuities between GaAs and AlAs are then

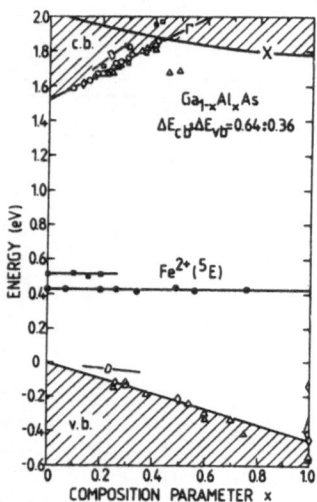

Fig. 2. $Ga_{1-x}Al_xAs$ band edges measured relative to the Fe-acceptor level (■ - [19] and ● - [20]). The BEO's were obtained either from MQW optical properties (□ and D), or HJ's properties ○ - C(V) by Kroemer method [11], ◇ - 2-Dim. properties of HJ, △ - J(V,T), ◁ - tunneling and photoemission for x = 1 [2], [8].

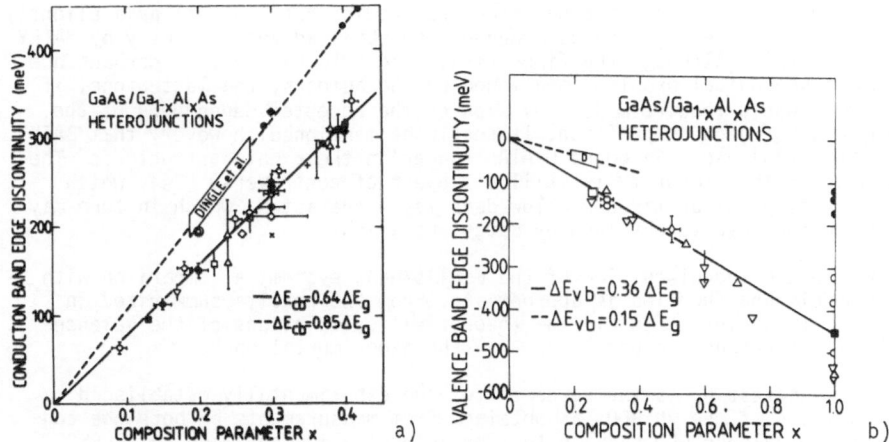

Fig. 3. The conduction- (a) and valence- (b) band-edge discontinuities in GaAs/Ga$_{1-x}$Al$_x$As HJ's. The solid line is the prediction of our model. For the original data sources see Ref. [8].

Table 2. Valence-band discontinuity (ΔE_{vb}) in GaAs/AlAs heterojunctions.

E_{vb}[eV]	Method	Year	Ref.
0.04	LCAO parametric, bulk	1977	[25]
0	pseudopotential, bulk	1977	[26]
0.25	SC, first principle, HJ	1978	[27]
0.28	SC, first principle, HJ	1982	[28]
0.35	SC, first principle, bulk	1984	[29]
0.4	revised Ref. [30]	1985	[6]
0.6 ± 0.2	UPS of AlAs/Ge and GaAs/Ge HJ	1985	[31]
0.48	Al/AlAs and Al/GaAs	1985	[32]
0.45 ± 0.01	TM energy levels in GaAlAs	1985	[8]
0.45	exp. data on ΔE_{cb} and ΔE_{vb}	summarized in Fig. 3	

$\Delta E_{cb} = (0.79 \pm 0.01) \text{ eV} = 64\% \ \Delta Eg,$

$\Delta E_{vb} = (0.45 \pm 0.01) \text{ eV} = 36\% \ \Delta Eg.$

Using these predictions, we constructed a full Ga$_{1-x}$Al$_x$As band diagram, in which the Fe^{2+} level was taken as the reference level independent of x, as shown in Fig. 2. All the data available to us are grouped here by the measurement technique used. The excellent fit of our model predictions is clearly seen. To facilitate a closer look at it we have also grouped all the data into two separate diagrams (Fig. 3) for the conduction- and valence- -band discontinuities, respectively. From these diagrams it is clear that in the direct gap region of GaAlAs there is essentially no discrepancy between our model and the data for both the conduction and valence bands. Quite puzzling however, are the very recent [12] results on ΔE_{cb} for x ≒ 0.5, strongly deviating from the band structure predictions, as well as a slightly larger slope of the ΔE_{vb} versus x dependence obtained very recently by BATEY and WRIGHT [23]. Although the first result seems to be erratic, perhaps due to the technological problems mentioned by the authors, the latter one, if confirmed, would require major revision of the accepted dependence of the energy gap $E_g(x)$ on composition. It should be mentioned, however, that BATEY and WRIGHT [23] use a Si doped GaAlAs spacer in their heterostructures. The behaviour of this impurity is still a subject of controversy [24], which reflects its peculiar mixed shallow-deep level character, which in turn may influence the results obtained by Batey and Wright.

To complete the discussion of the GaAlAs/GaAs system, a comparison with other models and theories is appropriate. From the results summarized in Table 2, it is clear that all the theoretical calculations of the valence-band offset produce values lower than the experimental ones.

Quite interesting is the agreement of the experimentally established 0.45 eV value of the vb BEO (as obtained from measurements on both the conduction and valence BEO's) with the Fermi-level alignment model of TERSOFF [29] extensively used by WANG and STERN [32].

The model, which will be discussed in the last section, proposes the use of the Schottky-barrier-height measurement for vb alignment. It assumes that the metal Fermi level may serve as the reference level in the same fashion as TM energy levels used by us. The recent measurements by WANG [33] on the Schottky-barrier heights of Al/p-type GaAs and Al/p-type AlAs junctions, as well as the older values of the Schottky-barrier height of Au contacts on n-type GaAs and n-type AlAs yield the vb offset between GaAs and AlAs equal to 0.48 eV and 0.42 eV, respectively. It should be noted, however, that the measurements of BEST [34] of the Schottky-barrier heights in the Au/n-type $Ga_{1-x}Al_xAs$ junctions for x < 0.8 give a larger vb offset of about 0.6 eV.

3.2 Lattice matched GaInAsP/InP Heterojunctions

The next class of HJ's, of rapidly growing importance for semiconductor optoelectronics, consists of lattice matched GaInAsP/InP HJ's. The spectral range of the light-emitting diodes and lasers made of them covers the region around 1.5 μm, a low-loss and low-dispersion region of contemporary light-guides. In contrast to GaAlAs/GaAs, there are far fewer studies of HJ band offsets [35-37]. As could be expected, their results do not agree. For the end member of this family, i.e. in the $Ga_{0.47}In_{0.53}As/InP$ HJ, the $\Delta E_{cb}:\Delta E_{vb}$ estimations vary from 60:40 [35] through 50:50 [36] to 40:60 [37].

The straightforward approach of our method is not possible, since there are no extensive measurements of TM's either in the GaInAsP quaternaries, or even in InAs. The only data available to us are on Mn in $Ga_{0.47}In_{0.53}As$, where the Mn acceptor level is 0.052 eV above the vb edge [38]. A comparison of this value with the 0.22 eV Mn binding energy [39] in InP and 0.12 eV in GaAs for the same quantity, yields $E_{vb} < 0.22$ eV for the $Ga_{0.47}In_{0.53}As/InP$ lattice matched system - the criterion passed by all the cited HJ discontinuity measurements. It should be remembered, however, that Mn in $Ga_{0.47}In_{0.53}As$ is most probably an effective-mass shallow acceptor, and thus the procedure outlined above gives only the lower boundary of the vb offset. This hydrogenic effective-mass contribution to the Mn acceptor states in GaAs and in GaAlAs with low Al content is most likely the reason for the composition dependence of the Mn binding energy in GaAlAs being weaker than that of Fe and Cu, thus making this impurity not very suitable for BEO estimates.

The data of FORREST et al. [35] and BRUNEMEIER et al. [37] for the whole accessible composition range can, however, be used in cross-checking the values of vb energies summarized in Table 1. The lattice-matching condition for the $Ga_xIn_{1-x}As_yP_{1-y}/InP$ HJ is fulfilled only when $x = 0.453y (1 + 0.03y)$ [37]. In the linear approximation, the vb offset between $Ga_xIn_{1-x}As_yP_{1-y}$ and InP is given by the relation

$$E_{vb}(x,y) = E_{vb}^{o}(GaAs)xy + E_{vb}^{o}(InAs)y(1-x)$$

$$+ E_{vb}^{o}(GaP)x(1-y) - E_{vb}^{o}(InP) (x+y-xy).$$

A least-squares fit of the above equation to the experimental data [35,37] with $E_{vb}^{o}(GaAs)$ and $E_{vb}^{o}(InAs)$ treated as the fitting parameters yields the sets of these parameters given in Table 3.

Table 3. Valence-band-edge energies of InAs and GaAs relative to GaP as obtained from the BEO trends in the lattice-matched GaInAsP/InP HJ's

Compound	Data of [35]	Data of [37]	TM Model
InAs	(0.66 ± 0.02) eV	(0.48 ± 0.01) eV	>0.45 eV
GaAs	(0.38 ± 0.07) eV	(0.33 ± 0.04) eV	0.33 eV

The data of BRUNEMEIER et al. [37] are closer not only to our model, but also to the value of BEO in InAs/GaAs HJ measured by the XPS technique [40] and equal to (0.17 ± 0.07) eV.

3.3. HgCdTe/CdTe Heterojunctions

The third family of lattice matched HJ's consists of narrow gap, or even zero gap HgCdTe (or HgCdSe) grown on CdTe (or CdSe) substrates. Two contradictory values of BEO's have been reported. One result, obtained from conventional J(V) or internal photoemission measurements for HgSe/CdSe [41] and HgTe/CdTe [42] HJ's, indicates the vb offset to be approximately equal to half the gap of CdSe or CdTe. The second, derived from the optical properties of MQW's, indicates an almost perfect alignment [43] of the vb's, at least in Hg-rich HgCdTe/CdTe HJ's.

The available data on the TM's in these compounds suggest the vb offset to be close to zero. The only study of the properties of the TM impurities in Hg-based II-VI compounds is that of JONES et al. [44], who found that a Cu-related defect is pinned to the vb of $Hg_{1-x}Cd_xTe$ in the open-gap region (0.25 x 0.45). The other studies deal with the position of either Mn or Fe "3d-like levels" in the mixed compounds, in which both Mn and Fe are host constituents, not mere impurities. Studying the optical and electrical transport in HgCdTeSe crystals, for a Fe concentration not exceeding 15%, MYCIELSKI et al. [45] have found that Fe^{2+} in HgSe produces a resonant state about 0.2 eV above the bottom of the cb. In $Cd_{0.5}Hg_{0.5}Se$, this level is about 0.5 eV above the vb, as in CdSe [45]. Since resonance with the continuum of the cb states may produce downward shifts, these results would indicate, at most, a +0.3 eV vb offset for the HgSe/CdSe HJ. A very small offset can also be inferred from a comparative study [46] of the UPS of ZnMnSe, CdMnSe and HgMnSe, in which a similar Mn^{2+} related feature was found (3.5 ± 0.1) eV below the top of the vb in all three compounds. Unfortunately, the same study has not been performed, as yet, for Te-based compounds. Trends in E_{vb}^{o} for the other Te, S and Se compounds (see Table 1) allow us to predict a quite similar result for HgCdTe and, therefore, also a small, if any, vb offset in HgCdTe/CdTe HJ's.

4. Summary and Perspectives

The spectacular agreement between the prediction of the BEO's based upon the TM impurity levels in GaAs/AlAs HJ's and the experimental data, as well as agreement of our predictions with the most reliable determinations of this quantity in lattice matched HgCdTe/CdTe and GaInAsP/InP HJ's gives some credit to our procedure. It must be stressed again, however, that our model is strictly <u>heuristic</u>, and until now has no sound theoretical justification. The contemporary theory of TM's in semiconductors still cannot reproduce with sufficient accuracy (say 0.1 - 0.2 eV) the energy spectra of these impurities. Therefore, the "pinning" behaviour of TM impurities in semiconductors remains an unsolved puzzle. The model approach of CALDAS et al. [47] gives only a guideline for further work, and can hardly be treated as quantitative. Equally good would be a justification based on noting that the peculiar electronic structure of TM's must derive from the whole band structure. Therefore, the constant reference level should rather reflect some average band energy level. Before our proposal, several authors [48, 9] had suggested a pinning of TM's to the vacuum level. Fig. 4, which summarizes the relative positions of the vb in the III-V and II-VI compounds, as obtained from our TM averaging procedure and from the differences in the electron work functions, clearly shows that the latter gives differences between the vb positions which are too large by about 25%.

It is quite tempting to extend our band-alignment procedure to other types of junctions, especially to metal-semiconductor (MS) structures. Such a proposal is an immediate consequence of the suggestion made by TERSOFF [29], and then extended by WANG and STERN [32], that vb alignment in HJ's and MS junctions have a common origin. He has postulated that the metal-Fermi level

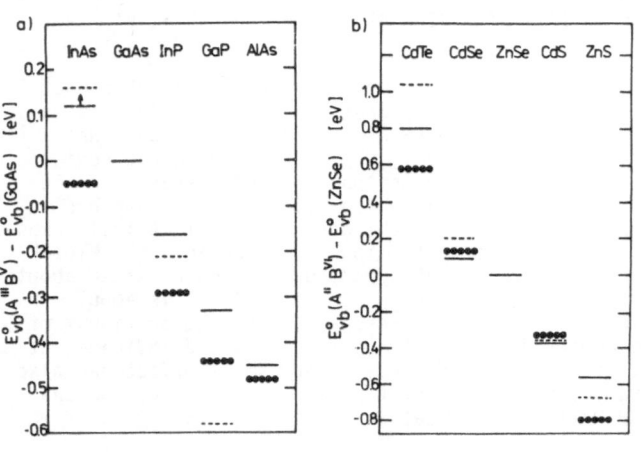

Fig. 4. The relative vb position in the III-V compounds, as obtained from TM averaging (—), differences in external work functions (---) and Au-Schottky junctions (···). For the latter the barrier heights were taken from [49].

may be used to align the vb's in a HJ. (This means that $\Delta E_{vb} = \Delta\phi_p$ and $\Delta E_{cb} = \Delta\phi_n$, where $\Delta\phi_{p,n}$ is the difference in the barrier heights for the same metal, for a pair of n- or p-type semiconductors forming a HJ.) This proposal works surprisingly well for GaAs/GaAlAs HJ, if the data of BEST [34] on the Au/GaAlAs MS junctions are disregarded (see Sect. 3.1 for discussion). Fig. 4 also summarizes the relative vb positions as obtained from this alignment procedure, but employing the Au-MS barrier height data tabulated by SZE [49]. Although the chemical trends are surprisingly well preserved, the quantitative correlation between the TM data and the Au-MS data does not allow, in our opinion, any firm conclusions to be drawn as to what kind of states cause the Fermi-level pinning in MS junctions. Moreover, at least for some II-VI compounds, the whole procedure is hardly justifiable, since it assumes that the MS barrier height is independent of the metal, which is not the case for most of them [49]. If the metal-induced gap states [50] were the source of the observed pinning in MS junctions, alternative use of MS or HJ barrier data would be justified. If these pinning states were of defect type [51], the success of this procedure would imply either the same type of pinning in the HJ as in the MS junctions, which is not likely according to most of the data on HJ's, or the chemical trends of the pinning defects should reflect some average energy level in a semiconductor, which is again not likely for sp^3 bonded type defects [52], considered to be responsible for the pinning behaviour in the III-V compounds [53]. The current debate on the origin of this effect is quite vigorous and full of opposing opinions, hence a more thorough study of the possible threefold correlation between the TM energy levels, HJ band offsets and MS barrier height may provide the key to the intricacies of the physics of these systems.

Acknowledgements

The authors sincerely acknowledge discussions with many colleagues whose names are listed in our original papers [8, 19]. There we would like to thank especially I. Lindau and J. Tersoff for sending us preprints relevant to the discussion on the MS barrier origin. The editorial help of K. Lesniak and E. Wieckowska is sincerely acknowledged. One of us (H. Heinrich) would like to acknowledge the support by the Austrian Ministry for Wissenschaft und Forschung.

References:

1. R.S. Bauer and H.W. Sang Jr.: Surf. Sci. 132, 479 (1983)
2. J.R. Waldrop, S. Kowalczyk, R.W. Grant, E.A. Kraut and D.C. Miller: J. Vac. Sci. Technol. 19, 573 (1981)
3. W.I. Wang, T.S. Kuan, E.E. Mendez and L. Esaki: Phys. Rev. B31, 6890 (1985)
4. R.W. Grant, J.R. Waldrop, S.P. Kowalczyk and E.A. Kraut: J. Vac. Sci. Technol. B3, 1295 (1985)
5. R.L. Anderson: Solid State Electron. 5, 341 (1962)
6. J.Van Vechten: J. Vac. Sci. Technol. B3, 1240 (1985)
7. A.D. Katnani and G. Margaritondo: Phys. Rev. B28, 1944 (1983)
8. J.M. Langer and H. Heinrich: Phys. Rev. Lett. 55, 1414 (1985)
9. A. Zunger: proposed during 2nd Brazilian School on Semiconductors, 1985 and in Solid State Physics (submitted for publication)
10. R.D. Dingle, A.C. Gossard and W. Wiegmann: Phys. Rev. Lett. 34, 1327 (1975), see also R. Dingle in Festkörperprobleme, edited by H.J. Queisser, Vol. 15 (Vieweg, Braunschweig, 1975), p. 21

11. H. Kroemer, W.-Y. Chen, J.S. Harris and D.D. Edwall: Appl. Phys. Lett. 36, 295 (1980); H. Kroemer, ibid 46, 505 (1985), and D.I. Babic and H. Kroemer: Solid State Electron. 28, 1015 (1985)
12. M.I. Heiblum, M.I. Nathan and M. Eizenberg: Appl. Phys. Lett. 47, 503 (1985)
13. a) W. Pötz and D.K. Ferry: Phys. Rev. B32, 3863 (1985)
    b) W. Porod, W. Pötz and D.K. Ferry: J. Vac. Sci. Technol. B3, 1290 (1985)
14. R.C. Miller, A.C. Gossard and D.A. Kleinmann: Phys. Rev B32, 5443 (1985)
15. H. Kroemer: Surf. Sci. 132, 543 (1983), and in Proceedings of the NATO Advanced Study Institute on Molecular Beam Epitaxy and Heterostructures, Erice, Sicily, 1983, edited by L.L. Chang and K. Ploog (Martinus Nijhoff, The Netherlands, 1984), p. 331, and J. Vac. Sci. Technol. B2, 433 (1984)
16. G. Duggan: J. Vac. Sci. Technol. B3, 1224 (1985)
17. H. Heinrich and J.M. Langer: Festkörperprobleme, edited by P. Grosse (Vieweg, Braunschweig, 1986) (to be published)
18. P. Vogl: in Festkörperprobleme, edited by P. Grosse (Vieweg, Braunschweig 1985), Vol. 25, p. 563
19. J.M. Langer and H. Heinrich: Physica 134B, 44 (1985) and Phys. Rev. (to be published)
20. F. Capasso: Surf. Sci. 132, 527 (1983), and ibid 142, 513 (1984)
21. D.V. Lang, R.A. Logan and L.C. Kimerling: in Physics of Semiconductors, Proc. of the 13th Int. Conf., Rome 1976, edited by F.G. Fumi (Tipografia Marres, Rome 1974), p. 615
22. Z.G. Wang, L.A. Ledebo and H.G. Grimmeiss: J. Appl. Phys. 56, 2762 (1984)
23. J. Batey and S.L. Wright: J. Appl. Phys. 59, 200 (1986)
24. N. Chand, T. Henderson, J. Klem, W.T. Masselink, R. Fisher, Y. Chang and H. Morkoc: Phys. Rev. B30, 4481 (1984); E.F. Schubert and K. Ploog, ibid B30, 7021 (1984), and Appl. Phys. A33, 63 (1984)
25. W.A. Harrison: J. Vac. Sci. Technol. 14, 1016 (1977), and ibid B3, 1231 (1985); see also E.A. Kraut, ibid B2, 486 (1984)
26. W.R. Frensley and H. Kroemer: Phys. Rev. B16, 2642 (1977)
27. W.E. Pickett, S.G. Louie and M.L. Cohen: Phys. Rev. B17, 815 (1978)
28. J. Sánchez-Dehesa and C. Tejedor: Phys. Rev. B26, 5824 (1982)
29. J. Tersoff: Phys. Rev. B30, 4874 (1984)
30. J. Van Vechten: Phys. Rev. 182, 891 (1969), and ibid 187, 1007 (1969)
31. M.K. Kelly, D.W. Nites, E. Colavita, G. Margaritondo and M. Henzler: Appl. Phys. Lett. 46, 768 (1985)
32. W.I. Wang and F. Stern: J. Vac. Sci. Technol. B3, 1280 (1985)
33. W.I. Wang: J. Vac. Sci. Technol. B1, 574 (1985)
34. J.S. Best: Appl. Phys. Lett 34, 522 (1979)
35. S.R. Forrest, P.H. Schmidt, R.B. Wilson and M.L. Kaplan: Appl. Phys. Lett. 45, 1199 (1984)
36. H. Temkin, M.B. Panish, P.M. Petroff, R.A. Hamm, J.M. Vandenberg and S. Sumski: Appl. Phys. Lett. 47, 394 (1985)
37. P.E. Brunemeier, D.G. Deppe and N. Holonyak, Jr.: Appl. Phys. Lett. 46, 755 (1985)
38. P.S. Whitney and C. Fonstad: J. Appl. Phys. 57, 4663 (1985)
39. L. Eaves, A.W. Smith, M. Skolnick and B. Cockayne: J. Appl. Phys. 53, 4955 (1982)
40. S.P. Kowalczyk, W.J. Schaffer, E.A. Kraut and R.W. Grant: J. Vac. Sci. Technol. 20, 705 (1982)
41. J.S. Best and J.O. McCaldin: J. Vac. Sci. Technol. 16, 1130 (1979)
42. T.F. Kuech and J.O. McCaldin: J. Appl. Phys. 53, 3121 (1982)
43. Y. Guldner, G. Bastard, J.P. Vieren, M. Voss, J.P. Faurie and A. Millon: Phys. Rev. Lett. 51, 907 (1983); D.J. Olego, J.P. Faurie and P.M. Racah: ibid 55, 328 (1985)

44. C.E. Jones, V. Nair, J. Lindquist and K.L. Polla: J. Vac. Sci. Technol. 21, 187 (1982)
45. A. Mycielski, P. Dzwonkowski, B. Kowalski, B.A. Orlowski, M. Dobrowolska, M. Arciszewska, W. Dobrowolski and J.M. Baranowski: J. Phys. C (in press)
46. A. Franciosi, S. Chang, C. Caprile, R. Reifenberger and U. Debska: J. Vac. Sci. Technol. A3, 926 (1985), and A. Franciosi, G. Chang, R. Reifenberger, U. Debska and R. Riedel: Phys. Rev. B32, 6682 (1985)
47. M.J. Caldas, A. Fazzio and A. Zunger: Appl. Phys. Lett.45, 671 (1984)
48. L.A. Ledebo and B. Ridley: J. Phys. C 15, L961 (1982)
49. S.M. Sze: Physics of Semiconductor Devices (Wiley, New York, 1982)
50. J. Tersoff: J. Vac. Sci. Technol. B3, 1157 (1985) and Phys. Rev. Lett. 52, 465 (1984)
51. W.E. Spicer, N. Newman, T. Kendelewicz, W.G. Petro, M.D. Williams, C.E. McCants and I. Lindau: J. Vac. Sci. Technol. B3, 1178 (1985); A. Zunger: Thin Solid Films 104, 301 (1983); T. Kendelewicz and I. Lindau (to be published)
52. P. Vogl: in Festkörperprobleme, edited by J. Treusch (Vieweg, Braunschweig, 1981), Vol. 21, p. 191
53. O.F. Sankey, R.E. Allen, S.-F. Ren and J.D. Dow: J. Vac. Sci. Technol. 133, 1162 (1985); C.B. Duke and C. Mailhiot, ibid, p. 1170

Part III

Resonant Tunnelling, Multi-Quantum-Well and Superlattice Structures

Quantum Tunnelling of Electrons Through III-V Heterostructure Barriers

L. Eaves[1], *D.C. Taylor*[1], *J.C. Portal*[2], *and L. Dmowski* [2*]

[1]Department of Physics, University of Nottingham,
 Nottingham, NG7 2RD, England
[2]LPS, INSA, F-31077 Toulouse, France and
 SNCI-CNRS, F-38042 Grenoble, France

Abstract

The effective mass theory of tunnelling through heterostructure barriers and its limitations is outlined. Experimental investigations of the effect of hydrostatic pressure (up to 15 kilobar) and magnetic field (up to 11T) on the low temperature J(V) characteristics of single barrier $n^+GaAs/(AlGa)As/n^-GaAs/n^+GaAs$ tunnelling structures are reported. The pressure dependence is accurately described by the effective mass/WKB model up to 10 kilobar. At higher pressure the observed breakdown of the model indicates the onset of band structure effects associated with the higher (X) conduction band minima. The reduction of the tunnelling current in an applied magnetic field is discussed in terms of the effect of the diamagnetic energy in increasing the height of the potential barrier.

1. Introduction

The rapid advances in Molecular Beam Epitaxy (MBE) and Metal-Organic Chemical Vapour Deposition (MOCVD) technology have led to a resurgence of interest in superlattices and related devices [1,2]. In addition, several new types of three-terminal (transistor) device have emerged recently which rely on quantum mechanical tunnelling through only a limited number of barriers. For example, in the tunnelling hot electron transfer amplifier [3] based on GaAs/(AlGa)As, electrons are injected into a very short base region (300 Å wide) by tunnelling through a thin (120 Å) (AlGa)As region from an n^+GaAs emitter contact. In the Stark effect transistor [4], the emitter-collector current is switched on when the bias voltage V_{CE} brings the electron energies in the n^+GaAs emitter region into resonance with a bound state of a narrow (100 Å) quantum well which forms the collector of the device. The application of a voltage to the base region can drag the device off resonance and produce a transistor action.

These developments emphasise the need for a thorough understanding of the physics of quantum mechanical tunnelling through semiconductor heterostructure barriers. The model which is widely employed is the so-called effective mass theory of tunnelling coupled with the WKB approximation. In this article, we describe a series of experiments that we have carried out to investigate this theory. The experiments examine the effect of large hydrostatic pressures and high magnetic fields on the current-voltage (J(V)) characteristics of single barrier GaAs/(AlGa)As/GaAs heterostructures. The article is arranged as

---

* On leave from High-Pressure Research Center, Polish Academy of Sciences, 01-142, Warsaw, Poland.

follows: section 2 describes the main features of the effective mass theory of tunnelling and its limitations; the results of the high pressure and high magnetic field experiments are described in sections 3 and 4 respectively; section 5 draws some general conclusions from the work.

2. The Effective Mass Theory of Quantum Tunnelling and its Limitations

The basic idea of this model can be understood by reference to Figure 1a. Consider a beam of electrons of kinetic energy ε incident on a rectangular potential barrier of height V_0 and width b. According to the laws of classical mechanics the electrons cannot cross the barrier if $\varepsilon < V_0$. However, they can traverse the barrier, though with attenuation, by quantum mechanically tunnelling. The forms of the wave function in the three distinct regions shown in Figure 1a are given by the well-known equations

$$x < 0, \qquad \psi = e^{ikx} + re^{-ikx} ;$$

$$0 < x < b, \qquad \psi = de^{\kappa x} + ge^{-\kappa x} ; \qquad\qquad (1)$$

$$x > b, \qquad \psi = te^{ikx} ;$$

where $k = \dfrac{(2m\varepsilon)^{\frac{1}{2}}}{\hbar}$ and $\kappa = \dfrac{[2m(V_0 - \varepsilon)]^{\frac{1}{2}}}{\hbar}$

are the wave vector and decay constant of the wavefunction in the classically allowed and classically forbidden regions respectively and r, d, g and t are appropriate amplitudes. By satisfying the boundary conditions for ψ and $\frac{\partial \psi}{\partial x}$ at the interfaces $x = 0$ and $x = b$, it is easy to show that for low transmission barriers ($\kappa b \gg 1$) the tunnelling probability T is given by

$$T = t^* t \sim \frac{16\varepsilon}{V_0} \left[1 - \frac{\varepsilon}{V_0} \right] \exp(-2\kappa b) . \qquad\qquad (2)$$

In the effective mass theory of tunnelling through a semiconductor heterostructure barrier, (1) and (2) are employed to calculate the tunnelling probability with V_0 set equal to the conduction band discontinuity and with the free electron mass m_e replaced by an appropriate effective mass m^* for the tunnelling electron. The tunnel current is then calculated by an integration over the density of states. The value of m^* depends on the composition of the barrier. In the case of an (AlGa)As barrier, if the Al-mole fraction $x < 0.4$, then the (AlGa)As layer is a direct gap semiconductor. The electron effective mass ratio of the direct (Γ) conduction band minimum in $Al_xGa_{1-x}As$ is rather low and is given by $m^*(\Gamma)/m = 0.067 + 0.08x$ [5]. For $x > 0.4$ the indirect X-minima are lower in energy than the Γ-minima. These X-minima have a significantly larger effective mass ($m^*(X)/m \sim 0.5$). Provided that the Al-fraction is low and that the energy of the tunnelling electron is not too far below the edge of the direct conduction band minimum, it is a reasonable approximation to equate the tunnelling effective mass m^* to that of the Γ-minimum conduction band of the (AlGa)As. Hickmott and coworkers [6] and ourselves [7] have used this approach in analysing the current-voltage characteristics of single barrier GaAs/(AlGa)As/GaAs tunnelling structures, and have found that coupled with the WKB approximation it gives an accurate estimate of the tunnel current.

The basic idea of this simple "one band" model of tunnelling is shown in Figure 1b. Consider an electron tunnelling elastically from a travelling wave state of energy ε' in the conduction band of GaAs into the (AlGa)As using this

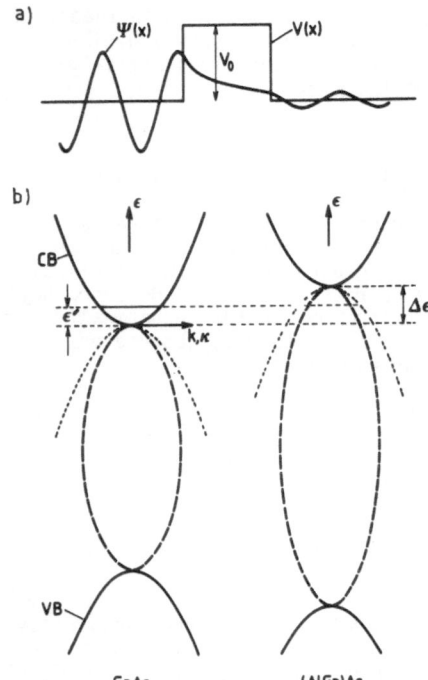

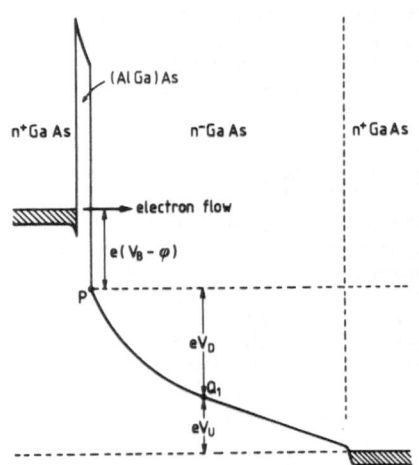

Figure 2 Schematic diagram (not to scale) showing the variation of the conduction band edge through the device under reverse bias conditions.

Figure 1(a) Tunnelling through a simple rectangular barrier (b) the form of the decay constant $\kappa(\varepsilon)$ in the "one band" (light dotted curve) and "two band" (heavy dashed curve) models for GaAs and (AlGa)As.

"one band" model. If it tunnels at an energy ε below the (AlGa)As edge, its decay constant $\kappa(\varepsilon) = (2m^*\varepsilon)^{\frac{1}{2}}/\hbar$ is a "mirror image" of the (AlGa)As conduction band ε-k curve for which $\kappa = (2m^*\varepsilon)^{\frac{1}{2}}/\hbar$. This is shown in the light dotted curve in the Figure. A more realistic picture of the form of $\kappa(\varepsilon)$ in the forbidden gap of a semiconductor should include the valence band (see for example [8]). Since propagating wave states occur in the valence band, the decay constant $\kappa(\varepsilon)$ which emerges from the edge of the conduction band must connect with the edge of the valence band as shown by the heavy dashed curve in Figure 1b. This type of "two band" model leads to a "non-parabolicity" of the decay constant $\kappa(\varepsilon)$. At a particular energy below the band edge, the decay constant is less than what would be expected from taking the tunnelling mass to be equal to the band edge mass. More precise models of tunnelling through a semiconductor barrier take into account not only the valence and conduction bands but the full band structure of the semiconductor. In the case of the GaAs/(AlGa)As system the evanescent waves associated with the high indirect conduction band minima near the L and X points of the Brillouin zone need to be considered. They decay much faster than those associated with the Γ-minimum and are hence most important for thin barriers. Descriptions of these band structure effects and the means of calculating them can be found in the recent literature [9, 10].

3. High-Pressure Studies of Electron Tunnelling through GaAs/(AlGa)As/GaAs Barriers

The structures used in our study were grown by Professor K.E. Singer (UMIST) using MBE. Some of their electrical properties (particularly those associated with hot electron effects) are described elsewhere [11-14]. Briefly, they consist of the following layers: 200 μm thick n$^+$GaAs substrate doped to 2 x 10^{18} cm^{-3}; 1 μm n$^+$GaAs buffer (collector), 2 x 10^{18} cm^{-3}; 1 μm n$^-$GaAs layer, 2 x 10^{15} cm^{-3}; 168 ± 10 Å of undoped (AlGa)As; 1 μm n$^+$GaAs top (emitter) layer, 2 x 10^{18} cm^{-3}. The Al concentration varies from 37% (substrate side) to 32 ± 2%. The structures were processed into the form of mesas with diameters of 200 or 800 μm. The variation of the conduction band edge through the device is shown schematically in Figure 2 under reverse bias conditions. The experiments were carried out at low temperatures, usually at liquid helium temperatures (2 to 4.2K) so that thermionic emission of electrons over the (AlGa)As barrier is negligible and the measured current J(V) is due only to quantum tunnelling of electrons. The large electric field in the depleted n$^-$GaAs region beyond the barrier ensures that effectively all of the tunnelling electrons are collected in the right-hand n$^+$GaAs contact.

The J(V) characteristics in reverse bias are shown in Figure 3 for a series of different pressures at a temperature of 4.2K. In the inset of the figure, the low reverse bias characteristics are shown in more detail. As we reported recently [11], for a given reverse bias V and pressures less than 10 kilobar (= 1 GPa), the current falls exponentially as the pressure is increased according to the relation

$$J(V, P) = J(V, 0)e^{-\gamma P}, \qquad (3)$$

where γ is a weakly dependent function of V. This behaviour is shown in Figure 4 which plots logJ versus P for various values of reverse bias and extends our previous measurement range up to 15 kilobar. It can be seen that for pressures above about 10 kilobar the linear relationship between logJ and

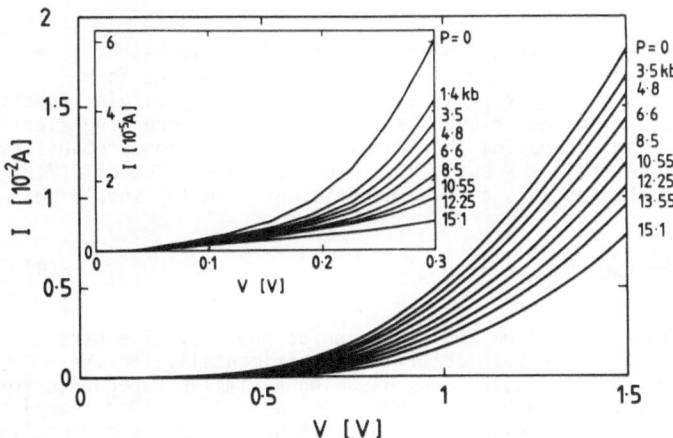

Figure 3 Reverse bias current-voltage characteristics of the single barrier structures (mesa diameter = 800 μm) for various applied hydrostatic pressures (T = 4K).

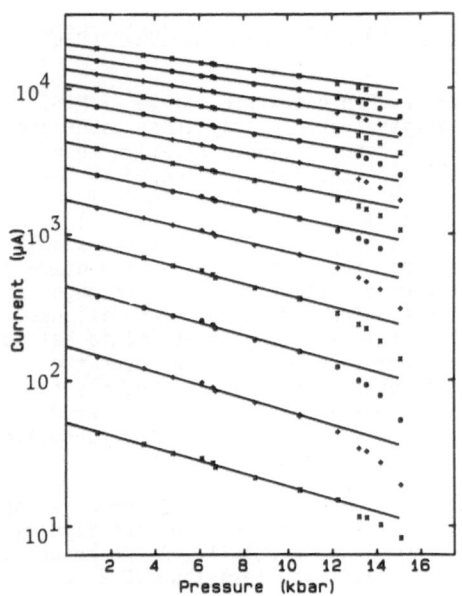

Figure 4 Logarithmic plot of current versus pressure for various fixed reverse bias voltages, T = 4K.

The reverse bias voltages are, in order of increasing current, 0.3 V to 1.5 V in 0.1 V increments.

P breaks down. At a pressure of 15 kilobar, the current is about a factor of 2 lower than what would be expected from a linear extrapolation of the low-pressure dependence.

The low-pressure dependence of $J(V,P)$ can be understood in terms of simple effective mass theory. In the WKB approximation, the tunnelling probability is given by

$$P = \exp[-2 \int_{0}^{b} \kappa(x)dx]$$

where $\kappa(x)$ is the decay constant in the barrier whose potential height is a function of x. Thus the current is given by a relationship of the form $J(V) \sim \exp[-\alpha f(V)]$ where $\alpha = 2b(2m^{*}V_{0}/\hbar^{2})^{\frac{1}{2}}$ and $f(V)$ is a slowly varying function of bias voltage. This function describes the effect that an increasing electric field in the barrier has in increasing the tunnelling transmission probability. At a fixed applied voltage $f(V)$ is a constant. Hence the variation of $J(V,P)$ with pressure is due to changes in α and the coefficient γ in (3) should be given by

$$. \; \gamma = \frac{-d}{dP}(\ell n \; J) = \frac{\alpha f(V)}{2}(\frac{d}{dP} \ell n \; m^{*} + \frac{d}{dP}\ell n \; V_{0}). \tag{4}$$

The main contribution to γ comes from the variation of the effective mass. The pressure dependence of m^{*} has been determined experimentally for the conduction band minimum of GaAs [15,16] and, assuming a similar dependence for the m^{*} appropriate to the tunnel barrier, gives $d\ell n m^{*}/dP = 7.4 \times 10^{-3}$ kbar^{-1}. The pressure dependence of the barrier height has not been determined directly but can be inferred from data on the variation of the energy gap E_{g1} in GaAs [16,17] and E_{g2} in (AlGa)As [17] if a constant band offset of 65% [12] is assumed i.e. $V_{0} = 0.65 (E_{g2} - E_{g1}) - E_{F}$ where E_{F} is the Fermi energy of the electrons in the n+ emitter contact. This gives $d\ell n V_{0}/dP \simeq -1.1 \times 10^{-3}$ kbar^{-1}.

Since $\alpha \simeq 27$ and $f < 1$, this leads to a theoretical value of the pressure coefficient $\gamma \simeq 0.0\overline{9}$ kbar^{-1} in reasonable agreement with the observed value. Note that as the bias voltage is increased, $f(V)$ decreases and slowly leads to the observed gradual change with bias in the slopes of the low pressure, linear sections of the plots in Figure 4.

We believe that the departure of the experimental data shown in Figure 4 from the linear dependence of logJ on P is evidence for the breakdown of the straightforward effective mass theory of tunnelling. This breakdown arises because the tunnelling phenomenon cannot be explained simply in terms of a wavefunction described by a single effective mass parameter m^*. The importance of band structure effects on tunnelling is described briefly in section 2. The application of large hydrostatic pressure is a particularly useful means of studying band structure effects on tunnelling since, in addition to increasing the direct energy gap and hence the Γ conduction band effective mass, it also decreases the separation of the Γ-X conduction band minima in both the GaAs and (AlGa)As layers. GaAs remains direct gap up to pressures of around 30 kilobar. However, for the composition Al = 35%, the (AlGa)As barrier becomes indirect gap at around 10 kilobar (for very recent measurements of this cross-over see [18]). This is almost precisely the pressure at which the data in Figure 4 deviates from the relationship described by (4). As the X-minima move towards the Fermi energy of the electrons in the n$^+$ emitter layer (essentially the energy of the tunnelling electrons), one can expect an increasing contribution of the X-minima both to the predominantly Γ-like travelling waves in the n$^+$ emitter region which are incident on the barrier and to the evanescent wave in the barrier region itself. The experimental data shown in Figure 4 indicate that the effect of the increasing influence of the X-minima is to decrease the transmission of the (AlGa)As barrier for tunnelling electrons. This could occur either through an increased reflection of the incident travelling waves or a faster decay of the evanescent waves in the barrier region, or indeed to a combination of these two effects. We are considering these possibilities in more detail.

4. The Effect of a Transverse Magnetic Field on the Tunnel Current

Just as for electrons moving freely in the conduction band, one can expect that a large magnetic field will have a significant effect on the tunnelling electrons. We observe that the effect is largest when the magnetic field $\underline{B}$ is applied perpendicular to the current direction $\underline{J}$ (the so-called transverse geometry). This has a classical analogue. When the electron is moving perpendicular to B, the Lorentz force $-e(\underline{v} \wedge \underline{B})$ is a maximum. Figure 5 shows the effect of a transverse magnetic field on the reverse bias current J for several different values of reverse bias voltage. There is a marked decrease of the tunnelling current with increasing B at all values of bias voltage.

As in the case of the pressure dependence of J(V), the WKB approximation offers a simple and obvious approach for interpreting the data shown in Figure 5. For an electron moving in a magnetic field B parallel to the z-axis the Hamiltonian is given by

$$H = \frac{(\underline{p} + e\underline{A})^2}{2m^*} + V_0(x) - eEx \quad . \tag{5}$$

In this equation $\underline{A} = B(0, x, 0)$ is the magnetic vector potential, $V_0(x)$ represents the step-like barrier potential and $\underline{E}$ is the electric field in the barrier region. The equation can be written in the form

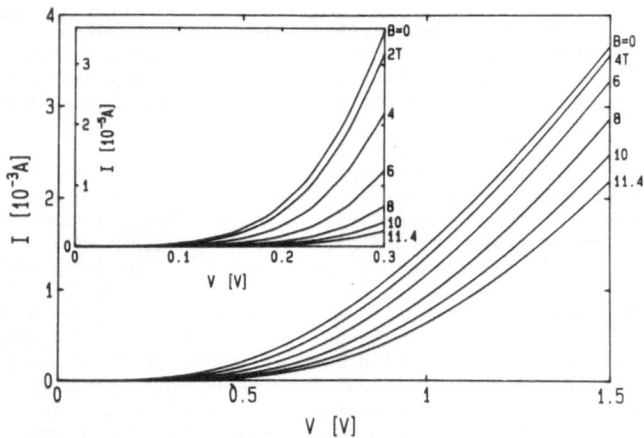

Figure 5 Reverse bias current-voltage characteristics of the single barrier structure (mesa diameter = 200 µm) for various applied magnetic fields , T=4K.

$$H = \frac{p^2}{2m^{*2}} + V_o(x) + \frac{e^2B^2x^2}{2m^*} + \frac{P_y eBx}{m^*} - eEx. \tag{6}$$

The first two terms give the Hamiltonian in the absence of external fields. The last three terms can be thought of as modifying the form of the barrier potential. It is convenient to choose the origin of co-ordinates $x = 0$ to be the interface between the left-hand n^+-emitter layer and the (AlGa)As barrier. By setting $A = 0$ for $x \leq 0$ and $A_y = Bx$ for $0 < x < b$, the dynamic effects of the magnetic field can be ignored in the heavily doped n^+-region. This approximation is justified by the smearing out of Landau level effects due to ionised impurity scattering in this region. In addition, this choice of origin (i.e. of gauge) provides a range of p_y-values for the electrons incident on the barrier which is symmetric about $p_y = 0$. Both A and the diamagnetic term are, of course, assumed to vary continuously at the (AlGa)As/n^-GaAs interface and into the n^-region.

In evaluating the tunnelling probability T in the presence of a transverse magnetic field, one would expect that the principal effect is in the exponential term, rather than in the pre-factor. The WKB approximation provides a simple means of calculating the exponential term. The effect of the magnetic field in reducing the tunnelling current can be understood qualitatively in terms of an increase in the effective height of the potential barrier due to the diamagnetic term in (6). There is an analogy here with the ballistic motion of a classical electron in a transverse magnetic field. Both classically and quantum mechanically, the combined effect of the two magnetic terms in (6) is to reduce the kinetic energy of motion along the x-axis and to increase it in the direction parallel to the y-axis. This reduction of the energy of the tunnelling electron along the x-axis is equivalent to an increase in the height of the barrier as given by the third term in (6). By performing a Taylor expansion of the exponential term containing p_y and then integrating over p_y, it can be shown that the tunnelling probability varies as

$$T = T_o \exp \left\{ - \frac{2\kappa b}{4V_o} \left(-eEb + \frac{e^2 B^2 b^2}{3m^*} \right) \right\} \tag{7}$$

Hence for a fixed voltage on the barrier the tunnel current should decrease with increasing B according to the relation

$$J = \exp(-\beta B^2) , \tag{8}$$

where $\beta = \dfrac{e^2 b^3 \kappa}{6m^* V_o}$.

Similarly, the additional voltage $\Delta V_B = -b\Delta E$ that must be applied to the barrier in order to offset the effect of B and maintain a constant tunnelling current is given by

$$e\Delta V_B = e^2 B^2 b^2 / 3m^* .$$

The analysis of the magnetic field dependence of tunnelling in our structures is complicated by the fact that the experiment measures the total voltage V applied to the device rather than the barrier voltage V_B. The total voltage is the sum of the voltages dropped across the barrier, the depletion layer and the undepleted region as shown in Figure 2. Assuming that the n⁻ layer is not fully depleted (V ≤ 1V) and that the voltage drop V_u across the depletion layer is small, we can write to a good approximation,

$$V_B = b \left[\frac{2N_D e}{\varepsilon \varepsilon_o} \right]^{\frac{1}{2}} (V - V_f)^{\frac{1}{2}} ,$$

where N_D is the ionised donor concentration in the n-layer and V_f is the flat band bias (V_f corresponds to a small forward bias voltage due to the negative space charge in the barrier). The following expression for the additional bias voltage ΔV required to offset a magnetic field B and maintain a constant current

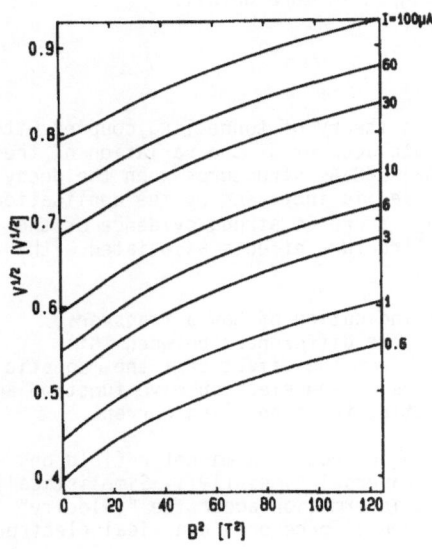

Figure 6 Plot of $V^{1/2}$ versus B^2 for a range of constant currents (reverse bias) through the device. The slopes of the curves are linear in the low magnetic field limit, as predicted by the model discussed in section 4 ,T=4K.

is then obtained:

$$\Delta V^{\frac{1}{2}} \sim b \left(\frac{\varepsilon\varepsilon_0}{2N_D e}\right)^{\frac{1}{2}} \left(\frac{eB^2}{3m^*}\right) \quad . \tag{9}$$

This relation is tested in Figure 6 which plots the square root of the applied reverse bias voltage against B^2. Note that the initial slope $(B \to 0)$ of the curves does not vary greatly over the wide range of tunnelling currents which are shown in the Figure: between 0.6 and 100 μA the slope decreases from 2.6 to 1.6×10^{-3} $V^{\frac{1}{2}}$ T^{-2}; at 1 mA it is 1.0×10^{-3} $V^{\frac{1}{2}}$ T^{-2}.

In comparison, (9) gives a slope of 4×10^{-3} $V^{\frac{1}{2}}$ T^{-2} for $N_D = 2 \times 10^{15}$ cm^{-3} and $m^* = 0.09$ m_e, the Γ conduction band effective mass of the (AlGa)As barrier. Considering the approximation made in deriving the equation, it can be seen that the model gives at least a qualitative indication of the effect of the magnetic field on the tunnelling current. Note that the model gives a closer fit if the tunnelling effective mass m^* is assumed to be smaller than that at the (AlGa)As conduction band minimum, as would be expected in any deviation from simple one band effective mass theory. We are presently extending the WKB model for arbitrary values of E and B to see if it can account for the curvature of the $V^{\frac{1}{2}}$ versus B^2 plots at high magnetic fields (see Figure 6).

Note that in Figures 5 and 6, the bias voltages (and hence field in the barrier) are large enough to ensure that the electrons tunnel into continuum states in the conduction band in the n⁻ region. At low enough reverse bias voltages and for large enough B the magnetic terms in (6) increase the effective energy of the edge of the conduction band in the n⁻ region to a level above the Fermi energy in the emitter contact. At this point there are no travelling wave states into which electrons can tunnel so that the current is effectively extinguished. Again, this can be understood by a classical analogue. If $e^2B^2b^2/2m^* > \varepsilon$, the energy of the electron as it enters the barrier, then the magnetic field deflects the electron motion so that all of its kinetic energy is parallel to the plane of the barrier. In low bias we see behaviour in qualitative agreement with this prediction (see insert of Figure 5),and we are currently investigating it in more detail.

5. Conclusions

We have shown that the simple effective mass theory of tunnelling,coupled with the WKB approximation, gives a quite accurate account of the variation of the tunnel current in single barrier GaAs/(AlGa)As/GaAs structures when the decay constant of the evanescent wave in the barrier is increased by the application of pressure. At pressures above 10 kilobar, there is strong evidence of the tunnelling current being affected by band structure effects associated with the high X-valley conduction bands.

The WKB theory also gives a qualitative indication of how a transverse magnetic field affects the tunnel current. The differences between this simple theory and experiment may well arise from the effect that the magnetic field has on the (non-exponential) prefactors of the electron wave-function and the densities of states involved in calculating the tunnelling current.

One of the motivations of our study of the effect of a magnetic field on tunnelling was the recent work on barrier traversal times [19]. Simplistically one might expect the magnetic field to give information about the "velocity" of an electron as it tunnels just as the Lorentz force on a classical electron give a measure of its velocity. (Indeed one naturally poses the question "Is it possible to measure the Hall voltage associated with the tunnel current?") Whilst our analysis appears to show that the changes of the J(V) characteristics in a transverse magnetic field tell us little about the tunnel velocity, they may encourage the theoreticians to calculate the effects of a magnetic field on the barrier traversal time.

This work is supported by SERC (UK) and CNRS (France). We are very grateful to Professor K.E.Singer for growing and supplying the layers used in this study and to Dr G.Hill for processing them. The work described here has been done in close collaboration with Drs F.W.Sheard and G.A.Toombs and with Professor K.W.H.Stevens, whose help we gratefully acknowledge.

References

1. L. Esaki, Proc. Int. Conf. on Phys. of Semiconductors, San Francisco, pp.473-83 publ. Springer (1984).

2. T.C.L.G. Sollner, P.E. Tannenwald, D.C. Peck and W.D. Goodhue, Appl. Phys. Lett. 45, 1319 (1984).

3. M. Heiblum, M.I. Nathan, D.C. Thomas and C.M. Knoedler, Phys. Rev. Lett., 55, 2200 (1985).

4. A.R. Bonnefoi, D.H. Chow and T.C. McGill, Appl. Phys. Lett., 47 888 (1985).

5. Landolt-Bornstein 17 Semiconductors III-V Compounds p.334 (1984).

6. T.W. Hickmott, P.M. Solomon, R. Fischer and H. Markoc, Appl. Phys. Lett. 44, 90 (1984).

7. D.C. Taylor, P.S.S. Guimaraes, B.R. Snell, L. Eaves, F.W. Sheard, G.A. Toombs and K.E. Singer, 4th Int. Conf. on Hot Electrons in Semiconductors, Innsbruck, Physica B 134, 12 (1985) see also G.A. Toombs, F.W. Sheard and L. Eaves, Proc. 2nd Int. Conf. on Phonon Physics, Budapest, 1985 (in press).

8. E.O. Kane, J. Appl. Phys. 32, 83 (1961).

9. M. Jaros, Rep. Prog. Phys. 48, 1091 (1985).

10. A.C. Marsh and J.C. Inkson, J. Phys. C: Sol. St. Phys. 17, 6561 (1984).

11. D.C. Taylor, P.S.S. Guimaraes, B.R. Snell, F.W. Sheard, L. Eaves, G.A. Toombs, J.C. Portal, L. Dmowski, K.E. Singer, G. Hill and M.A. Pate, Proc. Int. Conf. on Modulated Semiconductor Structures, Kyoto (Surf. Science 1986).

12. P.S.S. Guimaraes, D.C. Taylor, B.R. Snell, L. Eaves, K.E. Singer, G. Hill, M.A. Pate, G.A. Toombs and F.W. Sheard, J. Phys. C: Sol. St. Phys. 18, L605-9 (1985).

13. L. Eaves, P.S.S. Guimaraes, F.W. Sheard, B.R. Snell, D.C. Taylor, G.A. Toombs and K.E. Singer, J. Phys. C: Sol. St. Phys. 18, L885-9 (1985).

14. L. Eaves, P.S.S. Guimaraes, B.R. Snell, D.C. Taylor, K.E. Singer, Phys. Rev. Lett. 53, 262 (1985).

15. G.D. Pitt, J. Lees, R.A. Hoult and R.A. Stradling, J. Phys. C 6 (1973), 3282.

16. L.G. Shantharama, A.R. Adams, C.N. Ahmad and R.J. Nicholas, J. Phys. C 17, (1984), 4429.

17. N. Lifshitz, A. Jayaraman, R.A. Logan and R.C. Maines, Phys. Rev. B20, (1979), 2398.

18. Shubnikov-de Haas experiments on n^+(AlGa)As layers (Al = 0.3) (J.C. Portal, private communication) show a dramatic decrease in electron concentration in the Γ minimum at 11 kbar, indicating a Γ-X crossover.

19. K.W.H. Stevens, Eur. J. Phys. 1, (1980), 98.

Recent Results on III-V Superlattices and Quantum Well Structures

P. Voisin and M. Voos

Groupe de Physique des Solides de l'Ecole Normale Supérieure,
24 rue Lhomond, F-75005 Paris, France

We present recent studies of four different III-V quantum well structures, which illustrate the diversification of to-day's material science. We examine the controversal problem of the band offsets in the $GaAs-Ga_{1-x}Al_xAs$ system from the excitation spectroscopy of double step quantum wells. For this system, we discuss also the Stokes shift of a few meV's which is often observed between the luminescence and absorption peaks in terms of exciton trapping on individual interface fluctuations. In GaSb-AlSb strained layer superlattices, we have investigated the in-plane dispersion relations of the valence subbands from interband magneto-absorption experiments. Our results exhibit large qualitative differences with respect to the $GaAs-Al_xGa_{1-x}As$ case. The InAs-GaAs superlattice grown on an InP substrate is another interesting system in which the properties of large built-in strains and very thin layers come into play. We have performed calculations of the band structure of these pseudo alloy superlattices, which we compare to our low-temperature luminescence data. We have also studied theoretically a new consequence of the spatial separation of the electron and hole wave functions in InAs-GaSb superlattices, that is a macroscopic transient photovoltage which presents analogies with the piezo electric effect.

The increasing control of the epitaxial growth and the general understanding of the main electronic properties of two-dimensional systems are pushing the physics of III-V semiconductor heterostructures towards more and more sophisticated studies. We shall illustrate these trends with our recent studies on the following four different systems : $GaAs-Al_xGa_{1-x}As$, GaSb-AlSb, InAs-GaAs and InAs-GaSb. In the most well-known and best controlled system, i.e. $GaAs-Al_xGa_{1-x}As$, the Dingle determination of the sharing of the band gap difference between the conduction and valence bands was recently questioned. We have examined this problem in our group from the excitation spectroscopy of "separate confinement heterostructures" (or "double-step quantum wells"). The excitation spectra show transitions between states confined in the narrower and the thicker parts of the heterostructures, respectively. The calculated energies of these transitions are quite sensitive to the ratio of the conduction (ΔE_c) and valence (ΔE_v) band offsets. If ΔEg is the difference between the band gaps of GaAs and $Al_xGa_{1-x}As$, our analysis gives $\Delta E_c = 0.59 \Delta E_g$, in good agreement with other recent determinations. Besides the luminescence of $GaAs-Al_xGa_{1-x}As$ quantum wells is dominated by intrinsic excitonic recombination. However, a Stokes shift of a few meV's is often observed between the luminescence and the excitation (or absorption) peaks. This effect was attributed to trapping of free excitons on localized interface fluctuations. Calculations of the binding energy of an exciton on such defects and of the phonon-assisted hopping between adjacent defects support the interpretation of the luminescence linewidth and lineposition in terms of non-thermalized trapped excitons.

A remarkable aspect of quantum well structures is the in-layer dispersion relations of the valence subbands, which result from their strong $\vec{k}.\vec{p}$ coupling at finite in-layer wavevector $k_\perp$. GaSb-AlSb strained layer superlattices are interesting in this respect, as the accomodation of the lattice mismatch by elastic deformations may induce a reversal of the energy positions of the heavy and light hole subbands. The interaction between these subbands and the resulting in-plane dispersion relations are thus qualitatively different from the $Al_xGa_{1-x}As-GaAs$ case. We have measured the transmission of circular polarized light in a GaSb-AlSb superlattice under strong magnetic fields at low temperature. The spectra show many transmission minima corresponding to transitions between the Landau levels of the first light and heavy hole and conduction subbands. We find a very heavy mass, $0.8 \, m_0$, for the first light hole subband, which is the ground valence state in this sample, and a rather light mass, $0.11 \, m_0$, for the first heavy hole subband.

New questions arise also from the investigation of new materials. The InAs-GaAs superlattice grown on an InP substrate is an exemplary system, in which the properties of large built-in strains and very thin layers come into play. The low temperature luminescence of such structures, having equal layer thicknesses ranging from 10 to 20 Å, consists in single lines centered at 750 meV about. The main characteristic is that the line position almost does not vary with the individual layer thickness, as expected intuitively in the very thin layer of pseudo-alloy regime. Calculations of the band structure of these superlattices yield a good agreement with experiment for a conduction band offset of ~ 550 meV. However, we show that the band structure of such pseudo-alloy superlattices differs always appreciably from that of the corresponding bulk $In_xGa_{1-x}As$ ternary alloy.

Finally, new properties of well-known systems are sometimes discovered. For example, in InAs-GaSb superlattices, the electron and hole wavefunctions are spatially separated, due to the particular "type II" band-edge line-up of the host materials. Many consequences of this situation have already been investigated. However, the possible existence of a macroscopic transient photovoltage across the superlattice had not been considered. We shall describe this new effect, which presents analogies with the piezo-electricity of ionic cristals.

1 AlGaAs-GaAs Quantum Wells

1-1 Band offsets

The transmission and excitation spectra of $Al_xGa_{1-x}GaAs$ quantum wells (QW) exhibit excitonic peaks associated with transitions between the mth valence state (heavy hole HH_m or light hole LH_m) and the nth conduction state E_n. The energy position of these E_n, LH_m and HH_m states are sensitive to the band offsets ΔE_c and ΔE_v, because of the relative penetration of the electronic wave functions in the $Al_xGa_{1-x}As$ barrier. However, the energies of the predominent n-m = 0 optical transitions are not very sensitive to the ratio of these offsets in the case of a simple quantum well. Indeed, the increase of the confinement energy E_n, which is obtained when ΔE_c is increased, is compensated in part by the corresponding decreases of the confinement energies LH_n and HH_n. By fitting the weak parity allowed HH_3-E_1 transition observed in the excitation spectra, Miller et al [1] have obtained recently $\Delta E_c = 0.57 \Delta E_g$ for an aluminium concentration x $\sim$ 0.3. This value differs strongly from the previously admitted determination of Dingle et al [2], $\Delta E_c = 0.85 \Delta E_g$

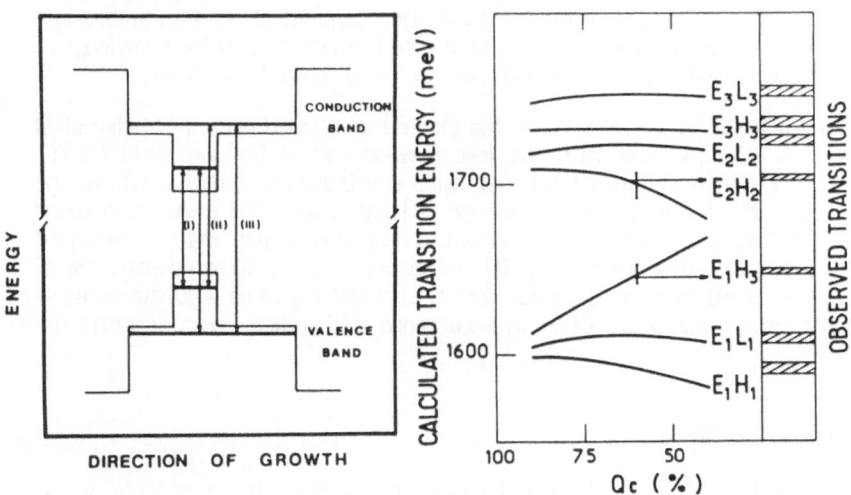

Fig. 1 : Scheme of the QWSCH structures investigated here. Also shown are the three types of optical transitions described in the text (left panel). Calculated dependence of the transition energies on the conduction-band discontinuity $Q_c = \Delta E_c/\Delta E_g$ for a 45 Å GaAs QW embedded in a 245 Å $Ga_{1-x}Al_xAs$ one, with x $\sim$ 0.13. The observed transitions, corrected for the binding energies of the excitons, are indicated in the right side of the figure (right panel).

This problem was also examined [3] by studing structures consisting of a GaAs QW embedded in a $Al_xGa_{1-x}As$ thicker one (hereafter called the barrier), the latter being clad between thick $Al_yGa_{1-y}As$ layers with y > x, as sketched in Fig. 1. The excitation spectra of these "double step" or SCH QW's [3] show actually three distinct kinds of transitions. Namely, these transitions may involve quantized levels (i) both essentially localized within the narrow GaAs well, (ii) with one state (initial or final) confined within the GaAs QW and the other one (final or initial) in the $Al_xGa_{1-x}As$ barrier and (iii) both in this barrier. The energy levels

109

have been calculated within the three-band envelope-function formalism [4] using values of $1519.2 + 1247x$(meV) for the $Ga_{1-x}Al_xAs$ band gap ; $341-66x$(meV) for the spin-orbit energy ; $0.067 m_0$, $(0.48 + 0.31x)m_0$ and $0.094 m_0$ for the electron, heavy-hole and light-hole effective masses. It was thus found that the transitions (i) are very sensitive to the width of the GaAs QW, the transitions (ii) to the band offset ratio and the transitions (iii) to the percentage of Aluminium x. This can be seen in Fig. 1 for a 45 Å GaAs embedded in a 245 Å $Ga_{0.867}Al_{0.133}As$ one. The transition E_1-HH_3, involving a state E_1 essentially localized in the GaAs QW, and a state HH_3 delocalized over the barrier, is strongly dependent on $Q_c = \Delta E_c/\Delta E_g$. The experimental (for the lower-lying transition [5,3]) and theoretical (for the other ones [6]) binding energies of quasi two-dimensional excitons are taken into account. Moreover, the excitation spectra of assymmetrical structures where the GaAs QW is not at the center of the $Al_xGa_{1-x}As$ one display also new (n-m) odd transitions which corroborate a value of $\Delta E_c = (0.59 \pm 0.03) \Delta Eg$ for the conduction band offset with an aluminium concentration of 0.13. The uncertainties due to the lack of precise knowledge of the parameters used in the calculation have been also shown to be small.

This value must be compared with the already mentioned value of Miller et al $(0.57, x \sim 0.3)$ [1] and with the less precise one of Dawson et al $(0.75, x \sim 0.3)$ [7], both obtained from excitation spectroscopy. Besides, fits of the measured sheet carrier density in p type [8] and n type [9] modulation-doped $Al_xGa_{1-x}As$-GaAs heterojunctions give values near 0.6, which is also consistent with C-V and I-V measurements [10]. In conclusion, the strong dependence of some optical transitions on Q_c in our double step QW's give an accurate value of the bands offsets for $x = 0.13$, in accordance with other recent results for $x \sim 0.3$.

1-2 Exciton trapping

The width of the excitonic peaks that appear in the excitation spectrum of the luminescence line has been correlated to interface fluctuations of a few hundred angströms in lateral size, i.e. larger than the 2D exciton diameter [11]. Less interest has been paid to the magnitude of the Stokes shift between absorption and luminescence peaks as well as to the thermalization of the recombining excitons. Figure 2 shows the low-temperature photoluminescence spectrum (solid line) and excitation spectrum (dashed line) of a 7 0 Å thick single GaAs quantum well grown by MOCVD. These experimental findings (luminescence linewidth, Stokes shift between photoluminescence and excitation spectra) have been interpreted (12) in terms of exciton trapping on interface defects. In a perfect QW, the exciton is free to move within the layer plane. Let us consider a protrusion of radius a and depth b of GaAs into the $Al_xGa_{1-x}As$ barrier ; because of the smaller energy gap of GaAs, this defect is a potential well which may trap the exciton, which is then in a defect-related bound state. A variational calculation [12] provides the dependence of the binding energy on the size of the defect, as shown in Fig. 2. This binding energy is typically a few meV's for a defect depth of a few monolayers when the lateral size becomes larger than the exciton radius, which is of the order of 100 Å. Using a gaussian distribution of defect sizes, the density of states of trapped

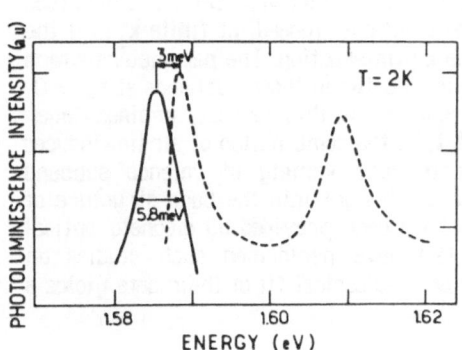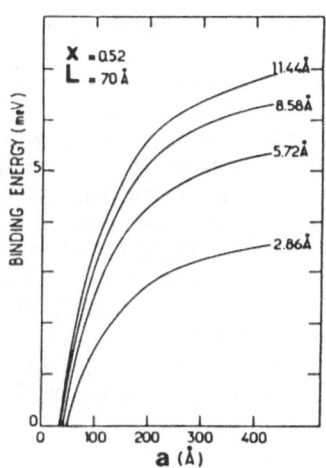

Fig. 2 : Photoluminescence spectrum (solide line) and excitation spectrum (dashed line) of a 70 Å-thick single GaAs quantum well at 2 K (left panel). Exciton binding energy on semi-Gaussian interface defects plotted versus the lateral size a for different values of the defect depth b in a GaAs quantum well of thickness L = 70 Å (right panel).

excitons, which is a peak about 3 meV's broad centered about 4 meV's below the edge of the exciton continuum was then derived.

These figures compare favorably with the experimental data of Fig. 2, i.e. a luminescence linewidth of 6 meV and a Stokes shift of 3 meV. The evaluation of the acoustical phonon-assisted trapping and hopping times shows that at low temperature, for defects with a binding energy of 3 meV separated by 1000 Å, the exciton has, before recombining, enough time to be trapped by a defect but not enough to jump to a deeper defect. These evaluations support the interpretation of the luminescence in terms of inhomogeneously broadened, non-thermalized trapped excitons. When the temperature is increased, the experimental Stokes shift decreases, as the luminescence also involves free excitons. A quantitative analysis of the temperature dependence [13] indicates a density of defects ranging between 10^{10} and 10^{11} cm^{-2}, which is consistent with the parameters used in the hopping time calculation.

In better quality samples [14,15], the luminescence and excitation linewidths are smaller and no Stokes shift is observed. At the opposite, in moderate quality samples, the binding of excitons on interface defects must be regarded in terms of the motion of an exciton in a random potential due to well-width fluctuations [16]. Below a mobility edge, excitons are localized, and above it they behave as extended states.

2 GaSb–AlSb Superlattices

Another important problem is that of the in-plane dispersion relations of the valence subbands. Altarelli [17] first showed that this complicated problem does not admit simple or intuitive theoretical solutions. Indeed, at finite $k_\perp$, all the valence subbands are coupled through the k.p interaction. The parameters which measure this interaction are the energy gaps between these subbands at $k_\perp = 0$, which depend strongly on the sample parameters. In this respect, strained–layer superlattices (SLS) are of special interest, as the combination of strain–induced and confinement–induced effects can produce a variety of valence subband ordering and spacing [18]. An experimental insight into the band structure of QW's or superlattices (SL) at finite $k_\perp$ have been provided by magneto-optical studies. In particular, Maan et al [19] have performed such studies on GaAs-Al$_x$Ga$_{1-x}$As quantum wells, and the semi-classical fit of their data yields a

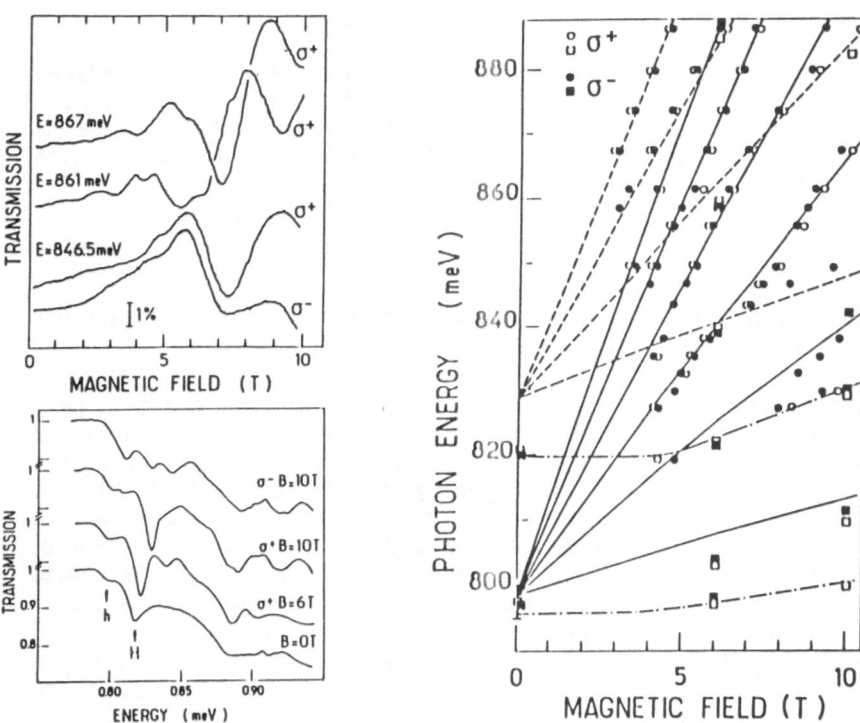

Fig.3 : Transmission spectra of a 181 Å – 452 Å GaSb-AlSb superlattice at various magnetic fields and for different light polarizations (lower left). Transmissions at fixed photon energies versus magnetic field. Note the small vertical scale (upper left). Plot of the transition energies (transmission minima) versus magnetic field; the solid and open circles correspond to spectra obtained at a fixed photon energy while the solid and open squares correspond to the transmission spectra at fixed B (right panel).

very heavy in-plane mass $m_H \sim m_0$ for the ground heavy hole subband and a lighter one, $m_h \sim 0.2\,m0$ for the first light hole subband. These results are in qualitative agreement with Altarelli's predictions, and brought strong contradiction to previous approaches [5,20]. Here, we present low-temperature magneto-optical absorption data [21] in a GaSb-AlSb SLS which is remarkable because its ground valence state is the first <u>light hole</u> subband [22].

The lowest spectrum in Fig.3 shows the zero field transmission at 2 K of our 10 period 181 Å – 452 Å GASb-ALSb SL, which was grown by MBE on a GaAs substrat. The absorption edge consists in the two structures noted h and H. H is a well-resolved exciton peak corresponding to the heavy hole exciton associated with the first heavy hole (HH_1) and conduction (E_1) subbands. The structure h, which appears at a lower energy and is less pronounced, corresponds to the absorption by the light hole exciton formed with the ground light hole (LH_1) and conduction (E_1) subbands. The next absorption step at ~ 880 meV corresponds to the unresolved $LH_2 \rightarrow E_2$ and $HH_2 \rightarrow E_2$ transition, and it will not be considered in the following.

The reversal of the ground heavy and light hole subbands, as well as the anomalously low energies of the corresponding absorption edges, results from the compensation of the strain- and confinement-induced effects-which occur when the small gap material experiences a biaxial tensile stress [17,20].

The other spectra in Fig.3 show the modification of the transmission spectrum when an increasing magnetic field is applied in the Faraday, B parallel to the SL axis, configuration. New structures progressively develop, which correspond to the quantization of the in-plane kinetic energy into Landau levels. In particular, a splitting of the h line and the enhancement of the excitonic feature H are observed in the σ^+ polarization, at 10 T. Both the H and h excitons are considerably weaker in the σ^- polarization.

An alternative experimental procedure consists in recording the transmission at a fixed photon energy when sweeping the magnetic field, which gives the other spectra shown in Fig.3 They exhibit an oscillatory behavior, characteristic of a series of transitions between the Landau levels associated with the LH_1, HH_1 and E_1 subbands. Some of the high field data exhibit a strong polarization-dependence, while the low field ($B < 7\,T$) data present only a very weak polarization-dependence.

All the observed transmission minima are reported in the usual transition energy vs magnetic field plot which is shown in Fig.3. This plot exhibits two distinct fan diagrams, eye-marked by the solid and dashed lines, which extrapolate towards $E_h = 799$ meV and $E_H = 829$ meV respectively. These two fan diagrams correspond to the $LH_1^N \rightarrow E_1^N$ and $HH_1^N \rightarrow E_1^N$ transitions between the Landau levels associated with the LH_1, HH_1 and E_1 subbands.

In addition, two dashed-and-dotted lines having a non-linear behavior have been drawn through the exciton data points. They extrapolate to ~ 795 meV and 820 meV for the light and heavy hole excitons respectively.

To analyse more quantitatively these data, we have : (i) discarded any spin effect, since the observed $\sigma^+ - \sigma^-$ splittings are generally small. (ii) evaluated the energies E_1^N from the semi-classical quantization rule $k_\perp^2 \to (2N+1)eB/\hbar$, using simplified in-plane dispersion relations [4] . (iii) evaluated the energies LH_1^N and HH_1^N in the same semi-classical approximation, using parabolic in-layer dispersion relations with the in-layer effective masses $m_{LH}^\perp$ and $m_{HH}^\perp$ as fitting parameters. This method leads to the fan diagrams shown in Fig.3, with $m_{LH}^\perp = 0.8\ m_0$ (solid line) and $m_{HH}^\perp = 0.11\ m_0$ (dashed lines), respectively.

We feel that the overall agreement witnesses that the involved hole subbands are not strongly non-parabolic in the energy range of interest, which in turn partly justifies our method. However, the present analysis cannot explain the polarization effects which are observed. More detailed models are thus required, which would take into account the finite g-factor of the conduction band and the complex nature of the valence subbands.

The other important point which has to be explained is the behavior of the exciton lines. They both show first a weak diamagnetic shift, and tend to follow the N = 0 band to band transition at higher magnetic field. This corresponds to the usual expectation [19]. Besides, the energy difference between the h and H excitons observed at B = 0 and the extrapolation E_h and E_H of the $LH_1^N \to E_1^N$ and $HH_1^N \to E_1^N$ transitions should measure the binding energies of the corresponding excitons. We would get ~ 3 meV and ~ 9 meV for the light- and heavy-hole excitons respectively, which is not consistent with the ratio of the corresponding reduced masses which we have obtained.

In fact, the calculated binding energies, using the experimentally determined reduced masses, are 6 meV and 5 meV for the h and H excitons, respectively [6]. The 4 meV discrepancy for the heavy hole exciton is probably due to a slight non-parabolicity of the heavy hole subband. Indeed, such a non-parabolicity would lower E_H by a few meV's without changing significantly the fit of the data. On the other hand, the data for the light hole exciton are by far too inacurate, due to its small oscillator strength. Actually, the ratio of the observed light and heavy exciton absorption at B = 0 is about 1/3, which is consistent with equivalent binding energies.

3 InAs-GaAs Superlattices

The demonstration that rather large lattice mismatch may be accomodated by elastic deformations in thin layer heterostructures [23] have opened rich perspectives for modern material science. For example, despite of the 6.8 % lattice mismatch in this system, periodic stackings of very thin alternate layers of InAs and GaAs have been grown successfully on InP substrates by Molecular Beam Epitaxy, and these samples show very nice structural properties [24]. One can expect such short period SL's to approach the band structure of a given bulk in $In_x Ga_{1-x}As$ alloy while they would be free from alloy scattering.

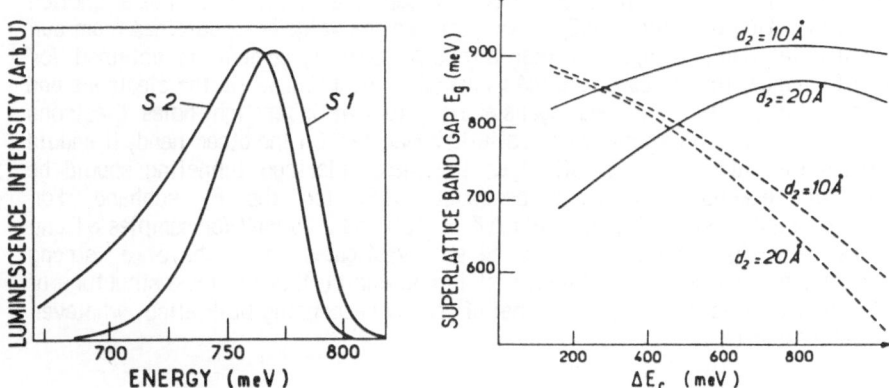

Fig.4 : Photoluminescence spectrum obtained in samples S1 et S2 at 2 K (left panel)
ant theoretical dependence of the InAs-GaAs superlattice band gap E as a function of
the conduction band offset ΔE_c for d_2 = 10 and 20 Å. Solid line : heavy-hole (HH1)
to conduction (E_1) band gap ; dashed line : light-hole (LH$_1$) to conduction (E_1) band
gap (right panel).

We have investigated the low temperature luminescence of structure consisting
in n period InAs-GaAs SL's with equal layer thicknesses d2, sandwitched between
thick Al$_{0.48}$In$_{0.52}$As buffer layers lattice matched to their InP substrates. For
samples S1, d2 = 10 Å and n = 15, while for sample S2, d2 = 20 Å and n = 10. The
luminescence spectra obtained at 2K with a 100 mW laser excitation are shown in
Fig.4. They consist in a single line, about 50 meV broad, exhibiting a somewhat
sample dependent lineshape and centered near 765 meV. From experiments as a
function of the excitation power, it is likely that the low energy part of the line is
due to recombination mechanisms involving impurities and that the rest of the
spectrum corresponds to band-to-band recombination. Since these two
recombination processes (extrinsic and intrinsic) are not resolved, the density of
impurities is certainly rather high. Anyhow, this indicates that the band gap of our
superlattices is roughly given by the energy position of the maximum of the
observed lines, i.e. 770 and 763 meV in samples S1 and S2, respectively, so that it
is not far from the band gap of bulk In$_{0.5}$GA$_{0.5}$As which is close to 800 meV.

In addition, we have calculated the band structure of the superlattices
investigated here in the effective mass approximation using conventional
values [25] of the band parameters of bulk InAs and GaAs. The calculations include
the effect of strains, the InAs and GaAs layers being both mismatched to the InP
substrate by 3.4 %. The GaAs(InAs) layers are under biaxial tension (compression)
which, taking into account the strain-induced coupling to the spin-orbit split-off
band, results in a band gap between the conduction and heavy-hole bands equal to
1389 meV (476 meV) and in a band gap between the conduction and light-hole bands
equal to 1033 meV (632 meV).

The calculated HH_1-E_1 (InAs-GaAs) SL band gaps are shown in Fig.4 as a function of the conduction band offset ΔE_c. If we compare the value of E_g obtained from our experimental data, it appears that a rather good agreement is obtained for $\Delta E_c = 550$ meV. In this case, the InAs layers are quantum wells for electrons and for heavy holes, while the GaAs layers are quantum wells for light holes. Electrons and light holes are thus somewhat spatially separated. On the other hand, it should be noted that, due to the small layer thickness, electron tunneling should be important, leading to a large bandwidth ΔE_1 for the E_1 subband. For $\Delta E_c = 550$ meV, the calculations yield $\Delta E_1 = 920$ and 273 meV for samples S1 and S2 respectively. Therefore, the SL's investigated here have a strong three-dimensional character. Finally, it is noteworthy that the band structure of such a SLS will never converge to that of the corresponding bulk alloy, whatever value of d2 is used.

4 InAs-GaSb Superlattices

A remarkable feature of InAs-GaSb superlattices [26,27] is the spatial separation of the electron and hole wavefunctions which arises from the fact that the InAs and GaSb layers are quantum wells for the conduction and valence states, respectively. This situation, which is sketched in Fig.5, results in quite original electronic structure and optical properties which have been extensively studied. in particular, due to the overlap of the InAs conduction band and GaSb valence band, the bandgap of the structure may range from a few meV's for a SL period d ~ 200 Å to a few hundreds of meV's for d ~ 60 Å. Though the periodic band bending caused by the spatial separation of the carriers [28] has been considered for a long time, until recently no attention was paid to the additive voltages [20] which can appear across the SL under pulsed carrier injection conditions. We describe here this new photovoltaic effect.

We consider the SL in the electric quantum limit, i.e. only the ground conduction and valence subbands are occupied. The corresponding envelope wavefunctions $\varphi_{e(h)}(z)$ (Fig.5), are centered in the InAs (GaSb) layers, respectively. this local polarization induces an overall potential difference ΔV across the SL. ΔV is equal to the sum of local potential differences δv between the planes $z = z_n$ and $z = z_{n+1}$ limiting the nth unit cell :

$$\Delta V = N \, \delta v \tag{1}$$

and

$$\delta V = e \, n_s / \varepsilon \, \varepsilon_0 \int_{z_n}^{z_{n+1}} dz \int_{z_n}^{z} dz' \, [|\varphi_e(z')|^2 - |\varphi_h(z')|^2] \tag{2}$$

N is the number of unit cells, ε the relative dielectric constant and n_s the areal density of injected carriers. The boundary conditions implicit in Eq.2 ensures that the electric field vanishes outside a two dimensional dipolar charge distribution.

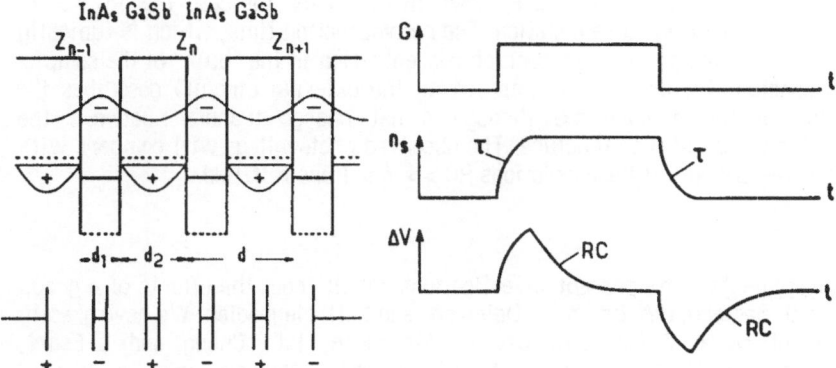

Fig.5 : Conduction and valence band edge profiles in a InAs–GaSb superlattice. The squared wavefunction amplitudes for the ground states are also shown (upper left). Series of capacitors, electrically equivalent to the InAs–GaSb superlattice (lower left). Right panel : time dependence of the pair generation rate G, photoinjected carrier density n_s and photovoltage ΔV when $T > RC >> \tau$.

Clearly, the system may be considered as a series of capacitors corresponding to each period, with the positive layer at the z-mean value of the hole wave function $<z_h>$ and the negative layer at the z-mean value fo the electron wavefunction, $<z_e>$. This is illustrated in Fig.5. Indeed, one verifies that if the wavefunctions are normalized in the z_n, z_{n+1} segment, the integral in Eq.2 may be written as :

$$\delta v = n_s \, Se \left[<z_h> - <z_e> \right] / (\varepsilon_0 \, \varepsilon \, S) = Q/C \qquad (3)$$

where S is the sample area. δv appears as the voltage across an individual capacitor of thickness $<z_h> - <z_e>$. For a 100 period 30 Å – 50 Å InAs–GaSb SL, with a density of injected carriers $n = n_s/d = 10^{17} \, cm^{-3}$, calculations of the wavefunctions yield $\Delta V = 200$ mV.

Up to now, we have discussed the quantum origin of this new photovoltaic effect, which arises from the wavefunction localization and lack of overall reflection symmetry. However, this effect cannot be observed under CW illumimation, due to the conductance paralell to the growth axis. In fact, we are dealing with a transient effect governed by the RC constant of the electric circuit and by the electron–hole recombination time τ.

If G is the pair generation rate, the carrier density n_s and the photovoltage ΔV obey the following rate equation :

$$dn_s/dt = G - n_s/\tau$$

$$d(\Delta V)dt) = 4\pi eN/\varepsilon \, \varepsilon_0) \, (<z_h> - <z_e>) \, dn_s/dt - \Delta V/RC \qquad (4)$$

The solutions of these equations are shown in Fig. 5 for the case T > RC >> τ, where T is the exciting pulse duration. The recombination time, which is typically of the order of 1 ns, governs the carrier concentration in the "bulk" of the sample. The RC constant (which may be imposed by the external circuit) describes the relaxation of the photovoltage through a net charge transfer between the terminating planes of the structure. The observed photovoltage will compare with the calculated ΔV only if the condictions RC >> τ > T are fulfilled.

Acknowledgements : The present investigations result from the efforts of a group including G. Bastard, J.A. Brum, C. Delalande and M.H. Meynadier. We have greatly benefited of cooperations with Dr. F. Alexandre, L.L. Chang and L. Esaki, P.J. Frijlink, J.K. Maan, J.Y. Marzin, R. Nahory and M. Tamargo.

References

1   R.C. Miller, D.A. Kleinman and A.C. Gossard, Phys. Rev. B 29, 7085 (1984).
2   R. Dingle in Festkörperprobleme, edited by H.J. Queisser, Advances in Solid State Physics, vol 15 (Pergamon/Wieweg, Braunschweig, 1975), p.21.
3   M.H. Meynadier, C. Delalande, G. Bastard, M. Voos, F. Alexandre and J.L. Lievin, Phys. Rev. B 31, 5539 (1985).
4   G. Bastard, Phys. Rev. B 24, 5693 (1981), Phys. Rev. B 25, 7584 (1982); see also G. Bastard in Molecular Beam Epitaxy and Heterostructures, edited by L.L. Chang and K. Ploog (Martinus Nishoff, 1985), p. 381.
5   R.C. Miller, D.A. Kleinman, W.T. Tsang and A.C. Gossard, Phys. Rev. B 24, 1134 (1981).
6   J.A. Brum and G. Bastard, J. Phys. C 18, L 789 (1985); J.A. Brum – Private Communication.
7   P. Dawson, C. Duggan, H.I. Ralph and K. Woodbridge, Superlattices Microstruct. 1, 173 (1985) and G. Duggan Proceedings of PCSI 12, to be published in J. Vac. Sci. Technol.
8   W.I. Wang, E.E. Mendez and F. Stern, Appl. Phys. Lett. 45, 639 (1984).
9   G. Bastard (Private Communication), from the data of J.C.M. Hwang, A. Kastalsky, H.L. Störmer and V.G. Keramidas, Appl. Phys. Lett. 44, 802 (1984).
10 H. Okumura, S. Misawa, S. Yoshida and S. Gonda, Appl. Phys. Lett. 46, 377 (1985) ; J. Batey, S.L. Wright and D.J. Di Maria, J. Appl. Phys. 57, 484 (1985).
11 C. Weisbuch, R. Dingle, A.C. Gossard and W. Wiegmann, J. Vac. Sci. Technol. 17, 1128 (1980).
12 G. Bastard, C. Delalande, M.H. Meynadier, P.M. Frijlink and M. Voos, Phys. Rev. B 29, 7042 (1984).
13 C. Delalande, M.H. Meynadier and M. Voos, Phys. Rev. B 31, 2497 (1986).
14 B. Deveaud, J.Y. Emery, A. Chomette, B. Lambert and M. Baude, Appl. Phys. Lett. 45, 1078 (1984).

15 H. Tanaka, H. Sakaki, J. Yoshino and T. Furuta, MSSII (Kyoto, sept. 1985) to be
    published in Surf. Science (1986).

16 J. Hegarty, M.D. Sturge, C. Weisbuch, A.C. Gossard and W. Wiegmann, Phys.
    Rev. Lett. $\underline{49}$, 930 (1982) ; J. Hegarty, L. Goldner, M.D. Sturge, Phys.
    Rev. B$\underline{30}$, 7346 (1985).

17 M. Altarelli, Lecture Notes in Physics (Springer Verlag, Berlin), Vol 177
    p. 174 ; Phys. Rev. B $\underline{28}$, 842 (1983).

18 P. Voisin, Surface Science 106 ... (1986).

19 J.K. Maan, G. Belle, A. Fasolino, M. Altarelli and K. Ploog, Phys. Rev. B $\underline{30}$,
    2253 (1984).

20 R.L. Green, K.K. Bajaj and D.E. Phelps, Phys. Rev. B $\underline{29}$, 1807 (1984).

21 P. Voisin, J.K. Maan, M. Voos, L.L. Chang and L. Esaki, Proc. EP2DS VI (Kyoto,
    Sept. 85), to be published in Surf. Science (1986).

22 P. Voisin, C. Delalande, M. Voos, L.L. Chang, A. Segmüller, C.A. Chang and
    L. Esaki, Phys. Rev. B $\underline{30}$, 2276 (1984).

23 J.W. Mattews and A.E. Blaskeslee, J. Cryst. Growth $\underline{27}$, 118 (1974) ; $\underline{29}$,
    273 (1975) ; $\underline{32}$, 265 (1976).

24 M.C. Tamargo, R. Hull, L.H. Greene, J.R. Hayes and A.Y. Cho, Appl. Phys. Lett.
    $\underline{46}$, 569 (1985).

25 "Landolt-Börstein Numerical Data and Functional Relationships in Science and
    Technology", edited by O. Madelung, Groupe III, Vol. 17, Springer, Berlin,
    1982.

26 L. Esaki and L.L. Chang, Journal of Magn. Magn. Mat. $\underline{11}$, 208 (1979).

27 P. Voisin, Thèse de Doctorat, Paris (1983) (unpublished).

28 L.L. Chang, N.J. Kawai, E.E. Mendez, C.A. Chang and L. Esaki, Appl. Phys. Lett.
    $\underline{38}$, 30 (1981).

29 F. Capasso, S. Luryi, W.T. Tsang, C.G. Bethea and B.F. Levine, Phys. Rev. Lett.
    $\underline{51}$, 2318 (1983).

Envelope Function Calculations for Superlattices

M. Kriechbaum

Institut für Theoretische Physik, Universität Graz, A-8010 Graz, Austria

1. Introduction

Envelope function calculations for heterostructures and superlattices have been performed with great success for some time
/1,2/. They allow in a simple way to make predictions on the
properties of a heterostructure by referring to the known properties of its constituents. In this article some light will be
thrown on these calculations. After an outline on the envelope
function approximimation (EFA) for bulk semiconductors we will
present its generalization to treat interfaces as well as discuss its limitations. No attempt is made to present a compre-
hensive survey of the vast amount of results obtained by EFA
for heterostructures.

2. Elementary derivation of the method

The envelope function calculation is a convenient means to
calculate the electron eigenstates in the vicinity of a certain point of interest k_o (Γ-point for most III-V and II-VI,
L-point for IV-VI semiconductors). This is achieved by making
the ansatz for the Schrödinger wavefunction

$$\Psi = \sum_n \Phi_{nk_o} f_n(r) \tag{1}$$

where the Φ are Blochfunctions for band n at point k_o. By
assuming the envelope functions f_n to vary slowly on the
range of a unit cell, the Schrödinger equation

$$H\Psi = E\Psi \tag{2}$$

can be transformed to the effective mass form (we include a
magnetic field, allow for strain and take into account a
slowly varying perturbation U)

$$(H - E)f =$$

$$\sum_{n=1}^{N} \left[(E_n - E + U)\delta_{mn} + P_{mn} \cdot k + \sum_{\alpha\beta} k_\alpha D_{mn}^{\alpha\beta} k_\beta + A_{mn} \cdot B + \sum_{\alpha\beta} C_{mn}^{\alpha\beta} \epsilon_{\alpha\beta} \right] f_n = 0 \tag{3}$$

Here $k = \frac{1}{i} \nabla - \frac{e}{c\hbar} A$, with A the vector potential to the magnetic field B, $\varepsilon_{\alpha\beta}$ is the strain tensor, α, β are Cartesian indices and $m = 1, 2, \ldots N$. N is the number of close lying bands.

Neither the periodic crystal potential nor the Blochfunctions Φ of the unstrained crystal appear in this set of coupled differential equations. They are hidden in the band edges E_n, the momentum matrix elements P_{mn}, the higher band contributions A_{mn} and D_{mn} and the deformation potentials C_{mn}. These quantities are treated as material parameters. A perturbation which is not slowly varying causes non diagonal matrix elements

$$U_{mn} = \int_{\Omega} d\tau \ \Phi_m^* \ U \Phi_n \qquad (4)$$

(Ω the unit cell) to appear in Equ.(3).

This effective mass formulation (Equ.3) is easily generalized to a heterostructure where the transition from layer A to layer B occurs gradually. Then the crystal potential will slowly change from crystal potential A to B and the material parameters in Equ.(3) will become position dependent. In order to retain the hermiticity of the coupled differential equations (3) the operators have to be symmetrized, i.e. P.k has to be replaced by (P.k+k.P)/2. In addition to the effective mass parameters which only specify band edge differences the band set up, i.e. the relative position of the band edges in A and B has to be stated. The envelope functions and their derivatives are of course still continuous.

Consider now the case of an abrupt transition from layer A to layer B at an interface in the (x,y) plane at z = 0. Then we have

$$z < 0 : H^A \ f^A = Ef^A$$
$$z > 0 : H^B \ f^B = Ef^B. \qquad (5)$$

The solutions f^A and f^B have to be joined at z = 0 thus determining the allowed energy values E. The abrupt change of the crystal potential at the interface will result in nondiagonal elements in the effective mass Hamiltonian (3) at the interface. However, their determination requires knowledge of the crystal potential and Blochfunctions at the interface and is beyond the scope of EFA. For sufficiently similar semiconductors A and B, we can assume in the sense of Schrödinger perturbation theory that the Blochfunctions in A and B are identical and the only effect of the changing crystal potential is to change the energy eigenvalues E_n.

The continuity of the Schrödinger function ψ of the heterostructure requires now, due to the orthogonality of the Φ_n, all envelope functions f_n to be continuous at the boundary:

$$f_n(z^-) = f_n(z^+) . \qquad (6)$$

The condition for joining the derivatives of the envelope functions is obtained in the spirit of the EFA, that is that only slowly varying changes are considered, in the following way. By integrating the effective mass equation (3) across the interface one obtains (only relevant terms considered)

$$\int\limits^z dz(E_m-E)f_m + \sum_n \left[P_{mn}f_n(z) + \sum_\alpha D_{mn}^{\alpha z}k_z f_n \right] = 0 .$$

(7)

The first term of (7) is certainly continuous and for P_{mn} identical in A and B we have the required condition

$$\sum_{\alpha n} D_{mn}^{\alpha z} \frac{\partial}{\partial z} f_n \qquad \text{continuous} \qquad (m = 1,2,\ldots, \alpha = x,y,z) .$$

(8)

Note that this condition is slightly different to the one obtained by Altarelli /2/, but it will be seen that this has no practical consequences. For a one band approximation (N=1), (8) reduces to

$$\frac{1}{m^*_A} \frac{\partial}{\partial z} f = \frac{1}{m^*_B} \frac{\partial}{\partial z} f \qquad \text{at} \qquad z = 0 .$$

(9)

It can easily be verified that conditions (6) and (8) guarantee the continuity of the current across the interface.

Now we are in the position to calculate the eigenstates of a heterostructure. For a superlattice we can invoke the Bloch theorem to characterise the solutions by observing

$$f_n(z+D) = e^{iKD}f_n(z)$$

(10)

with a (one dimensional) Bloch vector K and the superlattice period D.

The EFA is also very suitable for calculating effects of electrostatic potentials due to charge transfer in hetero-structures with staggered band set up or heavily doped structures. The potential appears in the diagonal of Equ.(3). Together with Poissons equation

$$\Delta U \approx - \frac{\rho}{\varepsilon \varepsilon_0}$$

(11)

it has to be determined self-consistently. The charge density ρ is obtained from the solutions of Equ.(3). In this article we will not pursue the question of charge transfer any further.

3. Some Results

We demonstrate a calculation for a IV-VI/IV-VI superlattice, like PbTe/PbSnTe. These semiconductors crystallize in rock salt structure with a direct small gap at the L-point. Only the

double degenerate conduction and valence band are considered.
The 4x4 EFA equation reads

$$
\begin{bmatrix}
(E_c-E+A_ck_\perp^2+B_ck_3^2)I_{22} & \begin{matrix} P_\parallel k_3 & P_\perp(k_1-ik_2) \\ P_\perp(k_1+ik_2) & -P_\parallel k_3 \end{matrix} \\[2em]
\begin{matrix} P_\parallel k_3 & P_\perp(k_1-ik_2) \\ P_\perp(k_1+ik_2) & -P_\parallel k_3 \end{matrix} & (E_v-E+A_vk_\perp^2+B_vk_3^2)I_{22}
\end{bmatrix}
\begin{bmatrix} f_1 \\ f_2 \\ f_3 \\ f_4 \end{bmatrix} = 0 \quad (12).
$$

1,2,3 denote a coordinate system with 3 the main valley axis
and I_{22} the 2x2 unit matrix. If we consider a superlattice
with perfect flat band edges the solutions to (12) are plane
waves in each layer A and B and the operators k are replaced
by wave-numbers. Furthermore, from the last two equations of
(12)

$$
\begin{bmatrix} f_3 \\ f_4 \end{bmatrix} = h \begin{bmatrix} f_1 \\ f_2 \end{bmatrix}
\qquad (13)
$$

with the meaning of h obvious. If the growth direction z does
not coincide with the main valley axis, k_1, k_2, k_3 are expressed by k_x, k_y, k_z. The in-plane momenta k_x and k_y enter as
free parameters into the calculation. We look now for all
possible solutions of Equ.(12) to a given set of E and k_x, k_y.
Four possible roots k_z are found by the zeros of the coefficient determinant in Equ.(12). Numerically it is immediately
found, that two values k_z are far out of the validity range of
EFA. The corresponding solutions therefore must not contribute
to the superlattice solutions for the EFA method to be reasonable. Now we have four independent degenerate solutions:

$$
A_1 \begin{bmatrix} 1 \\ 0 \\ f_3 \\ f_4 \end{bmatrix} e^{ik_z^1 z}, \quad
A_2 \begin{bmatrix} Q \\ 1 \\ f_3 \\ f_4 \end{bmatrix} e^{ik_z^1 z}, \quad
A_3 \begin{bmatrix} 1 \\ 0 \\ f_3 \\ f_4 \end{bmatrix} e^{ik_z^2 z}, \quad
A_4 \begin{bmatrix} 0 \\ 1 \\ f_3 \\ f_4 \end{bmatrix} e^{ik_z^2 z}.
$$

Each solution is multiplied by $\exp(ik_x x+ik_y y)$. The factors f_3
and f_4 are obtained by using Equ.(13). Similarly there are four
independent solutions in layer B. The eight amplitudes
$A_1,\dots,B_4$ are to be determined by the four continuity conditions
(6) at every interface. Together with the Bloch conditions (10)
this yields eight linear equations for the eight amplitudes. To

123

have a nontrivial solution the 8x8 coefficient determinant which is a function of the energy has to vanish. Thus the energy dispersion $E(k_x, k_y, K)$ is determined. In Ref./3/ energy dispersion curves are given. It should be noted, that the conditions (8) for the derivatives have not been used. Considering them would require one to take into account the other linear independent solutions, i.e. the plane wave solutions with the unreasonably large k_z wave vectors.

For in-plane momenta equal to zero, the dispersion relation can be given analytically /3/ as

$$\cos(KD) = -\frac{1}{2}\left[\frac{T}{S} + \frac{S}{T}\right] \sin(k_{zA}d_A)\sin(k_{zB}d_B)$$
$$+ \cos(k_{zA}d_A)\cos(k_{zB}d_B) \tag{14}$$

with d_A and d_B the thicknesses of layer A and B and

$$S = \frac{k_{3A}}{E_v^A + A_v k_\perp^2 + B_v k_{||}^2 - E}$$

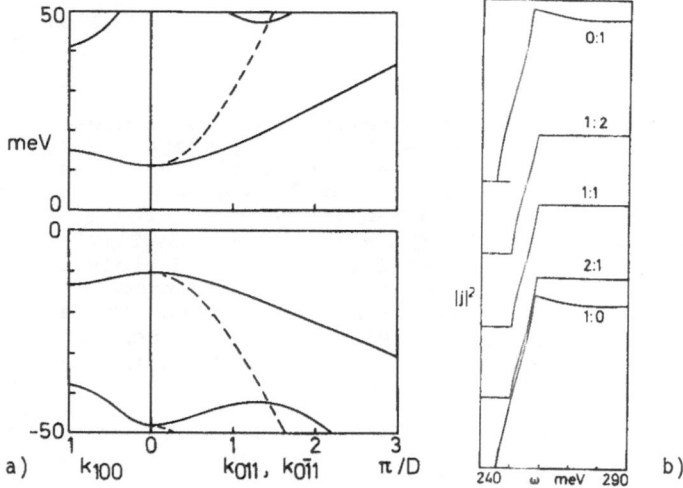

a)

b)

Fig. 1a. Energy dispersion curves for a type I PbTe/PbEuSeTe (20/7 nm) superlattice with equal conduction and valence band off set for valley [111] and growth direction [100]. Solid line: in-plane wave vector parallel [011]; dashed line:[0$\bar{1}$1]. The zeros of the energy scale denote the conduction and valence band edges of the PbTe layer.

Fig.1b. Optical interband transitions for different band alignments. The ratio of conduction to valence band off set is indicated. Note the independence of the leveling off energy of the band set up.

Table 1. Material parameters

	d	E_g	P	P̄	A_c	B_c	A_v	B_v
	(nm)	(meV)	(meVnm)			(meVnm^2)		
PbTe	20	229.3	144	476	586	78.2	-401	-44.6
PbEuSeTe	7	386.6	144	476	586	78.2	-401	-44.6

and T similar for B. The dependence of k_z on E is obtained from Equ.(12).

As the wavefunctions are known it is possible to calculate several physical quantities. In Fig. 1 the interband transition probability for a PbTe/PbEuSeTe superlattice grown in [100] direction is given for different band alignments. Table 1 shows the material parameters used (T = 100 K).

Fig. 1a gives the energy dispersion curve for the valley with main axis [111] for equal valence and conduction band offset. The wave vector k_{100} indicates the superlattice Bloch vector K of Equ.(10). Two electric subbands are within the energy range displayed in the figure in both conduction and valence band. The in-plane energy dispersion exhibits a strong orientational dependence. For the direction perpendicular to the valley axis the dispersion is rather bulk-like, whereas it is quite distorted for the other in-plane directions.

After calculating the energy dispersion for all values of K and in-plane momenta k_x, k_y, the electric dipole interband transition probability can be calculated by

$$|j|^2 = e^2 \int_{-\pi/D}^{\pi/D} dK \int dk_x dk_y |<f|v|i>|^2 \delta(E_f - E_i - \hbar\omega) \tag{15}$$

Here v is the velocity operator and i and f denote the initial and final state. Only the lowest subband contributes in the valence and conduction band in the range of transition energies displayed in Fig. 1b. The dependence of the oscillator strength is similar to that of a two dimensional electron gas. This is of course expected, as the band width for fixed in-plane momentum is only 15 meV. The onset of the transition depends on the band off-set. The transition energy, at which the maximum oscillator strength is reached is however independent of the band offset. It is therefore questionable to deduce the band offset from the frequency of a superlattice laser / 4 /.

In the case of an in-plane magnetic field the solution has to be obtained by solving the EFA equations numerically. We did this by the method of finite elements, or by the finite difference method, which results in a Hermitian eigenvalue problem. Fig. 2 shows the energy levels for a PbSnTe/PbTe superlattice

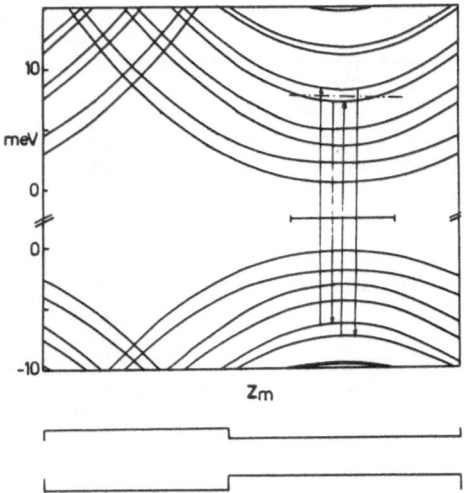

Table 2. Material parameters

	d (nm)	E_c (meV)	E_v (meV)
PbSnTe	54	0	-118
PbTe	46	28	-163

Other parameters as in Table 1.

Fig. 2. Energy versus cyclotron orbit center coordinate for a PbSnTe/PbTe (54/46) superlattice. Growth direction [111], valley axis [111], B [$\bar{1}$10] 1T. The zero of the energy scales is at the band edges. The dash-dotted curve indicates the Fermi level. Possible virtual transitions for a four wave mixing experiment for determination of the conduction band g-value are indicated. The bar indicates the magnetic length$\sqrt{\hbar c/eB}$. The band set up is shown.

for $B||[110]$ for the [111] valley and growth direction also [111]. The data are given in Table 2 (T = 4.2 K). Choosing the vector potential A in Landau gauge parallel to the super-lattice plane yields two constants of motion, viz. the wave-numbers parallel to B and to A. The latter is as usual inversely proportional to the cyclotron center coordinate Z_m. As Z_m is shifted along the superlattice growth direction the energy changes as indicated in Fig. 2. The selection rules for electric dipole transitions are similar as for bulk material. Due to the lifting of the degeneracy with Z_m no sharp transition energy is defined and neither a sharp inter- nor intra-band transition can be observed. However, the Landé factor is as seen from Fig. 2 rather independent of Z_m and can be measured in a four wave mixing experiment /5/. The possible transitions are indicated in the figure.

For III-V and II-VI semiconductors with a direct gap at the center of the Brillouin zone, the double degenerate Γ_6, the fourfould Γ_8 and the double degenerate Γ_7 band are to be considered. Without a magnetic field the 8x8 Hamiltonian may be reduced to two 4x4 equations. For quantization axis z, interface in the (y,z) plane and momentum k_z equal to zero and $\gamma_2 = \gamma_3$ /2/ the Hamiltonian reads for the Bloch functions

$$|1/2 \quad 1/2\rangle, \qquad |3/2 \quad 3/2\rangle, \qquad |3/2 \quad -1/2\rangle, \qquad |1/2 \quad -1/2\rangle$$

$$\begin{bmatrix} E_c + k^2/2 & -Pk_+/\sqrt{2} & Pk_-/\sqrt{6} & -Pk_-/\sqrt{3} \\ -Pk_-/\sqrt{2} & E_v - (\gamma_1+\gamma_2)k^2/2 & \sqrt{3}\gamma_2 k_-^2/2 & -\gamma_2 k_-^2\sqrt{3/2} \\ Pk_+/\sqrt{6} & \sqrt{3}\gamma_2 k_+^2/2 & E_v - (\gamma_1-\gamma_2)k^2/2 & \gamma_2 k^2/\sqrt{2} \\ -Pk_+/\sqrt{3} & -\gamma_2 k_+^2\sqrt{3/2} & \gamma_2 k^2/\sqrt{2} & E_v - \Delta - \gamma_1 k^2/2 \end{bmatrix}$$

$$(16)$$

$$k_\pm = k_x \pm ik_y.$$

A similar equation holds for the other four states. Fig. 3 shows the band structure for a CdTe/HgTe superlattice. HgTe has an inverted band structure compared to CdTe, i.e. the Γ_6 band is below the Γ_8 band. Table 3 gives the material parameters.

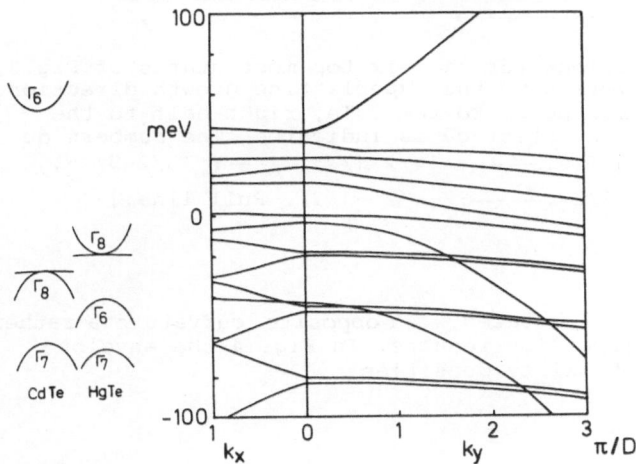

Fig. 3. Band structure of a CdTe/HgTe (15/15 nm) superlattice. k_x is the superlattice Bloch vector. The zero of energy is at the top of the CdTe valence band. The band set up is indicated schematically.

Table 3. Material parameters

	d (nm)	$E_{\Gamma 6}$ (meV)	$E_{\Gamma 8}$ (meV)	Δ	P (meVnm)	γ_1	γ_2
CdTe	15	1600	0	910	888	5.2	1.4
HgTe	15	-263	40	1000	828	5.2	1.4

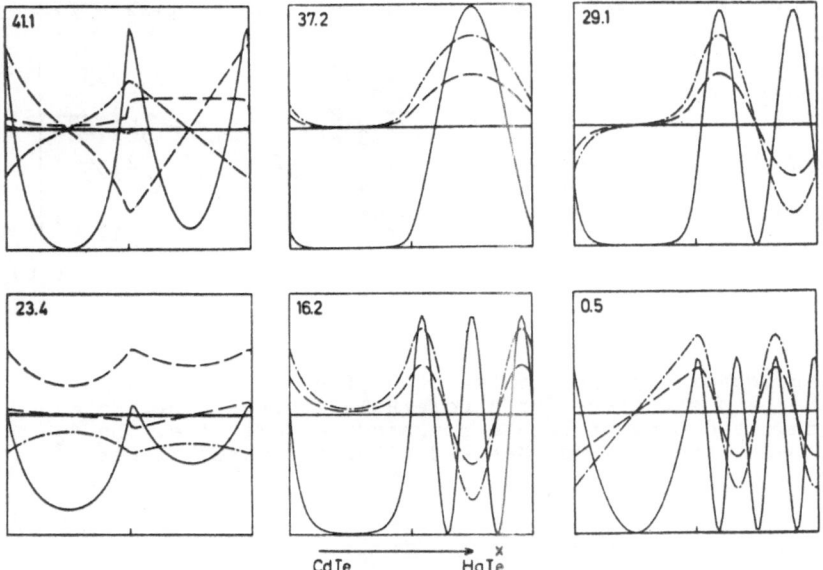

CdTe ──────→ x HgTe

Fig. 4. Envelope functions for the six top-most states of Fig.3
at $k_x = k_y = 0$ versus x, the superlattice growth direction.
Left half corresponds to the CdTe, right half to the
HgTe layer. The interface is indicated. The numbers de-
note the energies. ──── : $|1/2\ 1/2\rangle$, ──·── : $|3/2\ 3/2\rangle$,
── ── : $|3/2\ -1/2\rangle$, ──··── : $|1/2\ -1/2\rangle$. Full line: $|\Psi|^2$.

Due to the coupling of Γ_8 states with opposite curvature a rather
complicated energy dispersion results. In Fig. 4 the envelope
functions and the probability densities

$$|\psi|^2 = \sum_n |f_n|^2$$

are shown for six states with superlattice Bloch vector k_x and
in plane momentum k_y equal to zero. The two states with energy
41.1 meV and 23.4 meV are so-called interface states with charge
accumulation at the interface. This behaviour is expected by
joining one band solutions according to Equ.(9) with a positive
and a negative mass. However, as seen from Fig. 4, the magnitude of the Γ_6 state is also quite considerable for the interface states and a one band calculation is rather inadequate.
The four other states are the usual combination of oscillating
and exponentially decaying states. The amplitude of the Γ_6 and
Γ_7 states vanishes for these states. These results should be
compared to a calculation for a CdTe/HgTe/CdTe double heterostructure of Bastard /6/. Further calculations are required to
decide on the possible effects of the electrostatic potential
due to charge accumulation at the interface.

4. Outlook

The EFA proves to be an easy way to calculate the properties of a heterostructure of materials with similar band structure. It allows one to include magnetic field, strain, charge transfer and also shallow impurities. What is still missing is a formulation of continuity conditions at the interface for materials with different band structures like e.g. Ge/Si. A microscopic theory of the interface will be needed to give a quantitative estimate of the error made in neglecting the difference in Blochfunctions ϕ^A, ϕ^B. Numerical estimates show that in most cases the results for energy dispersion are within a certain range rather independent of the width of the transition region as long as the continuity condition (6) is fulfilled.

References

1. G. Bastard: Phys. Rev. B25, 7584 (1982); 24, 5693 (1981).
2. M. Altarelli: Lecture Notes in Physics 177, 174 (1982).
3. M. Kriechbaum, K.E. Ambrosch, E.J. Fantner, H. Clemens, G. Bauer: Phys. Rev. B30, 3394 (1984).
4. D. Partin: Superlattices and Microstructures 1, 131 (1985).
5. H. Pascher: P. Pichler, G. Bauer, H. Clemens, E.J. Fantner, M. Kriechbaum: Surface Science, in print.
6. G. Bastard: Surface Science, in print.

Optical and Electronic Properties of Si/SiGe Superlattices

G. Abstreiter, H. Brugger, T. Wolf*, R. Zachai, and Ch. Zeller*

Physik-Department E 16, Technische Universität München,
D-8046 Garching, Fed. Rep. of Germany

Abstract

Selectively doped Si/SiGe strained layer superlattices are
studied with various methods. Strong mobility enhancement is observed when the SiGe layers are doped with Sb. The two-dimensional nature of the electron gas is investigated by the angular
dependence of the Shubnikov-de Haas effect. Built-in strains are
determined by phonon Raman spectroscopy. The effects of strain
on the band structure are used together with self-consistent subband calculations to achieve a consistent picture of the band
ordering. In addition,superlattice effects are observed with
Raman scattering by folded acoustic phonons.

1. Introduction

Si/SiGe strained layer superlattices, quantum well structures, and single
heterostructures have received considerable attention recently /1,2,3/. In
the present article we concentrate on n-doped (Sb) superlattices grown by
molecular beam epitaxy on [001] oriented Si and SiGe substrates. Growth
properties and material preparation of the novel semiconductor structures
are reported in a separate paper of these proceedings /4/ and are not repeated here. In the following we concentrate on various aspects which are
related to the optic and electronic properties. There exists a large lattice mismatch between Si and Ge of about 4%. Below a critical thickness
this mismatch is accommodated by a lateral biaxial strain. The effect of
built-in strain on the band structure of Si and SiGe alloys,as well as the
measurements of the actual strain values by Raman scattering,are discussed
in the first part of section 2. It is followed by the description of the
experimental observation of mobility enhancement achieved by selective
doping and of the existence of a two-dimensional electron gas in the Si
layers for symmetrically strained superlattices. Self-consistent calculations of the potential wells, subband structure, and Fermi energy have been
performed in order to verify the assumptions on conduction and valence band
ordering as deduced from the experimental results. A model for the strain-
induced effects on the band structure and band offsets is extracted from
theory and experiments. It explains the results in a consistent way. Section 2 ends with the discussion of superlattice effects,which show up in
Raman scattering by folded acoustic phonons.

---

* Present address: Forschungslaboratorien der Siemens AG,
  D-8000 München 83, FRG

2. Discussion of Experimental and Theoretical Results

2.1. Effect of Built-in Strain on the Band Structure

The indirect band gaps of Si, Ge, and SiGe alloys have been studied by optical absorption measurements /5/. The conduction band ordering remains Si-like up to a Ge content of about 85%. The conduction band minimum is along the [100] (Δ)-direction and sixfold degenerate. In unstressed Ge-rich alloys the conduction band minimum is at the L-point of the Brillouin zone and consequently fourfold degenerate. The behavior is found to be equivalent to the application of hydrostatic pressure on Ge /6/.

Large biaxial lateral strain is present in samples where Si or SiGe layers are grown on substrates with different lattice constants, as it is the case in the superlattices discussed below. The built-in strain can be described as the sum of a hydrostatic pressure and a uniaxial stress normal to the layers, which is discussed in more detail in Sect. 2.2. It causes strong changes in the band structure /7,8/. The hydrostatic part basically shifts the valence and conduction band up or down depending on the sign of the pressure (compression or tension). It causes also small changes in the values of the indirect band gap energies. The uniaxial part of the strain is responsible for a splitting of the six-fold degenerate conduction bands and the heavy and light hole valence bands. In order to predict the changes of the band edges quantitatively, a precise knowledge of the deformation potentials is necessary. After a careful selection of the published data /9,10,11/ we used the following consistent set of deformation potentials in our calculations. Si: $\Xi_d(\Delta)$ = - 6 eV; $\Xi_u(\Delta)$ = 9 eV; a = 4.6 eV; b = - 1.9 eV; $\partial E_g/\partial p(\Delta)$ = - 1.6 meV/kbar; Ge: $\Xi_d(L)$ = - 12.3 eV; $\Xi_u(L)$ = 19.3 eV; a = 3.6 eV; b = - 2.2 eV; $\partial E_g/\partial p(L)$ = + 3 meV/kbar. For the L minimum of Si we used the same values as for Ge, for the Δ-minimum we used the Si-values.

In Fig. 1 we show the effect of biaxial strain on the band edges for a Si/Si$_{0.5}$Ge$_{0.5}$ superlattice grown on Si$_{0.75}$Ge$_{0.25}$ substrate. This results in a symmetrical strain distribution, tensile in Si and compressive in Si$_{0.5}$Ge$_{0.5}$. The hydrostatic part of the strain causes a shift of the Si-band edges to lower energies and a small increase of the energy gap. This is reversed for the Si$_{0.5}$Ge$_{0.5}$ layers. The uniaxial part of the strain on the other hand causes a splitting of the conduction and valence band. In Si the two-fold degenerate valley with the large mass normal to the layers is lowered in energy, and the light hole valence band is shifted upwards. The light hole valence band interacts with the spin-orbit split-off valence band. This is included in the calculation. The signs for the splitting of the Si$_{0.5}$Ge$_{0.5}$ layers are reversed. In order to obtain the band offsets we made the assumption that ΔE_c between Si and Si$_{0.5}$Ge$_{0.5}$ for the unstrained situation is slightly positive, i.e. the conduction band edge of Si$_{0.5}$Ge$_{0.5}$ is above the conduction band minimum of Si. This is justified from the experiments, as discussed below.

In Fig. 2 we show the strain-induced energy gaps of SiGe layers grown on different substrates. The energy gap for the unstrained situation (thick solid line) is taken from Ref. 5. The strain-induced band gaps are calculated by including the hydrostatic and the uniaxial pressure-induced shifts and splittings of the band edges. Results are shown for strained overlayers on Si, Si$_{0.75}$Ge$_{0.25}$, Si$_{0.5}$Ge$_{0.5}$, and Ge substrates. In all cases the energy gap is reduced for the whole concentration region from pure Si to pure Ge.

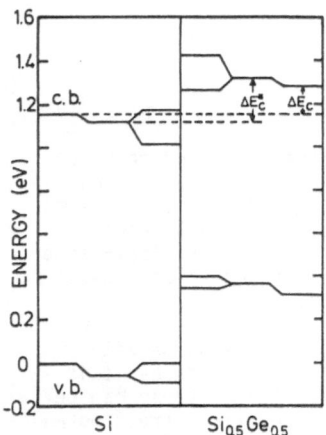

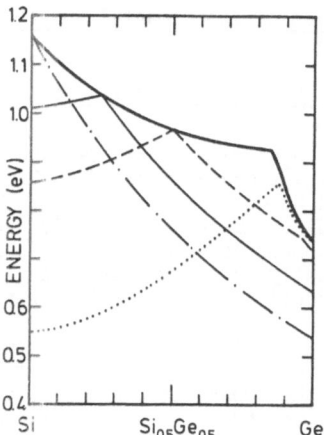

Fig. 1: The effect of hydrostatic and uniaxial stress on
the valence and conduction
band edges in Si and $Si_{0.5}Ge_{0.5}$

Fig. 2: Band gap energies in
SiGe for different biaxial
strain conditions: $-\cdot-$ Si , ——
$Si_{.75}Ge_{.25}$, ——$Si_{.5}Ge_{.5}$,..Ge subs.

For Si-rich substrates the band structure remains Si-like even for pure Ge.
The conduction band minima are along the Δ-direction. Due to the strain the
L-minima are higher in energy. They are only important for Ge-rich substrates and high Ge-content in the epitaxial layers. This can be seen by
the abrupt change of the slope in the region of high Ge-content.

2.2. Measurement of Built-in Strain by Raman Spectroscopy

In the following part we show that Raman scattering is a powerful tool for
studying the crystalline quality and the built-in strain in Si/SiGe superlattices. It is demonstrated that the entire lattice mismatch is accommodated as a homogeneous tetragonal strain in the superlattice layers.
By comparing peak positions in commensurate superlattices and single layers
with those from incommensurate thick layers of the same composition we can
obtain a quantitative determination of stress. From shape and linewidth of
the phonon modes we obtain qualitative information about inhomogeneous
strain, dislocation densities, and disorder.

In general, a strained superlattice is a periodic sequence consisting of
alternate layers of two materials (1,2) with lattice constants $a^{(1,2)}$ and
layer thicknesses $d^{(1,2)}$. The lattice mismatch Δa is elastically accommodated, i.e. the materials inside the superlattice are strained in order
to match their in-plane parameter a". This is called a pseudomorphic or
commensurate growth. Two types of superlattices have been used in our experiments. The layer sequence is shown schematically in the right part of
Fig. 3. They consist of alternate   layers of ten 60 Å Si and 60 Å
$Si_{0.5}Ge_{0.5}$ grown on two different nearly unstrained buffer layers of 2000 Å
thickness. The difference in lattice constant between Si and Ge varies
linearly with alloy composition. Consequently, in the case of Si buffer
layer (VS 94), it is expected that only the alloy layers $Si_{0.5}Ge_{0.5}$ are
strained ($\varepsilon \approx -2\%$). For $Si_{0.75}Ge_{0.25}$ buffer layers (VS 73) a medium lattice spacing between $Si_{0.5}Ge_{0.5}$ and pure Si is achieved. The buffer layer

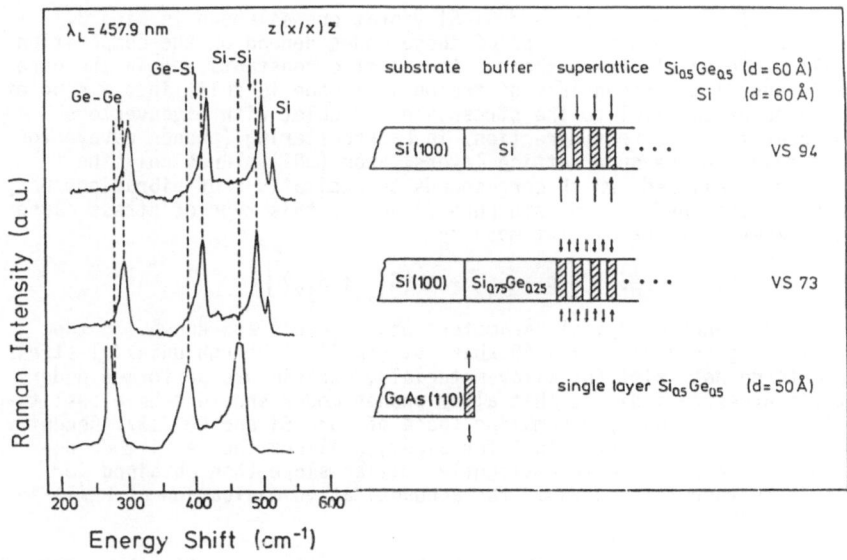

of 2000 Å is larger than the critical thickness of pseudomorphic growth.
This yields a relaxation of the film by producing misfit dislocations.
Therefore both layers should be strained symmetrically ($\varepsilon^{(1,2)} \simeq \pm 1\%$).

If one defines x $\parallel$ [100], y $\parallel$ [010], and z $\parallel$ [001] and takes into account
the cubic symmetry, then the strain tensor ε_{ij} has only diagonal components
($\varepsilon_{ij} = 0$, $i \neq j$): $\varepsilon_{xx} = \varepsilon_{yy} = (a'' - a)/a$, and the stress-strain relationship is given by the expression

$$\varepsilon_{ij} = \sum_{k,l} S_{ijkl}\sigma_{kl}.$$

Here the S_{ij} are the elastic compliance constants. The lattice is not
stressed in z-direction (free surface)$\sigma_{zz} = 0$, and therefore the extension
ε_{zz} perpendicular to the plane is given by $\varepsilon_{zz} = - 2S_{12}/(S_{11} + S_{12}) \cdot \varepsilon_{xx}$.
As already mentioned, the lattice mismatch is accommodated by lateral
biaxial strain, which can be described as a biaxial stress. This is equivalent to a hydrostatic pressure and a uniaxial stress of opposite sign
perpendicular to the interface. The stress tensor σ_{kl} can be written in
the following form:

$$\sigma_{kl} = \begin{pmatrix} \tau & & \\ & \tau & \\ & & 0 \end{pmatrix} = \tau \cdot \overline{\overline{1}} + \begin{pmatrix} 0 & & \\ & 0 & \\ & & -\tau \end{pmatrix}$$

τ is positive for tensile ($\varepsilon > 0$) and negative ($\varepsilon < 0$) for compressive
stress.

First order Raman spectra in Ge and Si exhibit one sharp line at about
301 cm^{-1} and 520 cm^{-1} respectively, which corresponds to the q $\simeq$ 0 triply
degenerate optical phonons ($\omega^{opt.}$). The alloy spectrum shows three main

peaks related to Ge-Ge, Si-Ge, and Si-Si vibrations as shown in Fig. 3. The frequency and intensity ratios of these modes depend on the composition of the alloy. Uniaxial stress changes the elastic constants, and in the case of pure Ge and Si the degeneracy of the optical mode is split into a singlet with eigenvector parallel to the stress and a doublet with eigenvectors perpendicular to the stress direction. In backscattering (phonon wavevector q perpendicular to the superlattice layers) from (001) planes only the singlet mode is allowed, which corresponds to excitations of vibrations perpendicular to the surface. In both pure Si and Ge this type of stress causes a shift in energy of the singlet mode by

$$\Delta\omega = \frac{1}{\omega_{opt.}} \cdot \left[p^* \cdot S_{12} + q^*(S_{11} + S_{12}) \right] \cdot \tau$$

p^* and q^* are phenomenological parameters whose values are known. In the case of Si one gets /12/: $\tau = 2.49$ kbar $\cdot \Delta\omega$ (cm^{-1}). Although uniaxial stress measurements do not exist for alloy materials, experiments performed under hydrostatic pressure indicate that alloy phonon modes should behave qualitatively in a manner closely resembling those of pure Si and Ge /13/. Recently Cerdeira et al. /3/ suggested that for Si_xGe_{1-x} alloys the Ge-Si mode depends linearly on strain with a slightly smaller slope than obtained for pure Si. It is therefore suitable for a quantitative evaluation of built-in strain in alloys.

Fig. 3 shows typical room-temperature Raman spectra from $Si_{0.5}Ge_{0.5}/Si$ strained superlattices and a single pseudomorphic film of $Si_{0.5}Ge_{0.5}$ on a GaAs (110) cleavage plane. In the latter case the spectrum is dominated by the allowed TO-phonon mode of GaAs. The penetration depth of the used laser light is small (< 1000 Å) and therefore information is only obtained of the superlattice layers. The expected positions of unstrained layers are marked by arrows.

The phonons of the alloy layers are shifted towards higher energies in the superlattice structure due to the compressive biaxial strain ($\varepsilon < 0$). The shift is nearly twice as large for sample VS 94 ($\varepsilon \simeq -2\%$) compared to sample VS 73 ($\varepsilon \simeq -1\%$). In the case of GaAs substrate with a lattice spacing nearly equal to Ge, the film is under tensile biaxial strain ($\varepsilon \simeq +2\%$), which results in a downward shift of about the same amount as in VS 94 in the opposite direction. In all cases the lineshapes do not differ significantly from unstrained bulk samples, i.e. the lattice mismatch is entirely accommodated by homogeneous tetragonal strain. The linewidth of the corresponding alloy modes of the superlattice spectra are nearly the same, and correspond to an excellent crystalline structure of the grown layers. On the other hand the alloy modes of the single film on GaAs are broadened. This may be due to phonon softening effects originating from the thin film and the free surface, but also due to a better quality found for layers embedded in a superlattice structure compared with single films. This effect is not yet understood.

In the superlattice VS 73 a downward shift of the Si phonon mode of about 7 cm^{-1} (0.87 meV) is observed. This corresponds to lateral tensile stress of approximately 17 kbar in the Si layers and a lateral compressive stress of about the same amount in the alloy layers. The in-plane strain distribution is schematically shown by arrows on the layers of the superlattices. An oscillating strain field ε occurs through the superlattice.

The situation in VS 94 is more complicated. Nominally the Si layers should be unstrained because of the Si buffer layer. On the other hand the Si mode shows a slight but measurable downward shift in energy, indicating a final tensile built-in stress. The reason might be that the critical superlattice thickness is already reached. Theoretical considerations and experiments show two critical thicknesses to apply to strained-layer superlattice growth, one relating to the thickness of individual layers $d_c^{1,2}$ and one relating to the overall thickness of the superlattice d_c^{SL}. If the total strain energy exceeds a critical value, then the superlattice will partially relax by formation of misfit dislocations close to the interface of buffer layer and superlattice. The crystal quality of the superlattice should be unchanged, in agreement with our Raman results. The strain field distribution is then altered: $\varepsilon^{(1)} = 0$, $\varepsilon^{(2)} = -2\%$ → $\varepsilon^{(1)} > 0$, $\varepsilon^{(2)} > -2\%$. The strain energy is essentially determined by the weighted stress difference $(d^{(1)}\sigma^{(1)} - d^{(2)}\sigma^{(2)})$ which is negligible in VS 73 (quasi equilibrium), but large in VS 94.

2.3. Mobility Enhancement and 2 DEG in Si

Recently Jorke and coworkers [14] developed a doping method denoted as secondary implantation, which allows an abrupt incorporation of Sb dopant atoms of more than 10^{19} cm^{-3}. This doping facility is extremely useful to achieve selectively doped Si/SiGe heterostructures and superlattices. In the samples studied, the position of the Sb doping spike was varied with respect to the Si/SiGe layer sequence. The mobilities of these samples were studied with Hall measurements using a standard van der Pauw method. Results of the temperature-dependent Hall mobility are shown in Fig. 4. In samples VS 81, 82, and 83 the Sb dopants are situated inside the SiGe layers. Their mobility is strongly enhanced at low temperatures. In VS 73, 75, and 76 the doping atoms are incorporated in the Si layers. The decreasing mobility at low temperatures of samples VS 75 and 76 is typical for highly doped n-type Si. In the other samples the mobility is much improved, especially at low temperatures. This is shown more clearly in Fig. 5, where the T = 20 K and T = 300 K mobilities of all six samples are plotted versus the position of the doping spike. The surprising result is that enhanced mobilities are observed when the Sb dopants are placed in the SiGe alloy layers. This is direct experimental evidence that the conduction band in Si is lower in energy than that of $Si_{0.5}Ge_{0.5}$ (see Fig. 1).

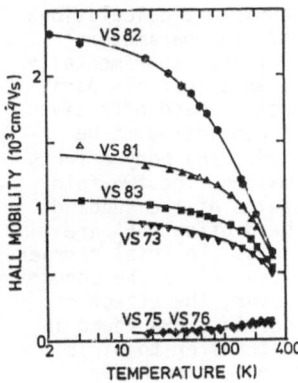

Fig. 4: Temperature dependence of Hall mobility for various samples

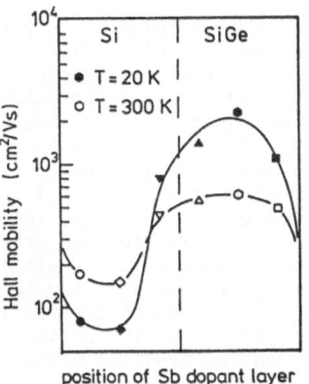

Fig. 5: Hall mobility at T = 300 K and T = 20 K versus position of the Sb dopants

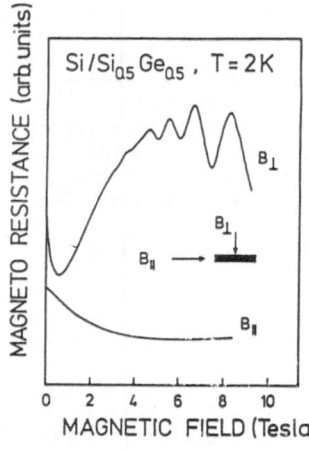

Fig. 6: Shubnikov-de Haas oscillations at T = 2 K for sample VS 82. The lack of oscillations in parallel field confirms the two-dimensional nature of the electron gas.

Direct confirmation of the existence of a two-dimensional electron gas is obtained from angular-dependent Shubnikov-de Haas oscillations. Fig. 6 shows the result of sample VS 82. The oscillation period with the magnetic field applied perpendicular to the layers leads to a two-dimensional electron density of 3.08×10^{12} cm^{-2} in reasonable agreement with the density per layer obtained from Hall measurements ($n \simeq 4 \times 10^{12}$ cm^{-2}). Shubnikov-de Haas oscillations with a similar period are also observed for samples VS 81 and VS 83. In the other samples the mobility is too low, such that $\omega_c \tau < 1$ up to B = 10 T. We also have studied a selectively doped superlattice grown directly on a Si substrate (VS 94). The Hall mobility is comparable to that observed for sample VS 83. However, only half the carriers are found to contribute to the high mobility. We therefore conclude that only half the carriers are localized in the Si layers occupying the two-fold degenerate valleys /15/.

2.4. Self-Consistent Subband Calculations

Recently Zeller and Abstreiter have performed self-consistent calculations of the subband structure of Si/SiGe superlattices /16/. The parameters used in this work were chosen to mimic the situation of the experimentally studied samples. The relevant band offset was used as an adjustable parameter. Some simplifications were made: The same conduction band effective mass was used for Si and SiGe. The doped material was considered to be degenerate. The donor binding energy has been neglected. Many-body effects were taken into account in the local density approximation. The two-fold and four-fold valleys were treated independently. Results of the subband energies and electron distribution versus effective band offset ΔE_c^* are shown in Fig. 7 for symmetrically strained superlattices. The total carrier concentration was fixed to $n_s = 4 \times 10^{12}$ cm^{-2} per layer. ΔE_c^* is the conduction band offset in the unsplit situation, where, however, the effect of hydrostatic pressure is included (see Fig. 1). Some carriers are found in the Si layers even for negative values of ΔE_c^*. Charge transfer to Si is made possible by the strain-induced splitting of the conduction band. A complete transfer from $Si_{0.5}Ge_{0.5}$ to Si is found for $\Delta E_c^* > + 20$ meV. In Fig. 8 we show similar results for samples grown directly on Si substrates. The splitting of $Si_{0.5}Ge_{0.5}$ conduction bands is doubled, while the Si band edge remains six-fold degenerate . A complete charge transfer

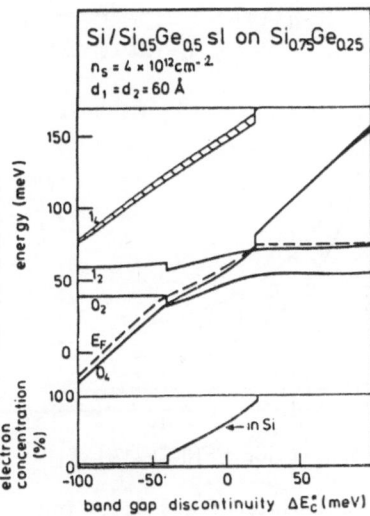

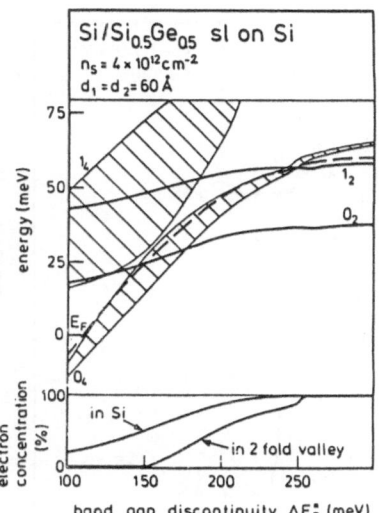

Fig. 7: Subband energies for the four-fold and two-fold degenerate valleys and carrier distribution versus ΔE_C^* (symmetrically strained superlattice)

Fig. 8: Like Fig. 7, but no strain in the Si layers, doubled strain in the $Si_{0.5}Ge_{0.5}$ layers

to the Si layers is only realized for $\Delta E_C^* > 250$ meV. At $\Delta E_C^* \simeq 200$ meV about half the carriers are quantized in the two-fold valleys in Si. The other part occupies the four-fold degenerate minibands which are not localized normal to the layers. Comparing the theoretical results with the transport experiments, we conclude that ΔE_C^* is about + 200 meV for this type of sample.

2.5. Folded Acoustic Phonons

The new periodicity d of artificial superlattices along the z-direction perpendicular to the layers causes a Brillouin-zone (BZ) folding and the appearance of gaps in the phonon spectrum for wavevectors satisfying the Bragg condition. Hence phonons propagating along z-direction with wavevectors $q = (2\pi/d)m$ (m = 1,2,3,...) within the extended zone are now mapped on the center of the folded Brillouin zone with the new reciprocal wavevector π/d. We have performed Raman measurements /17/ using backscattering geometry with the scattering wavevector $q_S \simeq 2 \cdot 2\pi n/\lambda_L \simeq 10^6$ cm^{-1} << π/a, where n is the medium refractive index and a the lattice constant of the sample, and λ_L is the used laser wavelength. In bulk crystals only phonons close to the zone center are observed. The scattering wavevector q_S is, however, comparable to the new Brillouin zone π/d and an experimental determination of the phonon dispersion curve $\omega(q)$ should be possible by changing the laser energy.

The energetic positions of the folded phonons vary sensitively with the layer thickness. This can be used to determine accurately the period $d = d^{(1)} + d^{(2)}$ of various Si/SiGe strained layer superlattices.

137

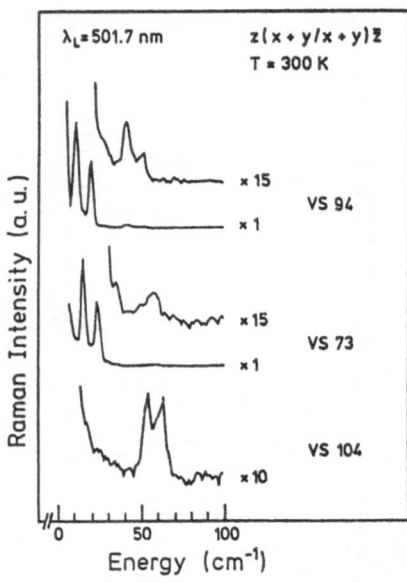

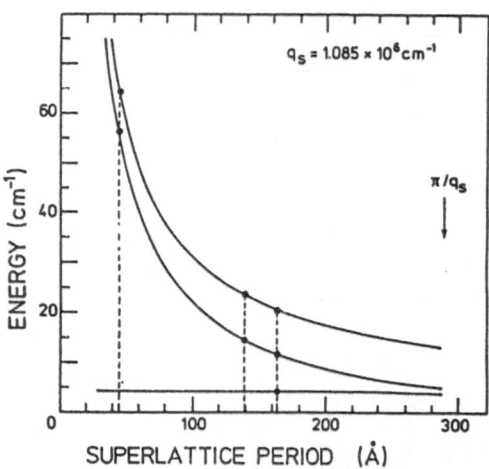

Fig. 9: Raman spectra of super-lattices with different periods d in the acoustical range

Fig. 10: Calculated frequencies for the first three eigenmodes of folded LA phonons as a function of superlattice period d. q_s is fixed by the used laser line (λ_L = 501.7 nm, n = 4.33). For large periods close to 29 nm q_s is equal to the new Brillouin zone boundary π/d.

In Fig. 9 the low-energy range of the phonon spectra of various superlattices is shown. Several sharp peaks are observed for all three samples. The two dominant ones of VS 73 and VS 94 are even stronger in intensity than the optical modes discussed above. The energetic positions of the doublets depend on layer thickness $d^{(1,2)}$ or superlattice period d.

The observed modes result from zone-folding of the longitudinal acoustic phonon branch as observed earlier in GaAs/AlAs superlattices /18/. The observed frequencies are explained in the elastic continuum limit. The circular frequency ω and the phonon wavevector q for equidistant layers $d^{(1)}$ = $d^{(2)}$ are related by /19/:

$$\cos(qd) = \cos\left(\frac{\omega d}{2v_1}\right)\cos\left(\frac{\omega d}{2v_2}\right) - \frac{1+k^2}{2k}\sin\left(\frac{\omega d}{2v_1}\right)\sin\left(\frac{\omega d}{2v_2}\right)$$

where $k = \rho_1 v_1/\rho_2 v_2$ and ρ_1, ρ_2 are the densities of different layers and v_1, v_2 are the sound velocities. We used the values v_1(Si) = $(c_{11}/\rho_1)^{1/2}$ = 8.433 x 10^5 cm/s, ρ_1(Si) = 2.33 g/cm^3, v_2(SiGe) = 7.5 x 10^5 cm/s, ρ_2(SiGe) = 3.83 g/cm^3. The sound velocity of the alloy material has been determined from the experimentally observed splitting $\Delta\omega = 2q_s v$ of the folded LA-phonon doublet /17/. It is the same for all samples investigated so far.

In Fig. 10 the calculated frequencies of the three low-energy modes are plotted versus superlattice period d for fixed q_s. The first mode at 4.6cm^{-1} is essentially independent of d and is determined by $\omega = q_s v$. All samples exhibit this mode, not shown in the Raman spectra of Fig. 9 due to the high amount of Rayleigh light. The other two modes represent the first strong doublet with constant splitting $\Delta\omega \simeq 9$ cm^{-1}, whose position shifts sensitively to higher energies with decreasing layer thickness. The experimental values are given by the dots. By comparing the positions with the calculated energies we determine the periods of the strained layer superlattices: d = 46 Å (VS 104), 138 Å (VS 73), and 163 Å (VS 94).

Acknowledgements: We thank E.Kasper,H.J.Herzog and H.Jorke from AEG Ulm for providing the Si/SiGe superlattices and for many stimulating discussions.

References:
1/ G.Abstreiter, H.Brugger, T.Wolf, H.Jorke, and H.J.Herzog, Phys.Rev.Lett. 54, 2441 (1985)
2/ R.People, J.C.Bean, B.V.Lang, A.M.Sergent, H.L.Störmer, K.W.Wecht, R.T. Lynch, and K.Baldwin, Appl.Phys.Lett. 45, 1231 (1984)
3/ F.Cerdeira, A.Pinczuk, J.C.Bean, B.Battlogg, B.A.Wilson, Appl.Phys.Lett. 45, 1138 (1984)
4/ E. Kasper, this volume; see also Surface Science (to be published)
5/ R.Braunstein et al., Phys.Rev. 109, 695 (1958)
6/ F.Bassani, D.Brust, Phys.Rev. 131, 1524 (1963)
7/ J.Balslev, Phys.Rev. 143, 636 (1966)
8/ R.People, Phys.Rev.B 32, 1405 (1985)
9/ Landoldt-Burnstein Bd. III/17a (1984)
10/ K.Murase et al., J.Phys.Soc.Jap. 29, 1248 (1970)
11/ J.A.Verges et al., phys.stat.sol.(b) 113, 519 (1982)
12/ Th.Englert et al., Solid State Electr. 23, 31 (1980)
13/ J.B.Renucci et al., Solid State Commun. 9, 1651 (1971)
14/ H.Jorke, H.Kibbel, Proc. 1st Int.Si-MBE Sympl, May 1985, Toronto, ed.: J.C.Bean, Electrochem. Soc.
15/ G.Abstreiter, H.Brugger, T.Wolf, H.Jorke, H.J. Herzog, Proc. 2nd Int. Conf. on "Modulated Semicond. Structures" Kyoto 1985, Surf.Sci.(in press)
16/ Ch. Zeller, G.Abstreiter, Z. Physik (to be published)
17/ H.Brugger et al., Phys.Rev. B, 1986 (in press)
18/ C.Colvard et al., Phys.Rev.Lett. 45, 298 (1980)
19/ S.M.Rytov, Sov.Phys.Acoust. 2, 68 (1956)

Resonant Tunneling Devices and Optoelectronic Ge/Si Superlattice Structures

S. Luryi and F. Capasso

AT & T Bell Laboratories, Murray Hill, NJ 07974, USA

This paper will review two classes of semiconductor devices: (1) resonant-tunneling diodes and transistors, and (2) infrared detectors on a silicon chip. The only connection between these topics is that both classes of devices employ heterojunctions.

1. Resonant Tunneling Devices

1.1 Introduction

Recently, a number of workers have demonstrated a negative differential resistance (NDR) and a microwave activity in double-barrier (DB) quantum-well (QW) structures, Fig. 1. Since the pioneering work of TSU, ESAKI, and CHANG [1] on tunneling in DBQW structures, the material quality has improved to the point that negative differential resistance (NDR) can be observed [2,3] directly in the current-voltage characteristics at 77 K, as opposed to the derivative of the

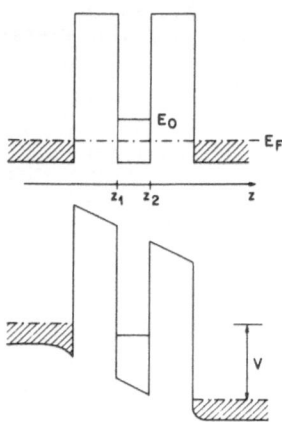

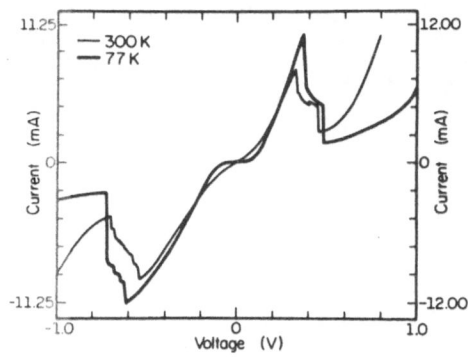

FIGURE 1: Band diagram of a double-barrier quantum-well diode in equilibrium (top) and under applied bias (bottom). Shaded regions indicate the Fermi sea of electrons in the degenerately doped emitter and collector layers.

FIGURE 2: The $I-V$ characteristics of a symmetric DBQW diode which contains an undoped 50 Å thick GaAs QW clad by two undoped 25 Å thick AlAs barrier layers and two n-doped GaAs layers [7]. Diode area is $\approx 2.8 \cdot 10^{-7} \text{ cm}^2$, corresponding to a peak current density of 30 kA/cm^2 at 300 K.

current as was the case with the first reports. The material quality has steadily improved, making it possible to observe the NDR at room temperature [4-7]. Recently, peak to valley (PTV) ratios in current of nearly 3:1 were obtained at room temperature [6,7] and 9:1 at 77 K [7], see Fig. 2.

As will be reviewed in the next Section, the NDR in DBQW diodes is a consequence of the dimensional confinement of states in a QW, and the conservation of energy and lateral momentum in tunneling. In addition to that, the operation of these structures has often been discussed in connection with a resonant tunneling effect analogous to that in a Fabry-Perot resonator. This effect is presumed to occur when the energies of incident electrons in the emitter match those of unoccupied states in the QW. Under such conditions, the amplitude of the resonant modes builds up in the QW to the extent that the electron waves leaking out in both directions cancel the reflected waves and enhance the transmitted ones. This physical picture has led to a design strategy intended to optimize the Fabry-Perot resonator conditions. In particular, RICCO and AZBEL [8] pointed out that achievement of a near-unity resonant transmission requires equal transmission coefficients for both barriers at the operating point — a condition not fulfilled for barriers designed to be symmetric in the absence of an applied field. To counter that, a resonant-tunneling structure was proposed [9] in which a symmetric DBQW was built in the base of a bipolar transistor, and the Fabry-Perot conditions were maintained through the use of minority-carrier injection.

High-frequency operation of DBQW diodes was recently considered [8,10] on the assumption that the underlying mechanism of NDR requires the Fabry-Perot resonant enhancement of the tunneling probability. It was found that the dominant delay results from the resonator charging time, which is of the order of the resonant-state lifetime. For a QW bounded by 50 Å-thick AlGaAs barriers, simple estimates [10] gave a frequency limit in the low gigahertz range. At higher frequencies, the amplitude of an electron wavefunction in the QW cannot readjust itself in response to an external field variation to provide resonant enhancement of the transmission coefficient. These estimates, contrasted with the experimental results [2] in which a DBQW structure was used as a detector and mixer of far-infrared radiation at 2.5 THz, have led one of us (SL) to the suggestion that the Fabry-Perot resonant transmission plays only a minor role (if any) in the operation of DBQW diodes. It has been shown [10,11] that the NDR can arise solely due to electron tunneling into a system of states of reduced dimensionality. In this picture, electrons subsequently leave the QW by tunneling through the second (collector) barrier, so that their transport through the entire DBQW structure is described by *sequential* rather than resonant tunneling.

1.2 Mechanism of Operation of Quantum Well Diodes

Let us review the mechanism of NDR in double-barrier QW structures — without invoking a resonant Fabry-Perot effect. Figure 3 illustrates the Fermi sea of electrons in a degenerately doped emitter. Assuming that the AlGaAs

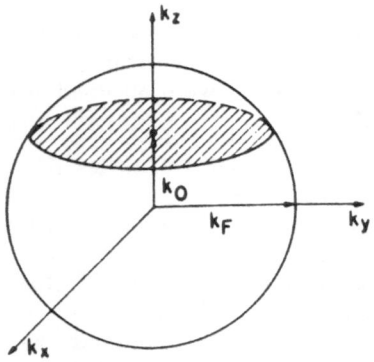

FIGURE 3: Illustration of the NDR mechanism in QW diodes. The shaded disk within the Fermi sphere corresponding to a degenerately doped emitter, describes all the electrons which can tunnel into the QW while conserving their lateral momenta. In an ideal diode at zero temperature the resonant tunneling occurs in a voltage range during which the shaded disk moves down from the pole to the equatorial plane of the emitter Fermi sphere. At higher biases resonant electrons no longer exist.

barrier is free of impurities and inhomogeneities, the lateral electron momentum (k_x, k_y) is conserved in tunneling. This means that for $E_C < E_0 < E_F$ (where E_C is the bottom of the conduction band in the emitter and E_0 is the bottom of the subband in the QW) tunneling is possible only for electrons whose momenta lie in a disk corresponding to $k_z = k_0$ (shaded disk in the figure), where $\hbar^2 k_0^2/2m = E_0 - E_C$. Only those electrons have isoenergetic states in the QW with the same k_x and k_y. This is a general feature of tunneling into a two-dimensional system of states. As the emitter-base potential rises, so does the number of electrons which can tunnel: the shaded disk moves downward to the equatorial plane of the Fermi sphere. For $k_0 = 0$ the number of tunneling electrons per unit area equals $mE_F/\pi\hbar^2$. When E_C rises above E_0, then at $T = 0$ there are no electrons in the emitter which can tunnel into the QW while conserving their lateral momentum. Therefore, one can expect an abrupt drop in the tunneling current. Extension of this picture to the case of several subbands in the QW is straightforward.

This sequential-tunneling mechanism of NDR is experimentally distinguishable from the Fabry-Perot model. In particular, it does not depend on the symmetry of transmission coefficients of the two barriers, and should not degrade, therefore, if the transparency of the second (collector) barrier is enhanced. Within the sequential-tunneling model, the terahertz results [2] can be explained — since rectification of an external signal by a DBQW diode requires the readjustment of only the phase of electronic wavefunctions and not their amplitude, so that the operation of a detector is not limited by a Fabry-Perot charging time [12]. The effect is conceptually similar to that in the Esaki diode. It should also be observable in various single-barrier structures in which tunneling occurs into a two-dimensional system of states.

Indeed, according to the described model, in DBQW structures the removal of electrons from the QW occurs via sequential tunneling, but other means of electron removal can also be contemplated, for example, *recombination*. REZEK et al. [13] studied electron tunneling through a single barrier into a QW located in a *p*-type quaternary material. In these experiments the diode current resulted from the subsequent recombination of tunneling electrons with holes in the direct-gap QW. The observed structure in the dependences of the current and the intensity of the recombination radiation on the applied bias can be explained in terms of the above picture based on the momentum conservation.

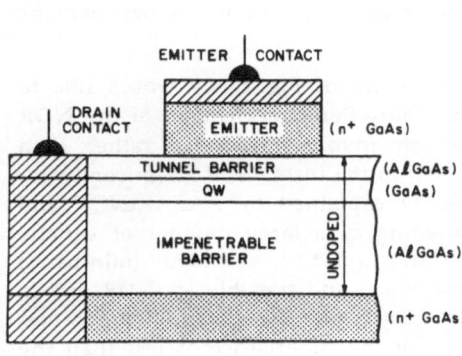

FIGURE 4: Illustration of a single-barrier QW structure which exhibits an NDR effect similar to that observed in DBQW diodes. The drain contact is assumed to be concentric with a cylindrical emitter electrode. The "impenetrable" collector barrier separating the QW from the conducting layer underneath (the latter is shorted to the drain) must be thin enough ($\lesssim 1000\,\text{Å}$), so that the emitter-to-QW potential could be effectively controlled by the emitter bias. In the experiments [7] the tunnel barrier and the QW thicknesses and composition were identical to those in the control double-barrier structure (Fig. 2), whereas the collector barrier represented $500\,\text{Å}$ thick $Al_x Ga_{1-x} As$ layer with $x = 0.3$.

The NDR effect of a similar nature can also be observed in a unipolar single-barrier structure, as was first proposed in [11] and demonstrated in [7]. Let the emitter be separated by a thin tunneling barrier from a QW which is confined on the other side by a thin but impenetrable (for tunneling) barrier, Fig. 4. The drain contact to the QW, located outside the emitter area, should be electrically connected to a conducting layer underneath. Application of a negative bias to the emitter will result in the tunneling of electrons into the QW and their subsequent drift laterally toward the drain contact. There will be no steady-state accumulation of electrons in the QW under the emitter if the drift resistance is made sufficiently small. Since the drain contact is shorted to the conducting layer underneath the collector barrier and its lateral distance from the edge of the emitter much exceeds the combined thicknesses of the two barriers and the QW, application of a drain-emitter voltage results in a nearly vertical electric field line under the emitter, which allows one to control by the applied voltage the potential difference between the emitter and the QW. Of course, this control is much less effective (by the lever rule) than it would be if the second barrier were as thin as the tunnel barrier. Experimentally, MORKOÇ et al. [7] were able to see a pronounced NDR already at room temperature and at 77 K the observed PTV ratio in current was more than 2:1. As expected, the NDR was seen only for a negative polarity of the emitter bias — with a peak current occuring at a voltage which is higher than that observed in a control symmetric DBQW structure by a factor given by the ratio of the barrier thicknesses (the lever rule).

1.3 Tunneling in Superlattices

We have established that all that is required for the NDR to occur in a resonant tunneling structure is the reduced dimensionality of electronic states in the tunneling range. One can relax this requirement — by replacing the QW by a superlattice with narrow minibands (a multiple QW structure, for which the tight-binding approximation is a good description). Clearly, we can expect the NDR effect in tunneling from a degenerately doped emitter into a superlattice. Recently, DAVIES et al. [14] reported similar effects in tunneling between the minibands of two coupled superlattices.

Returning to the double-barrier QW structure of Fig. 1, we would like to stress the essential difference between the Fabry-Perot mechanism of the NDR and the above-described mechanism, which involves *sequential* rather than resonant tunneling through the two barriers. In the instance of a semiconductor superlattice, this difference had been clearly explained by KAZARINOV and SURIS [15]. In an ideal superlattice consisting of a large number of equally spaced identical quantum wells, one can expect a resonant (miniband) transmission, analogous to the Fabry-Perot effect, and possibly an NDR due to the Bragg reflections, if the applied field is such that the potential difference, acquired by an electron over many periods of the superlattice, is less than the width of the lowest miniband. These effects, particularly the Bragg reflections, are extremely difficult to observe because of scattering and Zener tunneling between electron minibands [16].

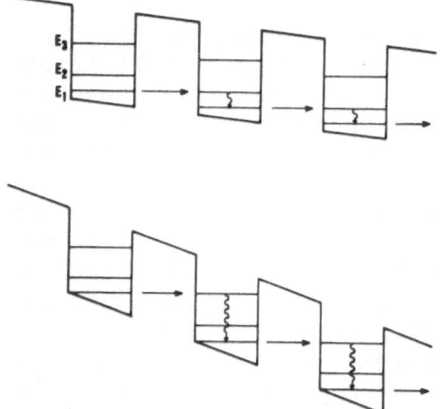

FIGURE 5: Schematic illustration of sequential tunneling of electrons for a potential energy drop across the superlattice period equal, respectively, to the energy difference between the first excited state and the ground state (a) and to the energy difference between the second excited state and the ground state of the wells (b).

In the opposite limit of a strong electric field an enhanced electron current will flow at sharply defined values of the external field, when the ground state in the n-th well is degenerate with the first or second excited state in the $(n+1)$-st well, as illustrated in Fig. 5. Under such conditions, the current is due to electron tunneling between the adjacent wells with a subsequent de-excitation in the $(n+1)$-st well. In other words, electron propagation through the entire superlattice involves again a sequential rather than resonant tunneling.

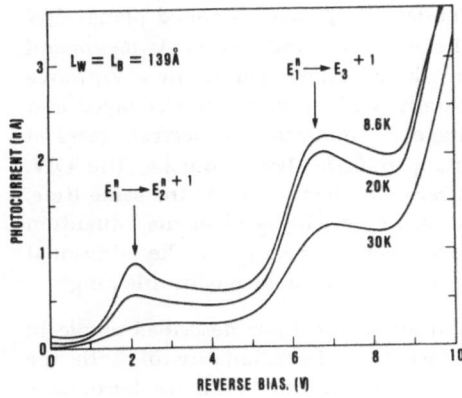

FIGURE 6: The photocurrent-voltage characteristics at $\lambda = 0.633\,\mu m$ (pure electron injection) for an $Al_{0.48}In_{0.52}As/Ga_{0.47}In_{0.53}As$ superlattice with 139 Å thick wells and barriers and 35 periods. The arrows indicate that the peaks correspond to resonances between the ground state of the n th well and the first two excited states of the $(n+1)$st well.

Experimental difficulties in studying this phenomenon are usually associated with the non-uniformity of the electric field across the superlattice and the instabilities generated by negative differential conductivity. To ensure a strictly controlled and spatially uniform electric field, CAPASSO et al. [17] placed the superlattice in the i region of a reverse-biased p^+-i-n^+ junction. This structure allowed for the first time to observe the sequential tunneling resonance predicted in [15]. Two NDR peaks observed in the photocurrent characteristics, Fig. 6, correspond to the resonances shown schematically in Fig. 5. For the sequential tunneling regime, Kazarinov and Suris had predicted the possibility of a laser action at the inter-miniband transition frequency — an effect not yet observed experimentally in superlattices.

1.4 Negative Transconductance Transistor

The preceding discussion has dealt with the bulk-carrier tunneling into a 2-D density of electronic states. Recently, we proposed a novel device structure [18] in which the QW is linear rather than planar and the tunneling if of 2-D electrons into a 1-D density of states. Figure 7 shows the schematic cross-section

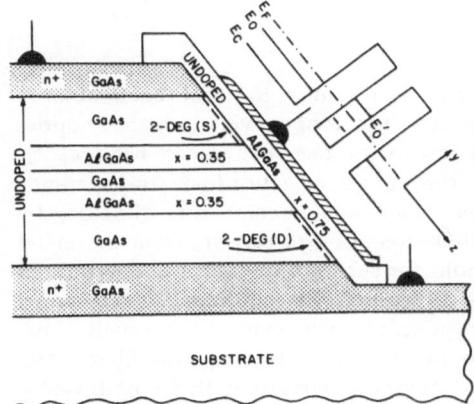

FIGURE 7: The negative differential transconductance device [18]. E_0 is the bottom of the 2D subband separated from the classical conduction band minimum by the energy of the zero-point motion in y-direction; E_0' is the bottom of the 1-D subband in the quantum wire, separated from E_0 by the confinement energy in the z-direction. In the operating regime the Fermi level E_F lies between E_0 and E_0'.

145

of the proposed device. It consists of an epitaxially grown undoped planar QW and a double AlGaAs barrier sandwiched between two undoped GaAs layers and heavily doped GaAs contact layers. The working surface defined by a V-groove etching is subsequently overgrown epitaxially with a thin AlGaAs layer and gated. Application of a positive gate voltage V_G induces 2-D electron gases at the two interfaces with the edges of undoped GaAs layers outside the QW. These gases will act as the source (S) and drain (D) electrodes. At the same time, there is a range of V_G in which electrons are not yet induced in the "quantum wire" region (which is the edge of the QW layer) — because of the additional dimensional quantization. The operating regime of our device is in this range.

Device characteristics can be understood along the lines described above in connection with Fig. 3. In the present case the dimensionality of both the emitter and the base is reduced by 1, so that the emitter Fermi sea becomes a disk and the shaded disk of Fig. 3 is replaced by a resonant segment. Application of a positive drain voltage V_D brings about the resonant tunneling condition and one expects an NDR in the dependence $I(V_D)$. What is more interesting, is that this condition is also controlled by V_G. The control is effected by fringing electric fields: in the operating regime an increasing $V_G > 0$ *lowers* the electrostatic potential energy in the base with respect to the emitter — nearly as effectively as does the increasing V_D (this has been confirmed [18] by solving the corresponding electrostatic problem exactly with the help of suitable conformal mappings). At a fixed V_G having established the peak of $I(V_D)$, we can then quench the tunneling current by increasing V_G. This implies the possibility of achieving the *negative transconductance* — an entirely novel feature in a unipolar device. A negative-transconductance transistor can perform the functions of a complementary device analogous to a p-channel transistor in the silicon CMOS logic. A circuit formed by a conventional n-channel field-effect transistor and our device can act like a low-power inverter in which a significant current flows only during switching. This feature can find applications in logic circuits.

2. Infrared Photodetectors on a Silicon Chip

2.1 Introduction

As is well known, the celebrated silicon technology has not been able to produce an on-chip infrared photodetector for long-wavelength fiber-optics communications. The obvious difficulty lies in the fact that silicon bandgap E_G is wider than the photon energy in the range of silica-fiber transparency ($\lambda = 1.3 - 1.55 \ \mu$m). Attempts have been made to overcome this difficulty by using Schottky-barrier structures with photoexcitation of carriers from the metal (or silicide) into silicon [19]. The threshold for such a photoeffect is determined by the Schottky-barrier height and can easily match the required infrared range; however, the quantum efficiency of absorption in such structures is usually low. So far, the only practical way of employing silicon technology for fiber-optics communications has been to combine Si integrated circuits with Ge or InGaAs-

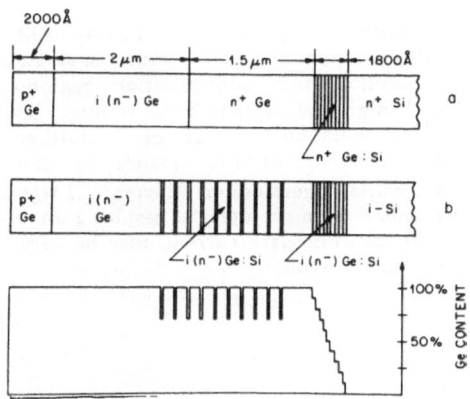

FIGURE 8: Schematic illustration of the composition of epitaxial layers in Ge/Si infrared photodetectors: (a) the original Ge *pin* structure on a silicon wafer [20]; and (b) the glitch-graded structure [21].

on-InP detectors on a separate chip. A different approach to this problem is based on growing single-crystal germanium *pin* junction on a silicon substrate, Fig. 8. Molecular beam epitaxy (MBE) grown diodes have been reported [20], which had a quantum efficiency $\eta \approx 40\%$ at $\lambda = 1.3\mu$m. However, the devices suffered from a relatively high reverse-bias parasitic leakage at room temperature. This leakage resulted from threading dislocations originating at the Ge/Si interface due to a large lattice mismatch and propagating through the germanium *pin* junction. In the subsequent work [21] the dislocation density was reduced by a novel MBE trick called the "glitch grading", which consists in inserting a Si_xGe_{1-x}/Ge superlattice in the epitaxially grown germanium film, Fig. 8b. It turns out that dislocations tend to be trapped in the strained superlattice region and do not propagate up into the working diode region (the region where photogenerated carriers are separated by the electric field). Ultimately, one should be able to produce on a silicon substrate Ge or even InGaAs layers comparable in quality to bulk samples. In this approach, no use is made of the electronic properties of the heterointerface, nor of the Si substrate itself. The latter merely serves as a carrier vehicle for an incommensurate growth of a useful single-crystal foreign semiconductor.

However, there is one property of silicon, which is very attractive for use in fiber-optics communications and whose utilization requires *commensurate* epitaxy. Silicon is an ideal material for avalanche multiplication of photogenerated signals. Neither Ge nor InGaAs are ideal avalanche photodetector (APD) materials from the point of view of the so-called *excess noise factor* F, which describes the stochastic nature of avalanche multiplication [22]. The F factor generally depends on the avalanche gain M and the ratio of the impact ionization coefficients $K = \alpha_n/\alpha_p$ for electron and holes. If $K \approx 1$ then $F \approx M$ and the total noise power scales $\propto M^3$. Such is the situation for Ge with $\alpha_p/\alpha_n \lesssim 2$ and InGaAs, where $\alpha_n/\alpha_p \lesssim 2$. On the other hand, if $K \gg 1$ or $K \ll 1$, then $F \approx 2$ even for $M \gg 1$, provided avalanche is initiated by the type of carrier with higher α. It is well established [22] that in Si at not too high electric fields ($\lesssim 3 \times 10^5$V/cm) the electron ionization coefficient is substantially greater than the hole ionization coefficient. Thus, properly designed Si APD's

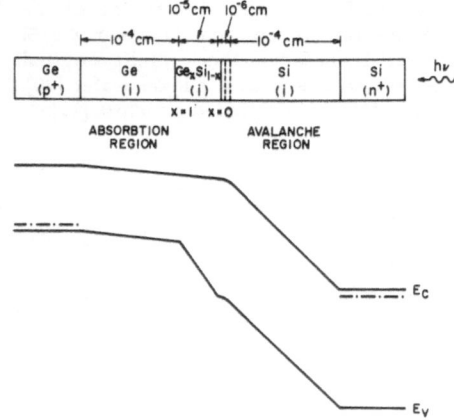

FIGURE 9: A possible Ge/Si hi-lo SAM APD structure [25] with separate absorption and multiplication regions and high-low electric field profile. Its implementation requires further improvement in the quality of the interfacial germanium layers. Large number of misfit defects, resulting in a high parasitic dark current, may be very difficult to avoid.

can have the noise performance near the theoretical minimum. At present, there are commercially available silicon devices with $K \approx 20 - 100$ (of course, these APD's do not operate in the range of interest for fiber-optical communications). It would be very attractive to implement a heterostructure device with *separate* absorption and multiplication regions (SAM APD),† in which electrons photogenerated in a Ge or InGaAs layer would subsequently avalanche in Si. An example of such a structure [25] is shown in Fig. 9. It combines Si multiplication with Ge absorption layers. It also contains a depleted layer of acceptors (charge sheet) built in silicon in the vicinity of the Ge interface, whose purpose is to separate the low-field region in the optically active Ge layer from a high-field Si layer where avalanche multiplication occurs. Similar SAM APD structures with hi-lo electric field profiles have been successfully fabricated in III-V compound semiconductors [26]; however, the implementation with Ge/Si heterojunctions requires far better material quality in the interfacial layers than that presently available with any crystal growth technique.

2.2 *Waveguide Infrared Detectors Based on Ge/Si Superlattices*

A novel infrared photodetector structure was recently proposed [27], which utilizes the silicon advantage. It represents a waveguide in which the core is a strained-layer Ge_xSi_{1-x}/Si superlattice (SLS) sandwiched between Si layers of a lower refractive index. Absorption of infrared radiation occurs in the core region due to interband electron transitions, and photogenerated carriers are collected in the Si cladding layers. Due to the recently discovered effect of bandgap narrowing by the strain in Ge_xSi_{1-x} alloy layers the fundamental absorption threshold of the SLS is shifted to longer wavelengths, so that the detector can be operated in the range of silica-fiber transparency. If the alloy

---

† Considerable research has been devoted to the use of III-IV compound-semiconductor SAM APD's for fiber-optics communications (see [23] and references therein). Excellent performance has been demonstrated by $InP/Ga_{0.47}In_{0.53}As$ APD's of this type [24].

absorption threshold in the Ge_xSi_{1-x}/Si SLS as a function of the Ge content in the alloy layers and found a good agreement with the theoretical predictions. At $x = 0.6$ the bandgap E_G is narrower than that of pure unstrained Ge, and for $x \geq 0.5$ one has $E_G \leq 0.8\,eV$. The absorption edge is thus brought down by the strain to below the photon energy at wavelengths of silica-fiber transparency.

2.1.1 Design of Waveguide Detectors and APD's

Consider first a waveguide-detector structure in which the core represents a single alloy layer, Fig. 10a. We assume that the Ge content in this layer is $x > 0.5$, and that the absorption coefficient at wavelengths of interest is $\alpha \approx 10^2\,cm^{-1}$ (as indicated by the preliminary results [28] at $\lambda = 1.3\,\mu m$). To be in the range of commensurate growth, the alloy thickness h must be less than the critical $h_c(0.5) = 100\,\text{Å}$. To a good approximation, the fraction Γ of the integrated intensity of the light wave which falls within the absorbing core, is given by:

$$\Gamma = 2\pi^2 \left[\frac{h}{\lambda}\right]^2 (n_{core}^2 - n_{clad}^2) .$$

For $x = 0.5$, $h = 100\,\text{Å}$, and $\lambda = 1.3\,\mu m$ one finds $\Gamma = 2.3 \times 10^{-3}$. The effective absorption coefficient of such a waveguide, $\alpha_{eff} = \alpha\Gamma \sim 0.2\,cm^{-1}$, is too low for a practical use (a detector would have to be several centimeters long and even the speed of light is not fast enough over such distances).

The use of a superlattice is thus imperative. Consider the structure illustrated in Fig. 10b. Ignoring in first approximation the influence of strain on the dielectric constant, the refractive index of an SLS can be estimated as an average of $n^2(x)$ and n_{Si}^2 over one period and the effective absorption coefficient of an

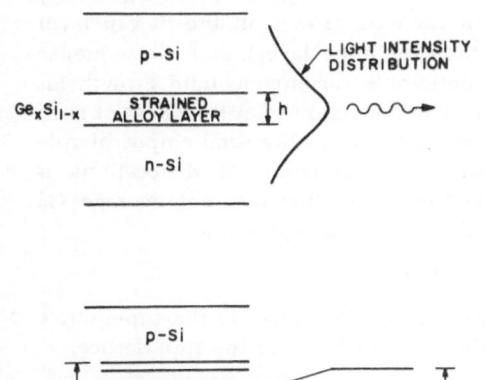

FIGURE 10: Schematic illustration of strained-layer Ge_xSi_{1-x} waveguide detectors [27]: (a) single-layer core; (b) SLS core.

layers are sufficiently thin, the SLS can be grown by MBE without nucleating dislocations. Experimentally, such structures were recently manufactured and tested. The first SLS waveguide *pin* diodes [28] showed an internal quantum efficiency of 40% at $\lambda = 1.3\,\mu m$ and a frequency bandwidth of close to 1 GHz. The first APD structure [29] showed an avalanche gain as high as $M = 50$ and a quantum efficiency of 100% at $M = 10$. The waveguide-detector approach is entirely compatible with the Si integrated circuit technology and offers the possibility of fabricating a complete receiver system for long-wavelength fiber-optics communications on a silicon chip. Following [27], we shall briefly discuss below the optimum composition of an SLS core, as determined by the trade-off between the confinement of radiation and the stability requirements for a Ge_xSi_{1-x}/Si SLS, as well as the design of a SAM APD waveguide structure, in which low-noise avalanche multiplication occurs in one of the Si cladding layers.

2.2.1 Material Properties of Strained-Layer Ge_xSi_{1-x}/Si Systems

Let us first discuss the questions of stability. The maximum thickness h_c of a single strained Ge_xSi_{1-x} alloy layer which can be grown pseudomorphically on Si depends on the germanium content, decreasing with x. PEOPLE and BEAN [30] have calculated $h_c(x)$ on the assumption that the film grows initially without dislocations, which are then generated at the interface, as the strain energy density per unit area of the film exceeds the areal energy density associated with an isolated dislocation. Their result, which implicitly gives $h_c(x)$ in [Å] by the equation

$$x^2 h_c = 13.3 \ln(h_c/4),$$

is in an excellent agreement with the empirical data.

Raman scattering studies [31] have shown that most of the strain in such structures resides in the alloy layer, with Si cladding layers being nearly unstrained. A second Ge_xSi_{1-x} layer can then be grown on the Si cap layer (provided the latter is 2-3 times thicker than the alloy layer), and the sequence can be repeated many times without a noticeable incommensurate growth (as many as 100 periods have been reported). The maximum total thickness of such strained layer superlattices (SLS) can be estimated from the semi-empirical rule [32] that the stability of the SLS against the formation of dislocations is equivalent to that of a single alloy layer of same thickness but average Ge content. This rule can be represented by the following expression:

$$h_{SLS}^{max}(x,r,T) \approx h_c(xr),$$

where $r \equiv h/T$ is the ratio of the thickness of the alloy layer to the superlattice period (i.e. the "duty cycle"), and h_{SLS} is the total thickness of the superlattice.

Next, we discuss the effects of strain on the *band structure* of an alloy layer. An important finding in this regard is the theoretical calculation of PEOPLE [33] who considered the bandgap narrowing in strained Ge_xSi_{1-x} alloys grown on Si (100) substrates, and found that the gap is substantially reduced in comparison with the unstrained alloy. LANG et al. [34] have measured the fundamental

SLS core is given by $\alpha_{eff} = r\Gamma\alpha$. Analysis [27] shows that α_{eff} is maximized by smaller r, which for $h_{SLS} = h_{SLS}^{max}$ implies maximizing the superlattice width, and that α_{eff} has its optimum value for those r which correspond to $\Gamma \approx 1/2$. Physically, as the superlattice is made thicker to absorb the wings of the light intensity distribution, the SLS requirement of decreasing r leads to less efficient absorption at the peak intensity, thus more than offsetting the gain. We can expect an optimum value $\alpha_{eff} \lesssim 0.2\alpha \approx 20\,\text{cm}^{-1}$ in an SLS consisting of 12 periods of $60\,\text{Å}$ $Ge_xSi_{1-x}/140\,\text{Å}$ Si. This means that the waveguide length must be of order 0.5 mm for high detector efficiency. If a *pin* detector represents a ridge waveduide of that length and the width $\leq 10\,\mu m$, then its capacitance is less than about 0.5 pF, assuming a typical depletion width of $1\,\mu m$. This value of the internal capacitance is acceptable and comparable to that of the conventional *pin* IR detectors. Note that the detector quantum efficiency grows with the optical path length, without degrading the speed of response.

The detector sensitivity will be further improved by an avalanche gain in Si cladding layers. To reduce an excess noise, the APD design should be guided by the following principles: *i*) since $K \equiv \alpha_n/\alpha_p \gg 1$ in Si, the multiplication should be initiated by electrons rather than holes; *ii*) since K decreases sharply when the electric field much exceeds the ionization threshold, E_i, the field in the avalanche layer should be near the threshold, $E \geq E_i \approx 3 \times 10^5\,\text{V/cm}$, and the thickness of that layer should be well above $\alpha_n^{-1}(E_i) \approx 0.5\,\mu m$; *iii*) the field in the SLS layers should not exceed $\sim 10^5\,\text{V/cm}$, the ionization threshold in Ge.

A possible waveguide APD structure is illustrated in Fig. 11. In addition to an undoped Ge_xSi_{1-x}/Si SLS of $x \geq 0.6$, $r \leq 0.3$, and thickness $h_{SLS} \geq 3000\,\text{Å}$, it contains an undoped Si avalanche layer of thickness $d \geq 2\,\mu m$ separated from the SLS by a thin ($\Delta \leq 10^{-6}\,\text{cm}$) p-type Si layer. In the operating regime, the Δ layer must be depleted by an applied reverse bias. The total surface density of charge in this layer should, therefore, be of order $\kappa E_i \approx 2 \times 10^{12}\,\text{e/cm}^2$. This will achieve the desirable hi-lo field separation of the absorption and multiplication layers and result in a low-noise SAM APD structure.

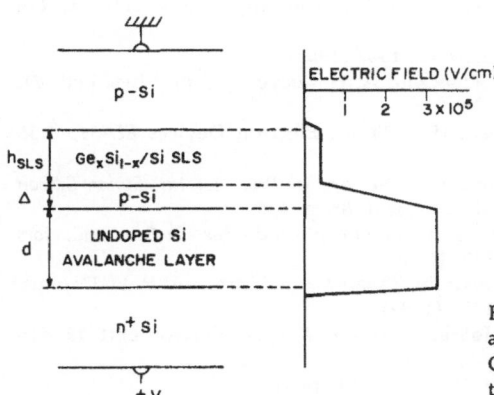

FIGURE 11: A waveguide APD structure and the electric field profile [27]. Conceptually, the structure is analogous to the SAM APD of Fig. 9.

2.3 Conclusion

Development of Si-based detectors for optical communications represents one of the most practical applications of silicon MBE research. The waveguide-detector approach is entirely compatible with the Si integrated circuit technology and offers the possibility of fabricating a complete receiver system for long-wavelength fiber-optics communications on a silicon chip.

REFERENCES

1. R. Tsu and L. Esaki, Appl. Phys. Lett. 22, 562 (1973); L. L. Chang, L. Esaki and R. Tsu, ibid. 24, 593 (1974).
2. T. C. L. G. Sollner, W. D. Goodhue, P. E. Tannenwald, C. D. Parker and D. D. Peck, Appl. Phys. Lett. 43, 588, (1983).
3. T. C. L. G. Sollner, P. E. Tannenwald, D. D. Peck, and W. D. Goodhue, Appl. Phys. Lett. 45, 1319, (1984).
4. T. Shewchuk, P. C. Chapin, P. D. Coleman, W. Kopp, R. Fischer and H. Morkoç, Appl. Phys. Lett. 46, 508 (1985).
5. M. Tsuchiya, H. Sakaki and J. Yashino, Jap. J. Appl. Phys. 24, L466 (1985).
6. T. Tsuchiya and H. Sakaki, Tech. Digest of IEEE Int. Electron Dev. Meeting, Dec. 2-4, Washington, DC (1985), p. 662.
7. H. Morkoç, J. Chen, U. K. Reddy, T. Henderson, P. D. Coleman, and S. Luryi, Appl. Phys. Lett., to be published.
8. B. Ricco and M. Ya. Azbel, Phys. Rev. B 29, 1970 (1984).
9. F. Capasso and R. A. Kiehl, J. Appl. Phys. 58, 1366 (1985).
10. S. Luryi, Appl. Phys. Lett. 47, 490 (1985).
11. S. Luryi, Technical Digest of the IEEE Int. Electron Dev. Meeting, Dec. 2-4, Washington, DC (1985), p. 666.
12. G. E. Derkits, to be published.
13. E. A. Rezek, N. Holonyak, Jr., B. A. Vojak, and H. Schichijo, Appl. Phys. Lett. 703 (1977).
14. R. A. Davies, M. J. Kelly, and T. M. Kerr, Phys. Rev. Lett. 55, 1114 (1985).
15. R. F. Kazarinov and R. A. Suris, Sov. Phys. - Semicond. 5, 707 (1971).
16. L. Esaki and L. L. Chang, Phys. Rev. Lett. 33, 495 (1974).
17. F. Capasso, K. Mohammed, and A. Y. Cho, Tech. Digest of IEEE Int. Electron Dev. Meeting, Dec. 2-4, Washington, DC (1985), p. 764; also Appl. Phys. Lett. 48, 478 (1986).
18. S. Luryi and F. Capasso, Appl. Phys. Lett. 47, 1347 (1985).
19. T. R. Harrison, A. M. Johnson, P. K. Tien, and A. H. Dayem, Appl. Phys. Lett. 41, 734 (1982).
20. S. Luryi, A. Kastalsky, and J. C. Bean, IEEE Trans. Electron Devices ED-31, 1135 (1984).
21. A. Kastalsky, S. Luryi, J. C. Bean, and T. T. Sheng, in Proc. 1st Int. Symp. Silicon MBE, ed. by J. C. Bean, Electrochem. Soc. Press, 1985, p. 406.
22. See, for example, F. Capasso, Physics of avalanche photodiodes, in Semiconductors and semimetals, vol. 22, part D, pp. 1-172, 1985.
23. G. E. Stillman, L. W. Cook, G. E. Bulman, N. Tabatbaie, R. Chin, and P. D. Dapkus, IEEE Trans. Electron Devices ED-29, 1355 (1982).
24. J. C. Campbell, A. G. Dentai, W. S. Holden, and B. L. Kasper, Electron Lett. 19, 818 (1983).
25. F. Capasso, A. Kastalsky, and S. Luryi, 1983 (unpublished).

26. F. Capasso, A. Y. Cho, and P. W. Foy, Electron. Lett. *20*, 635 (1984).
27. S. Luryi, T. P. Pearsall, H. Temkin, and J. C. Bean, IEEE Electron Device Lett. *EDL-7*, 104, (1986)
28. H. Temkin, T. P. Pearsall, J. C. Bean, R. A. Logan, and S. Luryi, Appl. Phys. Lett., to be published.
29. T. P. Pearsall, H. Temkin, J. C. Bean, and S. Luryi, to be published.
30. R. People and J. C. Bean, Appl. Phys. Lett. *47*, 322 (1985).
31. F. Cerdeira, A. Pinczuk, J. C. Bean, B. Batlogg, and B. A. Wilson, Appl. Phys. Lett. *45*, 1138 (1984).
32. R. Hull, J. C. Bean, F. Cerdeira, A. T. Fiory, and J. M. Gibson, Appl. Phys Lett. *48*, 56 (1986).
33. R. People, Phys. Rev. *B32*, 1405 (1985).
34. D. V. Lang, R. People, J. C. Bean, and A. M. Sergent, Appl. Phys. Lett. *47*, 1333 (1985).

Part IV

Bound States in Quantum Wells

Far Infrared Studies of Shallow Donors in GaAs-AlGaAs Quantum Wells

B.D. McCombe, N.C. Jarosik[+], and J.-M. Mercy

Department of Physics and Astronomy, SUNY at Buffalo,
Buffalo, NY 14260, USA

Calculations for GaAs-AlGaAs quantum wells show that impurity
energies depend strongly on the quantum well (QW) width and/or the
impurity position within the well. Recent experimental work in GaAs-
AlGaAs MQW structures is reviewed with emphasis on FIR spectroscopic
studies of donor impurities in magnetic fields up to 9.5T. Samples
nominally doped in narrow regions at the center and the edge of GaAs
wells have been studied for QW widths between 80Å and 450Å. Results
are generally in very good agreement with recent calculations.
Possible use of these measurements to determine the distribution of
impurities is discussed.

1. Introduction

Recent developments in materials growth technologies, in particular
Molecular Beam Epitaxy (MBE) and Organometallic Chemical Vapor Deposition
(OMCVD), have permitted the growth of semiconductor heterojunctions,
superlattices and quantum wells (QW), that can be selectively doped on the
scale of a few lattice constants. This has led to new devices such as the
high electron mobility transistor (HEMT) and QW lasers, as well as making
possible the observation of new physical phenomena, such as the Fractional
Quantum Hall Effect. In all cases, an important requirement of the
structures is controlled doping. Knowledge of the effect of confining
potential barriers on the electronic states of the impurities,
particularly binding energies, is becoming increasingly important.

The present experimental study focusses on far infrared (FIR)
spectroscopy of the electronic states of shallow (hydrogenic) donors in QW
structures. Donors are suitable for an initial study directed at
understanding the effects of confinement due to their rather simple
electronic structure in the materials of interest. Due to the high degree
of control of growth and doping, the MBE-grown GaAs-AlGaAs system was
chosen as the vehicle for these studies. Both absorption and
photoconductivity experiments have been performed.

Related experimental studies have been carried out by photoluminescence
[1,2] and Raman Scattering [3]. The FIR experiments allow a more precise
determination of the impurity energies and, in the presence of a magnetic
field, a detailed study of line profiles.

+ Present address: AT&T Bell Laboratories

2. Theoretical Background

Impurities in semiconductors have received considerable attention, both experimentally and theoretically, for a number of years. Many shallow impurities in bulk semiconductors can be well described by a simple hydrogenic model. For a donor, the extra electron is weakly bound to the net positively charged impurity center through the Coulomb attraction. The semiconductor host crystal is taken into account in the one-band effective mass approximation through the use of an effective mass for the electron and a Coulomb interaction screened by the appropriate dielectric constant of the host crystal, ε_s. For a simple parabolic and isotropic conduction band, the known results for the hydrogen atom can simply be taken over by replacing the free electron mass by the effective mass, m^* and $-e^2/r$ by $-e^2/\varepsilon_s r$. This yields the following well-known allowed energy states,

$$E_n = - \frac{m^* e^4}{2\varepsilon_s^2 \hbar^2 n^2} \quad , \quad n = 1, 2, 3, \ldots, \tag{1}$$

where the energy is measured with respect to the conduction band edge. The effective Bohr radius is

$$a_o^* = \frac{\hbar^2 \varepsilon_s}{e^2 m^*} . \tag{2}$$

For donors in GaAs the binding energy, $R_y^*(3D)$, is approximately 5.8 meV, and the effective Bohr radius is $\sim 100\text{\AA}$.

The effects of confinement are expected to become important when the impurity ion is within a few Bohr radii of some confining potential barrier. There are two limiting cases that provide a qualitative picture of these effects. In the first case, the width of the confining potential is allowed to go to zero, resulting in a 2 dimensional (2D) situation. The allowed energy eigenvalues of the 2D hydrogen atom are

$$E_n(2D) = \frac{m^* e^4}{2\varepsilon_s^2 \hbar^2 (n-\frac{1}{2})^2} \quad , \quad n = 1, 2, \ldots \tag{3}$$

The binding energy is $R_y^*(2D) = 4R_y^*(3D)$, and the effective Bohr radius is $a^*(2D) = a^*(3D)/2$. The other interesting limit is that of an impurity ion located at an infinite potential discontinuity. In this case the envelope wave function of the impurity is excluded completely from the half-space where the potential is infinite, and by symmetry, only a subset of the 3D hydrogen atom solutions are allowed in the other half-space (all s-like solutions are excluded). This results in the following allowed eigen-energies [4].

$$E_n(\text{edge}) = - \frac{m^* e^4}{2\varepsilon_s^2 \hbar^2 n^2} \quad , \quad n = 2, 3, \ldots \ , \tag{4}$$

where ε_s is the average dielectric constant of the two materials. In this case the binding energy is $R_y^*(\text{edge}) = R_y^*(3D)/4$. Thus it is clear that confinement leads to large changes in the binding energies (a factor of 16 between the 2 limits for infinite potential barriers).

Recently, several workers have considered the problem of a hydrogenic impurity in a semiconductor quantum well (formed by the energy band discontinuity) in the effective mass approximation. The Hamiltonian for this problem can be written

$$H = \frac{p^2}{2m_i^*} - \frac{e^2}{\varepsilon_{si}[\rho^2+(z-z_i)^2]^{1/2}} + V(z) , \qquad (5)$$

where 　　$V(z) = V_o$, i = 2, for $|z| > L/2$

　　　　　　 $= 0$, 　i = 1, for $|z| < L/2$,

$\rho^2 = x^2+y^2$; z_i is the coordinate of the impurity within the well; the origin of the coordinate system has been taken to be at the center of the QW with the z-axis parallel to the growth axis; L is the width of the well; $m_1^*(m_2^*)$ and $\varepsilon_{s1}(\varepsilon_{s2})$ are the electron effective mass and dielectric constant of the smaller (larger) gap semiconductor; V_o is the conduction band discontinuity ($\equiv$barrier height), and image contributions have been ignored. The Hamiltonia are not separable in general, and are not amenable to analytic solution. Theoretical work has made use of variational aproaches with infinite barrier height [5] and finite barrier heights [6,7] with parameters appropriate to the GaAs/AlGaAs system. These calculations differ in the complexity of the trial functions and the treatment of boundary conditions. However, all results are in qualitative agreement; and the finite barrier height calculations agree quantitatively. Substantial shifts in binding energy are found as a function of well width and of position of the impurity within the well. The results are qualitatively in accord with expectations from the limiting cases outlined above. The binding energy increases as the QW width is decreased, while it decreases as the impurity ion is moved from the center to the edge of the well at constant well-width.

An external magnetic field along the QW axis (the z-direction) has recently been included in the theoretical calculations, and the dependence on magnetic field of the ground state (1s) and the lowest lying excited states, 2p(m=±1) energies have been determined [8]. Here, the usual hydrogen atom notation for states in the low magnetic field limit, with m = azimuthal quantum number, has been used. The results are qualitatively very similar to results for hydrogenic impurities in bulk semiconductors.

The continuum states consist of quantized confinement subbands in the z-direction, with the x-y motion of each subband quantized into Landau levels, whose energies are given for parabolic bands by (neglecting spin)

$$E_n = (N+\tfrac{1}{2})\hbar\omega_c , \qquad (6)$$

where N is the Landau quantum number, and ω_c = eB/m*c, with B the magnetic induction, and m_c^* the cyclotron effective mass (m_1^* for the QW of Eq. (5)). In the low field (Zeeman) limit ($\gamma = \hbar\omega_c/2R_y^*(3D)$) the 2p(m = ±1) states are split apart linearly by the field ($E(2p^+) - E(2p^-) = \hbar\omega_c$), while the 1s state is unaffected in lowest order. At higher magnetic fields ($\gamma = 1$ and greater) the 1s state begins to move up in energy with slope approaching that of the N=0 Landau level ($\hbar\omega_c/z$) for $\gamma \to \infty$. The 2p$^-$ state, after initially decreasing, also begins to move up in energy with a similar slope. Both states in the high field limit, $\gamma \gg 1$, are bound states associated with the N=0 Landau level. The $2p^+$ state moves up more rapidly with a slope approaching that of the N=1 Landau level ($3\hbar\omega_c/2$); this becomes a bound state associated with the N=1 Landau level for $\gamma \gg 1$. The energy separation between $2p^+$ and $2p^-$ is $\hbar\omega_c$ at all values of B.

Electric dipole transitions are governed by the parity selection rule ($\Delta\ell = \pm 1$), with ℓ the orbital angular momentum quantum number) and angular momentum conservation along the magnetic field direction, $\Delta m = \pm 1$. This results in allowed transitions $1s \rightarrow 2p^+$ and $1s \rightarrow 2p^-$ for light propagation along the magnetic field with electric field vector polarized perpendicular to the magnetic field.

3. Experimental Details

The structure of the MBE-grown GaAs-AlGaAs samples is shown in Fig. 1. The GaAs QWs (widths between 80Å and 450Å) were selectively doped with Si donors at $5 \times 10^{15} cm^{-3}$ (for the 450Å well-width sample) or $1 \times 10^{16} cm^{-3}$ (for all others) over the central 1/3 of the well for the center-doped samples, or the "top" 1/3 of the wells for the edge-doped sample. The number of wells varied from 12 to 45; barrier widths were 125Å to 150Å to minimize interactions between adjacent QWs. The molar fraction of Al in the $Al_xGa_{1-x}As$ barriers was $x \approx .22$ (450Å wells) or $x \approx 0.3$ (all other samples). The composition and well-width were determined by photoluminescence measurements at NRL.

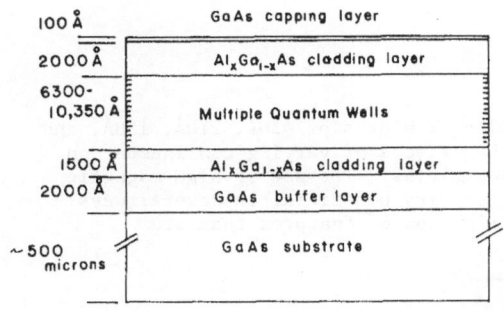

100 Å	GaAs capping layer
2000 Å	$Al_xGa_{1-x}As$ cladding layer
6300-10,350 Å	Multiple Quantum Wells
1500 Å	$Al_xGa_{1-x}As$ cladding layer
2000 Å	GaAs buffer layer
~500 microns	GaAs substrate

Fig. 1: Cross-section of structure of the multi-quantum-well samples.

The FIR measurements were carried out in a light pipe system in conjunction with a Fourier Transform Interferometric Spectrometer with various cryogenic detectors, and a 9T superconducting magnet system. Related measurements have been made with a FIR laser spectrometer [9]. Transmission data were obtained at low temperatures ratioed either to background spectra from a reference MQW sample with no intentional donor doping, or to backgrounds from the same sample immediately after cool down in the dark. In the latter case, the confined donor impurity absorption was not observable; donors in the wells are ionized under these circumstances. After illumination for a few seconds at low temperatures with a red LED, donors in the QWs are neutralized and clear intra-impurity absorption lines are observable as long as the sample is maintained at temperatures below ~140K. This persistent photoeffect is illustrated in Fig. 2. With the LED off after cool down, no structure is observed. With the LED on, both confined impurity and bulk impurity (from the GaAs buffer layer) $1s$-$2p^+$ transitions are observed. When the LED is turned off again,

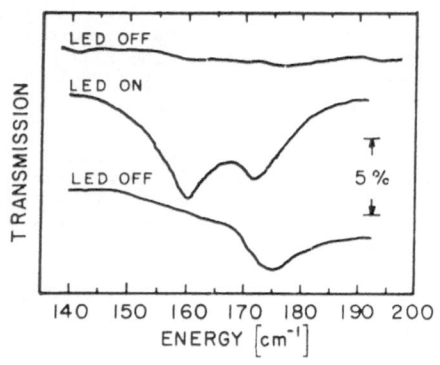

Fig. 2: Transmission spectra at 9T and 4.2K for a center-doped 210Å well-width MQW sample illustrating the effects of illumination by a red LED. Top-before: Center-during: Bottom-after

only the confined impurity 1s-2p$^+$ line remains. The assignment of the high-energy feature to confined impurities was verified by tilting the magnetic field. The confined impurity line position scales approximately with the normal component of the magnetic field, while the feature attributed to bulk impurities is independent of the angle between the magnetic field and the normal to the sample surface. Although of intrinsic scientific interest, the persistent photo-effect is not central to the present work, and it is used here simply as a convenient way of providing high quality background spectra. The data discussed in the following section were obtained in this manner for the most part.

4. Experimental Results

4.1. Center-Doped Donors

A series of 4 center-doped samples with QW widths of 450Å, 210Å, 138Å, and 80Å were investigated to determine the effects of varying confinement on the electronic states of the donor impurities. The use of high magnetic fields in these studies was found to be very beneficial in several ways: 1) It permits the unambiguous identification of features that are

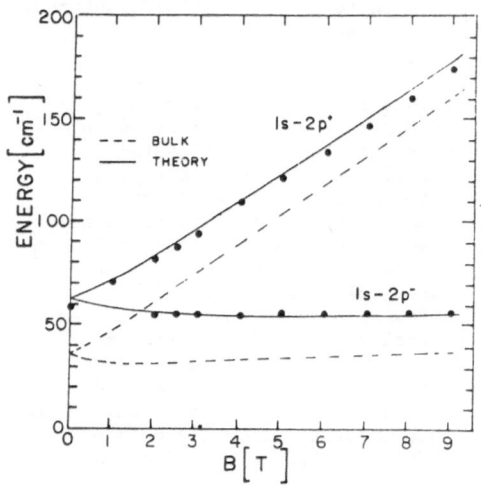

Fig. 3: Plot of transition energies vs. magnetic field for the sample of Fig. 2. Experimental results (●) are compared with theoretical calculations (solid lines) and the corresponding bulk transitions (dashed lines)

electronic in nature; 2) The lines become narrowed at high fields permitting much higher precision in the transition energy determinations; and 3) Tilting the magnetic fields permits the separation of confinement-related features from bulk features. An example of the results is shown in Fig. 3. Experimental points for both 1s-2p$^-$ are compared with theoretical calculations [8] (solid lines) and the corresponding transitions observed in a "bulk" MBE Si-doped epitaxial layer (dashed lines). Substantial shifts to higher energy for the confined impurity transitions are clearly evident, and the agreement with theory is extremely good over the whole range of magnetic fields investigated. Similar data obtained for the other samples were also in very good agreement with theory [10]. For the 80Å QW the observed 1s-2p transition energies show slightly more deviation (to lower energies) from the theoretical values. This may be a result of increased sensitivity of the calculations to well-width and barrier height for the narrower wells, combined with experimental uncertainties in these determinations, and/or a slight shift of the observed peak to lower energies resulting from the distribution of impurities away from the well center.

4.2. Edge-Doped Donors

The edge-doped sample used in these studies consisted of twelve 375Å GaAs QWs separated by 125Å barriers of $Al_{0.3}Ga_{0.7}As$. A transmission spectrum for this sample is shown in Fig. 4. Also shown are the calculated positions of the transition from the ground state with m=0 to the first excited state with m=+1 for an impurity at the edge of the QW (lower energy arrow) and at the center (higher energy arrow). These are all generally referred to as 1s-2p$^+$ transitions, although for edge impurities they more closely resemble 2p^o-3d$^+$ transitions. Note the strongly asymmetric line shape, broadened to higher energies. The experimental transmission minimum in this case occurs measureably below ($\sim$4cm^{-1}) the calculated transition energy for an impurity at the edge; there is no perceptible absorption at the energy corresponding to an impurity at the center of the well. A summary of the measured transition energies (transmission minima) vs. magnetic field for this sample is presented in Fig. 5. Also shown are the corresponding calculated transition energies vs. field for an edge impurity (dashed line) and the measured 1s-2p$^+$ transition energies for Si donors in bulk GaAs (solid line). The experimentally measured energies are consistently below the calculated curve. Since the calculations are insensitive to uncertainties in well-

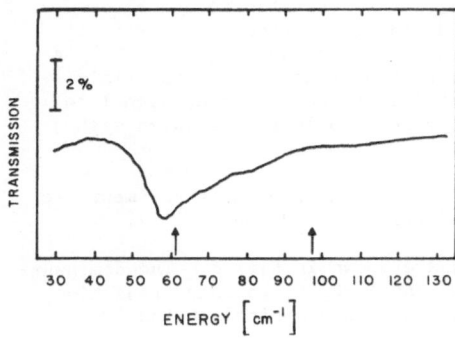

Fig. 4: Transmission spectra at 4T and 4.2K for an edge-doped, 375Å well-width MQW sample. The arrows indicate the calculated ground state with m-0 to 1st excited state with m=+1 transition energy for an impurity at the edge of the QW (lower energy) and at the center of the QW (higher energy)

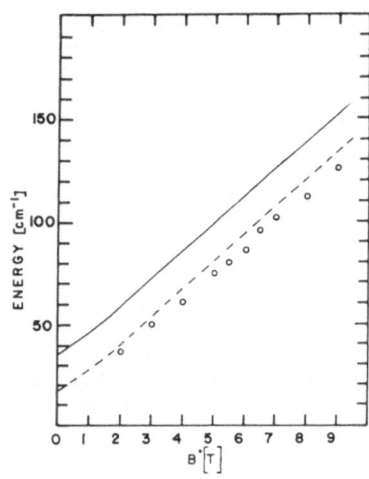

Fig. 5: Plot of the ground state (m=0) to 1st excited state (m=1) transition energies vs. magnetic field for the edge-loped MQW sample of Fig. 4. Experimental results (0) are compared with theoretical calculations (dashed line) and the corresponding bulk transition energies (solid line)

width for edge impurities in QWs of this width, it is necessary to look elsewhere for an explanation of the discrepancy. Two possibilities appear plausible: 1) The dopant distribution (1/3 of the well width) skews the line profile toward higher energies (Fig. 4) and may result in a shift of the transmission minimum of magnitude; 2) Calculations for impurities at the edge of the well are more sensitive to the choice of wave function matching conditions at the boundary (than impurities at the center); thus the discrepancy may result from the use of wave function matching conditions that are not adequate for this structure.

Finally, the lack of any absorption in this sample corresponding to impurities at the well center means that there is no substantial redistribution of impurities on the scale of ~60Å under these growth conditions.

5. Photoconductivity Measurements

Since the transition energies depend strongly on the position of the impurity ion in a QW, the line profile of a particular intra-impurity transition contains information about the distribution of impurities along the growth direction. Recent theoretical work has focussed on calculations of the absorption line profile in a magnetic field and its dependence on impurity distribution along the QW axis [11]. In order to investigate details of the line profile, it is desirable to use a technique that avoids the necessity of dividing two spectra to extract weak (typically 1-4%) absorption features. Photoconductivity has been a very useful technique in studying shallow impurities in bulk materials. This approach typically has enhanced sensitivity and improved signal to noise and does not suffer from the background problems associated with transmission measurements.

A capacitive coupling technique was used in the present experiments to study photothermal ionization spectra of the confined donors [12].

Results for two center-doped samples are shown in Fig. 6. The dominant feature at zero field in both samples is the 1s-2p hydrogenic transition, with a series of overlapping transitions to higher excited states

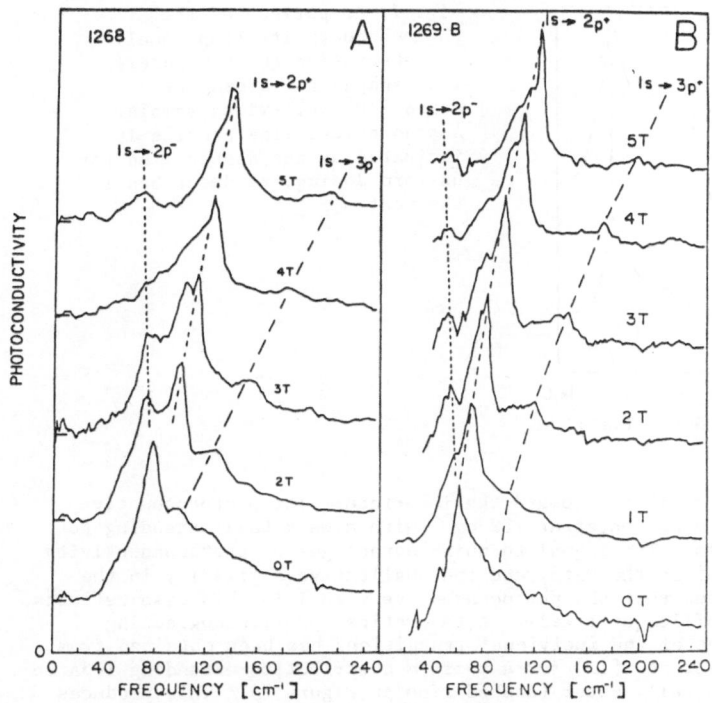

Fig. 6: Photoconductivity spectra at ∿8K and various magnetic fields for two center-doped MQW samples. A – 138Å wells; B – 210Å wells.

contributing to the high-frequency "tail". The evolution of these lines with magnetic field clearly shows the advantages of the photoconductive technique, as well as the beneficial effect of high magnetic fields in separating the various overlapping transitions into clear and well-defined lines. The various transitions are identified in the figure.

Although the photoconductive signal at zero magnetic field is of excellent quality, due to the large number of overlapping transitions, it is not suitable for a detailed lineshape analysis. On the other hand, the $1s \rightarrow sp^+$ transition at high field is well separated from other transitions and is free from such complications. Greene and Bajaj [11] have calculated the $1s \rightarrow 2p^-$ transition energy as a function of positive impurity ion position at high magnetic field for various well widths. These authors have also calculated the line profile for certain impurity distributions. For a uniform distribution the line profile is proportional to $d\Delta(z_i)/dz_i)^{-1}$ where $\Delta(z_i)$ is the $1s \rightarrow 2p^+$ transition energy, and z_i is the position of the impurity ion in the QW, with $z_i=0$ at the center of the well.

Figure 7 shows a comparison of the $1s \rightarrow 2p^+$ line in photoconductivity with the same line in absorption at a field of 7T. Both measurements exhibit similar lineshapes, a sharp peak corresponding to impurities near the center with an asymmetric tail extending to lower energies resulting

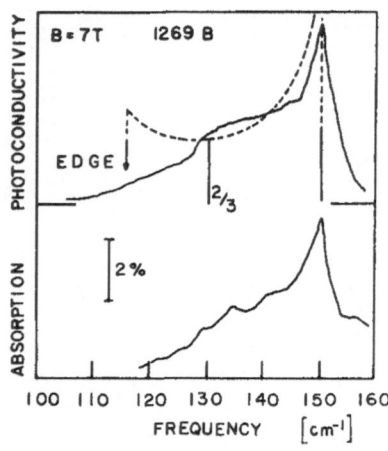

Fig. 7: Comparison of the photoconductivity (top panel) and absorption (bottom panel) line shapes for a center-doped 210Å well-width sample. A theoretical line profile is indicated by the dashed line for uniform doping, as described in the text

from impurities distributed toward the interface. The photoconductive signal exhibits a clear onset at 128 cm^{-1} with a weak tail extending to even lower energies. The signal to noise advantages of photoconductivity are clearly evident in the data, and the qualitative similarity in the lineshapes indicates that the photoconductive signal in this case reflects the absorption profile quite well. A theoretical profile neglecting lifetime broadening of the individual transitions has been obtained from the results of reference [11] for a uniform distribution extending 125Å to either side of the well center (dashed line in Figure 7). This produces an onset close to that observed,and otherwise reproduces the qualitative features of the line profile. This agreement provides evidence of impurity redistribution ~35 Å beyond the intended doping profile. The line profile in narrower wells should be even more sensitive to redistribution on this scale,since the transition energy changes more rapidly with position for impurities near the edge.

Acknowledgements

We are grateful to a number of collaborators who were involved in various aspects of this work; B.V. Shanabrook, R.J. Wagner and J. Furneaux of NRL for photoluminescence measurements, FIR laser magnetospectroscopy, and numerous enlightening discussions; G. Wicks and J. Ralston of Cornell University, and J. Comas of NRL for MBE samples. This work was supported in part by ONR through grant #N0001483K0219 and NSF through grant #ECS-8200312 to NRRFSS at Cornell University. Special thanks go to L. Cooper of ONR for continuing support.

References

1.   R.C. Miller, A.C. Gossard, W.T.Tsang, and O. Munteaunu, Phys. Rev. B 25, 3871 (1982).
2.   B.V. Shanabrook and J. Comas, Surf. Sci. 142, 504 (1984).
3.   B.V. Shanabrook, J. Comas, T.A. Perry and R. Merlin, Phys. Rev. B 29, 7096 (1984).
4.   J.D. Levine, Phys. Rev. 140, A568 (1965).

5.   G. Bastard, Phys. Rev. B $\underline{24}$, 4714 (1981).
6.   C. Mailhiot, Yia-Chung Chang, and T.C. McGill, Phys. Rev. B $\underline{26}$, 4449 (1982).
7.   R.L. Greene and K.K. Bajaj, Solid State Commun. $\underline{45}$, 825 (1983).
8.   R.L. Greene and K.K. Bajaj, Phys. Rev. B $\underline{31}$, 913 (1985).
9.   R.J. Wagner, B.V. Shanabrook, J.E. Furneaux, J. Comas, N.C. Jarosik, and B.D. McCombe, Proc. of the 11th Int'l Symposium on GaAs and Related Compounds, Brarritz, France, 1984 (Conference Series #74, Adam Hilger Ltd., Bristol and Boston, 1985) p. 315.
10. N.C. Jarosik, B.D. McCombe, B.V. Shanabrook, J. Comas, John Ralston, and G. Wicks, Phys. Rev. Letters $\underline{54}$, 1283 (1985).
11. R.L. Greene and K.K. Bajaj, Phys. Rev. B, to be published.
12. J.M. Mercy, N.C. Jarosik, B.D. McCombe, J. Ralston and G. Wicks, Proc. of PCSI-13, Pasadena, CA, Jan. 1986, to be published in JVST.

Magneto-Impurities and Quantum Wells

A. Raymond[1], *J.L. Robert*[1], *and W. Zawadzki*[2]

[1]Groupe d'Etudes des Semiconducteurs, (associé au CNRS, UA 357),
Université des Sciences et Techniques du Languedoc,
F-34060 Montpellier, France
[2]Institute of Physics, Polish Academy of Sciences,
PL-02668 Warsaw, Poland

Properties of impurities in quantum wells are reviewed, both theoretically and experimentally, emphasizing effects of an external magnetic
field. New results on donors in two-dimensional $GaAs-Ga_{1-x}Al_xAs$ heterostructures with a spacer are presented. Influence of an external magnetic field and a hydrostatic pressure on their properties
is discussed in terms of magnetic freeze-out effect and metal-nonmetal transition.

1. Introduction

Properties of impurities in quasi-two-dimensional structures have lately
become the subject of numerous investigations, because of their interesting physical properties as well as their important applications. As in
other cases of "man-made" quantum structures, the geometry of a given
situation is decisive for its behaviour. Until the present,the most frequently investigated cases were those of impurities in the quantum well,
either at its center or at the edge. This is a situation in which an impurity atom is constrained from the outside, i.e. the electron is kept
closer to the impurity center than it would be according to its "natural"
behaviour. In modulation-doped structures with a spacer a different situation is possible, in which the impurity atom is constrained from the inside,
i.e. the electron is kept farther from the impurity center than it would
be according to its behaviour without constraint.

It is well known from investigations on bulk semiconductors that a
presence of an external magnetic field has a significant influence on
shallow impurities. When the magnetic energy becomes comparable to the
coulomb energy, the wave function of a magneto-impurity shrinks considerably and, as a result, its binding energy increases. In narrow-gap semiconductors the presence of a magnetic field is a necessary condition for
observation of shallow impurities.

In the first part we will review basic properties of donors in quantum
wells, both theoretically and experimentally. Next we will discuss the
influence of an external magnetic field perpendicular to the interface on
such quasi two-dimensional impurities. In the second part we will review
new results on donors in heterostructures with a spacer. We once more em-
phasize the influence of an external magnetic field and, in addition, of a
hydrostatic pressure on their behavior. In both cases we mostly have in
mind donors in $GaAs-Ga_{1-x}Al_xAs$ heterostructures.

2. Shallow impurity in a quantum well

The advance of molecular beam epitaxy and of metal-organic chemical-vapor deposition techniques has made it possible to grow heterostructures in which the layer thickness may be arbitrarily varied. Let us consider the problem of an electron attracted by a coulombic impurity and confined between two planes $z = \pm L/2$ (BASTARD [1]). Let z_i denote the impurity position. The effective mass Hamiltonian for the problem reads :

$$H = \frac{p^2}{2m^*} - \frac{e^2}{\kappa} \frac{1}{[x^2+y^2 + (z-z_i)^2]^{1/2}} + V(z) \qquad (1)$$

with the boundary condition $\psi (\pm L/2) = 0$ for the eigenstates. One looks for a variational solution of eq.(1) with the one-parameter trial function :

$$\psi_\lambda (\vec{r}) = \cos (\frac{\pi}{L} \frac{z}{}) A (L,\lambda,z_i) \exp (-\frac{1}{\lambda} \left[x^2+y^2 - (z-z_i)^2\right]^{1/2}) \qquad (2)$$

if $|z| \leqslant L/2$ and $\psi_\lambda (\vec{r}) \equiv 0$ otherwise. A is a normalisation coefficient. If $E(L,z_i)$ is a binding energy between the ground impurity state and the first electric subband in the well, there is

$$E (L,z_i) = \frac{\hbar^2 \pi^2}{2m^* L^2} - \min_\lambda < \psi_\lambda |H| \psi_\lambda > \qquad (3)$$

The trial wave function (2) leads to exact results for $L = \infty$ (in this case $E = R^*$), and for $L = 0$ (then $E = 4R^*$), where the effective Rydberg $R^* = m^* e^4 / 2\kappa^2 \hbar^2$. In Fig. 1 we show variation of $E(L,0)$ (impurity at the well's center) and $E(L, \pm L/2)$ (impurity at the well's edge) with L. It is seen that the 2D character disappears very quickly : for $L/a^* = 1$ there is already $E(L,0) \cong 2.25 R^*$ (we denote $a^* = \hbar^2 \kappa /m^* e^2$ the Bohr radius, $a^* \cong 100 \overset{\circ}{A}$ for GaAs). For a fixed L the binding energy $E (L,z_i)$ depends on the impurity location z_i along the well. Thus the electron confinement lifts the usual degeneracy with respect to the impurity position. The binding energy is maximum for impurity at the center ($z_i = 0$) and minimum for impurity at the edge ($z_i = \pm L/2$) for fixed L.

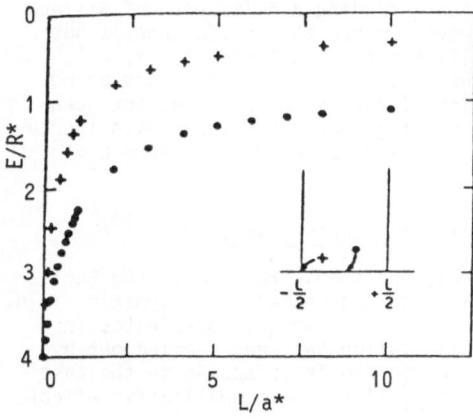

Fig.1 : Impurity binding energies E (L,0) and E(L, ± L/2) (in units of the effective Rydberg) versus well thickness (in units of the effective Bohr radius). After BASTARD [1].

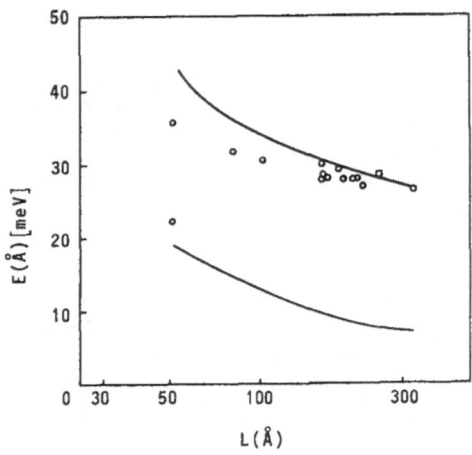

Fig. 2. The binding energy of neutral acceptors versus the GaAs well width L_z. The upper and lower curves represent the theoretical prediction of BASTARD for carbon acceptors in the center of the well and at the interface, respectively. After MILLER et al. [2]

Thus, if impurities are distributed randomly within the layer (i.e. within the well) one deals with an impurity band and not with an impurity level (like in 3D case).

Qualitative features obtained from the above simple model have been confirmed experimentally by MILLER et al [2]. Their results are shown in Fig.2.

The simple theoretical model presented above has been refined by various authors. GREENE and BAJAJ [3] took into account a finite size of the potential V_o at the GaAs-GaAlAs interface. As a consequence the impurity wave function does not have to vanish identically outside the potential well. It turns out that the donor binding energy decreases with the decreasing height of the interface potential. Also a new effect appears, as compared to results presented in Fig. 1, namely the binding energy has a maximum at low well widths. This is explained as follows : as the value of L is reduced, the wave function becomes more compressed in GaAs well, leading to more binding. However, beyond a certain value of L, more and more donor wave function is found in the surrounding GaAlAs. This leads, for L → 0, to the binding energy which is characteristic of a bulk donor in GaAlAs. For the case of the infinite potential barrier this does not occur and E (L , ± L/2) increases monotonically up to the 2D limit. MAILHIOT et al [4] performed a variational calculation including a difference of dielectric constants in GaAs and GaAlAs. However, since this difference is not large, the numerical results do not differ much from those of Ref. [3]. Finally, BRUM et al [5], considered theoretically an effect of screening by free carriers on the binding energies of donors in semiconductor quantum wells. For a given well width the binding energy decreases with increasing electron concentration until a saturation is reached at large n_s.

3. Effect of magnetic field on impurities in quantum wells

If an external magnetic field is present, one should replace in the Hamiltonian (1) $\vec{p} \rightarrow \vec{p} + e \vec{A}$, where $\vec{A}$ is the vector potential of a magnetic field. One usually takes $\vec{A} = (1/2) (\vec{B} \times \vec{r})$ and uses cylindrical coordinates for the trial functions. The variational calculation has been carried out by GREEN and BAJAJ [6] , for $\vec{B}$ perpendicular to the interface. As in the case of 3D magneto-impurities (cf. YAFET et al [7]) the main qualitative effect of a magnetic field is an increase of impurity-binding energy with increasing magnetic field. This is due to shrinking of the electron orbit (cf. Fig. 6), so that on average the electron is closer to the impurity center and the coulomb energy increases. A characteristic parameter for the problem is : $\gamma = \hbar\omega_c/2\ R^*$ where $\omega_c = e\ B/m^*$ is the cyclotron frequency. For $\gamma \ll 1$ the coulomb interaction dominates, for $\gamma \gg 1$ the magnetic interaction dominates. For an unscreened coulomb interaction for donors in GaAs there is $R^* = 6$ meV and at available magnetic fields one deals with $0 < \gamma < 2.5$. There are three magneto-impurity states of first experimental importance. In the hydrogen atom notation they are : 1s, $2p_-$ and $2p_+$. At high magnetic fields they correspond to states : (000), (0$\bar{1}$0) and (010), respectively, in the standard notation. 1s $\rightarrow$ $2p_-$ is a low-energy transition, while 1s $\rightarrow$ $2p_+$ is a higher energy transition, called at high fields an impurity-shifted cyclotron resonance.

Since, as mentioned above, the donor energies depend on their positions in the well, the latter have to be well defined by the growth procedure in order to produce well-resolved optical spectra.

JAROSIK et al |8| investigated Si-donors in the center of GaAs-GaAlAs quantum wells for the well widths between 80 and 450 Å. Both 1s - $2p_-$ and 1s - $2p_+$ transitions were observed. As a general rule, the well confinement increases magneto-optical energies as compared to the bulk case. The results are in good agreement with the theory of GREENE and BAJAJ |6|. The effects of well confinement become somewhat smaller at high magnetic fields, which is understandable since the magnetic field makes the electron orbits smaller and the outside limitations become less important. It is somewhat surprising that no screening effects were observed at doping levels of $1 \times 10^{16} cm^{-3}$.

The donors at the edge of GaAs-GaAlAs have also been observed |8|. In this case the agreement between the theory and experiment is somewhat worse, which may indicate that either the imposed boundary conditions (very important in this case) or the effective mass approximation are not equally valid for the edge impurities as they are for the center impurities.

4. Magneto-donors in heterostructures with a spacer

Now we turn to the second subject of our review and consider the case of the impurity atom outside the well, separated from the confined electron by a spacer. This situation has been very recently investigated by means of magneto-transport measurements. External magnetic field and hydrostatic pressure have been used as variable parameters. In particular, the pressure has been applied to reach sufficiently low surface-electron densities, at which the ultra quantum region occurs at available magnetic fields (only the lowest Landau level occupied). One can diminish the surface electron density n_s in GaAs-Ga$_{1-x}$Al$_x$As heterostructures by the effect of pressure on the Si donor in Ga$_{1-x}$Al$_x$Sb |9|, which shifts rapidly downward with respect to the Γ conduction-band minimum because of its deep-level character. As a consequence, donor deionization takes place and n_s decreases when the pressure is applied.

The situation is shown schematically in Fig. 3. The electrons from the quantum well in GaAs are transferred under pressure to the Si donors in Ga$_{1-x}$Al$_x$As. As a result, the potential well in GaAs becomes shallower.

At sufficiently high pressures, one can reduce n_s to values lower than 5×10^{10} cm^{-2} even for highly doped samples.

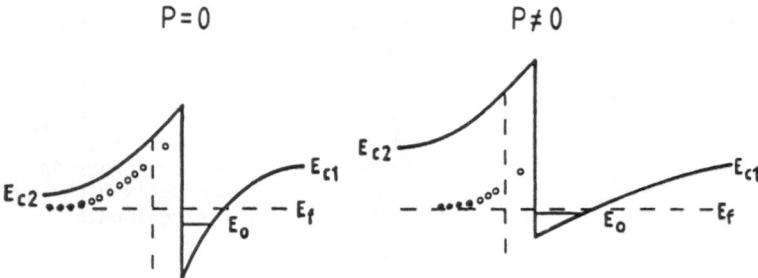

Fig. 3 : Variation of the energy diagram of GaAs-Ga$_{1-x}$Al$_x$As hetero-
rostructure under hydrostatic pressure.

Structures grown by molecular-beam epitaxy or metal-organic chemical
vapor deposition techniques, consisting of an undoped GaAs layer and undoped Ga$_{1-x}$Al$_x$As spacer (with thickness d varying between 60 and 250 Å),
and a Si-doped Ga$_{1-x}$Al$_x$As layer, have been studied by ROBERT et al [10].

In table I, the sample characteristics at 4.2 K (with and without pre-
sure) are presented. The values of n_s and the mobility are the Hall
values, measured at low magnetic field (B = 0.5 T). During the experiments, special care has been taken to stay within the ohmic regime. All
the measurements have been carried out for two current and magnetic field
directions in the temperature range 1.5 - 4.2 K. When the condition
$\omega_c \tau \gg 1$ is satisfied, n_s has been determined from the σ_{xy} component of
the conductivity tensor [11,12] :

$$n_s = \frac{1}{e} \frac{R_H B^2}{\rho_\perp^2 + R_H^2 B^2} \tag{4}$$

Table I : Sample characteristics, critical magnetic fields and
surface electron densities for MNMT in GaAs-Ga$_{1-x}$Al$_x$As heterostructures at different hydrostatic pressures. The last
column gives the value of a Mott-like criterion for the MNMT.
After ROBERT et al |10|

Sample	x	Spacer thickness (Å)	P = 0 T = 4.2 K		Hydrostatic Pressure T = 4.2 K						
			n_s 10^{10} cm^{-2}	μ 10^4 cm^2/Vs	p kbar	n_s 10^{10}cm^{-2}	μ 10^4cm^{2}Vs	B_c (T)	n_{sc} 10^{10}cm^{-2}	Fig.4	$\sqrt{n_{sc}}$ L_c
1	0.3	60	52	7.5	13.3	8.5	0.95	6	6		0.26
2	0.25	150	· 24	12.9	8.8	6.2	2.55	5	5.9	a	0.28
					8.8	5.7	2.36	4.2	5.6	b	0.29
					8.8	5.1	1.84	3	5	c	0.33
					8.8	4.9	1.73	2.8	4.8	d	0.34
3	0.27	250	20.8	5.2	5.9	12.5	2.0	10	11		0.27
4	0.3	250	35	41.9	13	6.5	6.82	8	6.5		0.23

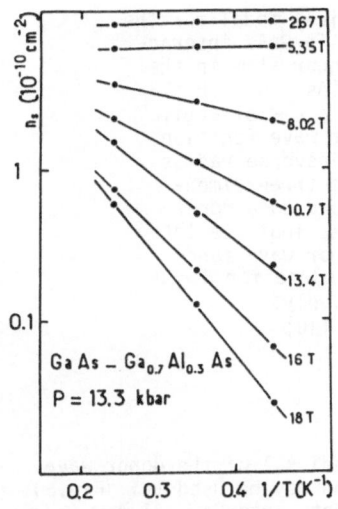

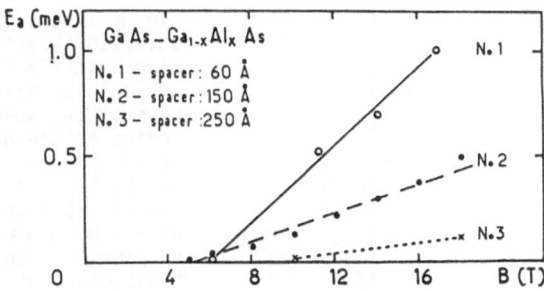

Fig. 5: Activation energies E_a of magneto-donors (the inversion electron is separated from the donor atom by a spacer) for samples 1, 2 and 3 versus magnetic field. After ROBERT et al. [10].

Fig. 4: Temperature dependence of the surface electron density for different magnetic fields. Sample 1 under pressure of 13.3 kbar. After ROBERT et al. [10].

Figure 4 indicates clearly the evidence of a MNMT : for magnetic fields higher than 6T the surface electron density n_s is thermally activated (magnetic freeze-out). This allows to determine the critical density n_{sc}, corresponding to the MNMT which occurs at the critical magnetic field B_c. The activation energy E_a can be directly determined from the slopes of the curves of $\ln(n_s)$ versus $1/T$. (Fig. 4). The activation energies determined for samples 1,2,3 with various spacer thickness (60, 150 and 250 Å) are presented in Fig. 5. For samples 1 and 2 the pressure has been chosen in such a way as to obtain approximately the same critical density n_{sc} (6 x 10^{10} cm^{-2}) and pratically the same critical magnetic field B_c. In these conditions the overlap of the donor wave functions and the screening of the donor potentials by surface electrons are the same for the three samples at the transition field.

Figure 5, nevertheless, shows that the magnetic field dependences of the activation energies are distinctly different for the three samples : the activation energy decreases with increasing spacer thickness. This result suggests that the observed localization effect should not be ascribed to the Wigner condensation of a dilute 2D electron gas, which, in the case of similar electron densities, would lead to the same activation energy value whatever the structures.

A qualitative explanation of the observed MNMT has been proposed by ROBERT et al [10]. They assume that the ionized donor atoms located in the doped $Ga_{1-x}Al_xAs$ layer at a distance d from $GaAs-Ga_{1-x}Al_xAs$ interface are responsible for the localization : when d decreases, the coulombic interaction between 2D electrons and ionized Si donors increases, and the binding energy E_a increases too. Let us be more precise and use the following simple model. We consider a perfect 2D electron at a distance d from an ionized

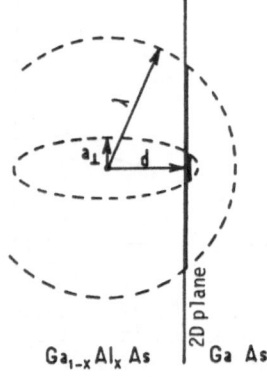

Fig. 6 : Semiclassical model for the inversion electron in GaAs interacting with the Si donor atom in the doped layer of GaAlAs : d - thickness of the spacer ; λ - effective radius of the donor wave function at B = 0 ; a - transverse radius of the orbit of the three-dimensional electron bound to a donor at B ≠ 0. Thick lines indicate intersections of the donor wave functions with the 2D plane for B = 0 and B ≠ 0, respectively. After ROBERT et al, [10]

$Ga_{1-x}Al_x As$ 2D plane   Ga As

donor atom. If d is smaller than the effective radius λ of the donor wave function, a semiclassical picture of the situation may be used (cf. Fig.6). Since the two-dimensional electron may not penetrate into $Ga_{1-x}Al_xAs$, in the bound donor state the electron wave function is given approximately by a disc, resulting from the cross-section of the Bohr sphere of the radius λ with the 2D plane. It has been shown by variational calculation that the donor binding energy E_a in this case is a small fraction of the bulk effective Rydberg R^*, and its effective radius λ is much larger than a^* ($a^* \cong 100$ A for GaAs)[13] . It is well known from the bulk investigations that in the presence of a magnetic field the donor wave function has a cigar shape,and its dimensions decrease with increasing magnetic field [7] . It is seen from Fig. 6 that the intersection of the magneto-donor wave function with the interface gives in this case a smaller disc, so that the electron is on an average closer to the donor atom than without magnetic field. As a result, the coulomb binding energy is enhanced by the presence of a magnetic field. This results in a magnetic freeze-out,which has been effectively observed experimentally [14,10]. It is also clear that in high-field conditions the binding energy of such a magneto-donor should be much smaller than the one in bulk GaAs, since in the latter case the electron is on an average much closer to the donor atom. This is fact what ROBERT et al [10] have observed : for B = 17 T they measure for sample 1 $E_a \cong 1$ meV, whereas for the bulk magneto-donor in GaAs at the same field one calculates $E_d \cong 12.5$ meV [7] . The above reasoning is qualitatively valid also for $\lambda < d$, although in this case one may not use the semiclassical picture.

In Figure 7 the activation energy E_a of magneto-donor is reported for sample 2. Measurements have been performed at the same hydrostatic pressure P = 8.8 kbar, and the surface density n_s has been varied by changing the speed of cooling of the sample (cf. Table I). In this case one deals with the same spacer thickness and it is seen that the activation energy increases with decreasing n_s. This can be interpreted in terms of a decrease of the screening.

On the other hand, the MNMT occurs at lower magnetic field when n_s is decreased.This can be associated with the metal-nonmetal Mott-type transition (change in the overlap of the impurity wave functions). As follows from Fig. 6, one expects that the surface electron moves on the orbit with radius $a_{\perp} \cong L = (\hbar/e B)^{1/2}$. On the other hand, the average distance bet-

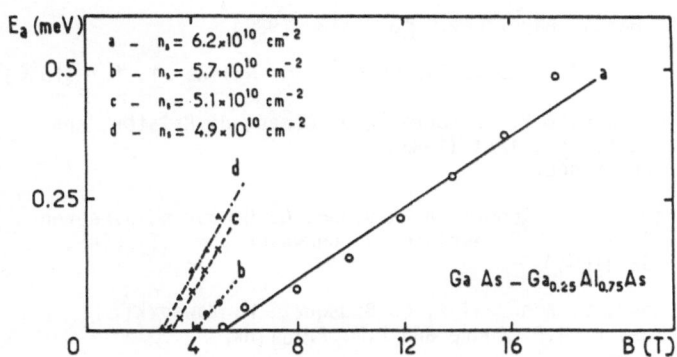

Fig. 7 : Magneto-donor activation energies E_a for sample 2 (under pressure of 8.8 kbar) versus magnetic field : The above lines are drawn to guide the eye. Different E_a (B) dependences correspond to various n_s values, obtained by different speeds of cooling the sample, cf. Table I. After ROBERT et al |10|.

ween surface electrons for the critical density is $n_{sc}^{-1/2}$. Thus the overlap condition for the Mott transition is approximately given by $2 L_c \cong n_{sc}^{-1/2}$, i.e. $n_{sc}^{1/2} L_c \cong 0.5$. As shown in table 1, the product $n_{sc}^{1/2} (\hbar/e B_c)^{1/2}$ is, in fact, close to that value.

Acknowledgements

The authors would like to thank Dr. J.P. ANDRE, P.M. FRIJLINK (LEP) F.ALEX-ANDRE and J.M. Masson (CNET) for providing the samples. The authors acknowledge the SNCI-CNRS and Dr. C. BOUSQUET and Dr. L. KONCZEWICZ for their participation in some experiments supported by the MIR and the CNRS. The authors are grateful to Dr. G. BASTARD for many valuable discussions.

References

1. G. Bastard, Phys. Rev. B24, 4714 (1981)
   G. Bastard, Surf. Science 113, 165 (1982)

2. R.C. Miller, A.C. Gossard, W.T. Tsang and O. Munteanu,
   Phys. Rev. B25, 3871 (1982)

3. RL.Greene and K.K. Bajaj
   Solid Stat. Comm. 45, 825 (1983)

4. C. Mailhiot, Yia-Chung Chang and T.C. Mc Gill
   Phys. Rev. B26, 4449 (1982)

5. J.A. Brum, G. Bastard and L. Guillemot
   Phys. Rev. B30, 905 (1984)

6. R.L. Greene and K.K. Bajaj, Phys. Rev. $\underline{B31}$, 913 (1985)

7. Y. Yafet, R.W. Keyes and E.N. Adams, J. Phys. Chem. Solids $\underline{1}$, 137 (1956)

8. N.C. Jarosik, B.D. McCombe, B.V. Shanabrook, J. Comas, J. Ralston, and
   G. Wicks, Phys. Rev. Lett. $\underline{54}$, 1283 (1985)
   B.D. McCombe, this conference.

9. J.M. Mercy, C. Bousquet, J.L. Robert, A. Raymond, G. Gregoris, J.Beerens,
   J.C. Portal, P.M. Frijlink, P. Delescluse, J. Chevrier and N.T. Linh,
   Surf. Science $\underline{142}$, 298 (1984)

10. J.L. Robert, A. Raymond, L. Konczewicz, C. Bousquet, W. Zawadzki,
    F. Alexandre, I.M. Masson, J.P. André and P.M. Frijlink.
    Phys. Rev. B. (in the press).

11. J.L. Robert, A. Raymond, R.L. Aulombard and C. Bousquet
    Phil. Mag. $\underline{B42}$, 1003 (1980)

12. R. Mansfield, J. Phys. C.Solid State Phys. $\underline{4}$, 2084 (1971)

13. G. Bastard, (private communication).

14. J.M. Mercy, C. Bousquet, A. Raymond, J.L. Robert, G. Gregoris,
    J. Beerens, J.C. Portal, P.M. Frijlink.
    Proc. 17th Int. Conf. Physics of Semicond. San Francisco, USA,
    Ed. J.D. Chadi and W.A. Harrisson, Springer-Verlag, New York,
    Berlin, Heidelberg, Tokyo, 1099 (1984)

The $\delta(z)$ Doping Layer:
Impurities in the 2-d World of Layered Systems

F. Koch, A. Zrenner, and M. Zachau

Physik-Department E 16, Technische Universität München,
D-8046 Garching, Fed. Rep. of Germany

Abstract

We consider a layer of dopant atoms embedded in a single atomic
plane during epitaxial growth of a GaAs crystal. It is shown
that the electronic states of this δ-layer are subbands of a V-
shaped potential well. The subband levels are studied in magneto-
transport and tunneling experiments. We discuss the case of a
periodic sequence of layers, the role of valley-orbit effects,
and central cell corrections.

I. Introduction

Many aspects of semiconductor physics have changed profoundly with the advent of techniques for the deliberate and controlled growth of layered
crystal structures. Even the dopant impurity is no longer what it once was,
namely a diffused or implanted, randomly situated ionic center with a bound
carrier circling about it in hydrogen-like orbits. In the world of the synthetic, epitaxial layer growth impurity states can be designed, the carrier
orbits shaped to suit a purpose. Carriers can be removed from their parent
atoms by a spacer layer and impenetrable band-offset barriers, the so-called
remote doping.

Many other impurity designs are possible and have been explored in
theory and practice. There is, for example, the half H-atom as in Fig. 1a.
The donor ion is embedded in the interface barrier and the electron is confined to the half-space of the lower conduction band material. When acted
on by an additional depletion layer field the orbit is pressed against the
interface. The electron binding energy can thus be tuned. The case of an
isolated Na^+ ion at the $Si-SiO_2$ interface has been described in detail in
the literature /1/. Another object of current interest is the impurity in
the center of the GaAs quantum well in Fig. 1b /2/. Its hydrogenic-orbit
style is crimped by AlGaAs barriers. In the 2-d limit of a pancake orbital
the binding energy is increased by a factor of 4. In Fig. 1c is yet another
impurity state model /3/ with the electronic orbit pulled in additionally
by a hypothetical, positive sheet of charge in the z = 0 plane.

Related to all of these is the $\delta(z)$-function doping layer in Fig. 1d
which is the central theme of this paper. Donor ions are statistically
distributed in the plane to simulate a continuous, positive sheet of charge.
A given hydrogenic orbit is pulled inward by the electrostatic force of
each of the many ions. In the limit of an ion areal density N_D greater than
the inverse of the Bohr radius squared, the electronic states are spread

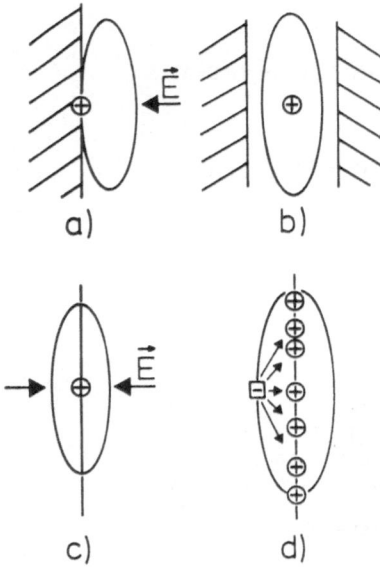

Fig. 1: Electronic configuration
for various custom-tailored
doping geometries

out over the donors that make
up the V-shaped potential well.
With Si atoms embedded in
GaAs the required N_D is of
order 10^{12} cm^{-2}.

It is the delocalization of
the carriers in the dopant
plane, and the fact that with
$N_D = N_S$ it is a many-electron
system that makes the δ-func-
tion in Fig. 1d different from
the others. Quantization of the
electron states into 2-d subbands was first demonstrated in
magnetotransport experiments
/4/. A MES-FET transistor has
been fabricated and described
in the literature /5/. Saw-
tooth-like, periodic structures
of alternate n- and p-type δ-
layers have been examined in
Ref. /6/. Such a periodic structure is also to be found in
Döhler's arsenal of n-i-p-i
variants /7/.

We here intend to review
what is known about the subband physics of the δ-doping
layer in GaAs and to speculate
about things yet to come.

II.   Samples and Their Subband Structure

A sharp and well-defined doping layer is most easily realized in epitaxially grown GaAs. Using molecular beam epitaxy (MBE), K. Ploog of the MPI für
Festkörperforschung has provided the Si-doped δ-layer samples for our experiments. Typically the GaAs growth is interrupted, while Si is deposited at
a rate of order $\sim 10^{11}$ atoms sec^{-1}. With the doping layer covered by several
μm of GaAs, most of the electrons remain in the δ-layer, i.e. $N_S \approx N_D$. For
the tunneling experiments samples are used where the covering layer is only
200 Å. Because of the pinning of the Fermi energy at the surface some electrons are transferred from the potential well to the surface. In that case
$N_S < N_D$.

Alternatively, P. Roentgen of the electrotechnical laboratory in the
RWTH Aachen has worked with metal-organic vapor-phase epitaxy (MOVPE) to
produce the desired doping structures. This process employs Se-doping during
continuous growth. His procedure leads to an expected spread over several
tens of Å in the dopant distribution.

The subbands of the doping layer are given by a straightforward, self-
consistent, one-dimensional potential calculation. The only complication
is conduction-band nonparabolicity, which needs to be included because of
the relatively high energies above the band edge. An example of the computed potential, the square of the wave-function, and the first three

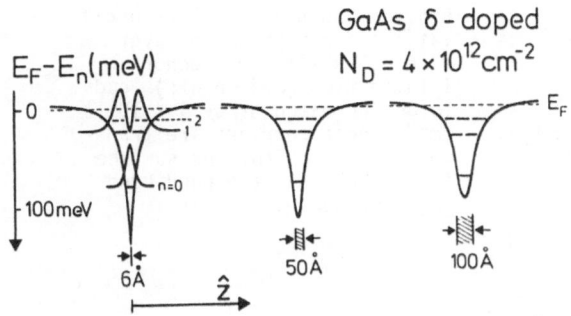

GaAs δ-doped

$N_D = 4 \times 10^{12} cm^{-2}$

$E_F - E_n$(meV)

Fig. 2: Subband levels in the δ-layer for various spreadings of the donor ions

levels is given in Fig. 2. The calculation assumes a light p-type background doping with depletion charge 1×10^{10} cm^{-2} and considers a modest $N_S = N_D = 4 \times 10^{12}$ Si cm^{-2}. For the sharp doping profile expected in MBE, where the Si donor is confined to 1-2 atomic layers, we use a model positive charge spread of 6 Å as in the left-hand side of the figure. Of the four occupied levels with energies 78, 26, 10, and 2.5 meV three are shown. The ratios between these numbers are typical for a quasi-accumulation potential with sharp doping profile. The characteristic binding length of the ground state electrons (chosen as the distance between the classical turning points in the potential well) is ∿ 50 Å. This is considerably less than the diameter of the hydrogenic 1s orbit which is ∿ 200 Å.

To examine the sensitivity of the level structure to a spread in the doping width we have repeated the calculation for a 50 and 100 Å smearing of the Si deposition. While the total depth of the well reacts sensitively, the level structure and the expected subband occupation numbers change only slowly. Typical error limits of 5 - 10% in the experimental determinations of energy would mean that a ∿ 50 Å spread could not be detected reliably. The case of 100 Å broadening of the doping profile, however, is easily identified by monitoring the ratio of the n = 0 and n = 1 energies.

III. Magnetotransport Experiments

The first evidence of 2-d subbands in a δ-layer came from Shubnikov-de Haas oscillations in a magnetic field perpendicular to the sample. In Ref. /4/ two distinct periods in 1/H were identified at fields up to ∿ 10 T. A calculation assuming that the layer had an N_S value equal to the design doping showed that the two periods were linked to the n = 1 and 2 subbands. The n = 0 ground state, it was speculated, would contain most of the remaining charge. Because of the relatively lower mobility in this band, conductivity oscillations were expected at higher fields. That this indeed is the case we show here in Fig. 3. For data on the very same sample as had been used before /4/, the missing n = 0 electron conductivity oscillations have now been identified at fields above 10 T. Fig. 3 shows a distinct third period in high fields. The sum of all three groups of electrons equals 97% of the total charge. The remainder is expected in the highest, n = 3 subband. The occupations, in particular the relative numbers in each of the bands, are well described in terms of a sharply defined doping layer.

A related magnetotransport effect, one that we have labelled the diamagnetic analogue of the conventional Shubnikov-de Haas oscillations, was

177

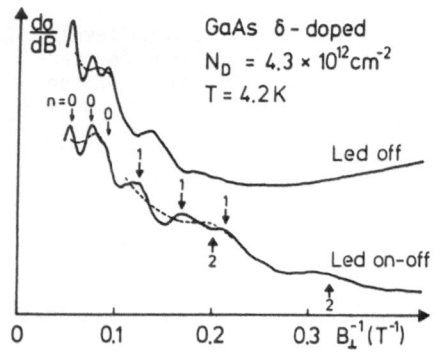

Fig. 3: Shubnikov-de Haas oscillations for a Si-doped layer with 4.3 x 10^{12} cm^{-2} donors. Illumination (Led on-off) leads to an increased N_S by 4 x 10^{11} cm^{-2}. These carriers are transferred mainly from the surface to the layer as the band bending is reduced.

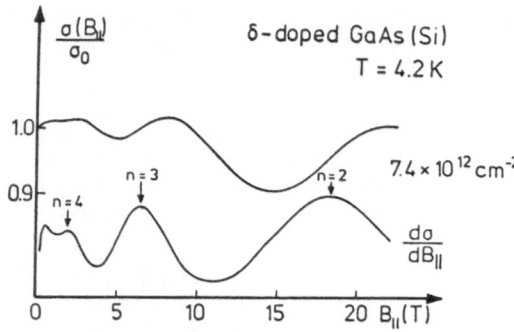

Fig. 4: The diamagnetic Shubnikov-de Haas effect oscillations in $\sigma(B_{\parallel})$ and $d\sigma/dB_{\parallel}$ for a sample with 7.4 x 10^{12} Si cm^{-2}

observed with the magnetic field in the sample plane /8/. The subbands in the V-shaped potential represent carriers that move symmetrically in $\hat{z}$ about the parallel field. For the lowest energy, i.e. the $k_{\parallel} = 0$ states of the subband, it is a closed cyclic motion in the $\hat{z}$-direction. Such a subband minimum will be shifted up by a diamagnetic energy in the field $B_{\parallel}$ until the level is emptied. This gives rise to the oscillations that are shown in Fig. 4. A calculation has allowed us to identify the various bands and has shown them to be levels of the doping layer potential.

IV. Tunneling Spectroscopy

Tunneling experiments serve to demonstrate the quantization into discrete subband levels. They allow one to measure directly the relevant energies. In the work of Ref. /9/ a tunneling structure involving a Schottky barrier has been devised. A Ag electrode is evaporated onto the GaAs surface and forms a barrier with height $e\emptyset_B \sim 900$ meV. The doping layer is 200 Å from the interface as in Fig. 5 below. Although the layer is doped with 8 x 10^{12} cm^{-2} the number of electrons N_S is considerably less. The calculation of the energy level structure as in the figure shows that 3.8 x 10^{12} electrons are transferred to the surface until the required barrier is achieved. With the applied voltage eV_g additional carriers are transferred. The potential and the subbands adjust accordingly. The square of the n = 0 and 1 wavefunctions is entered in the figure. The overall arrangement resembles that for an ohmic contact made by degenerately doping the semiconductor surface region.

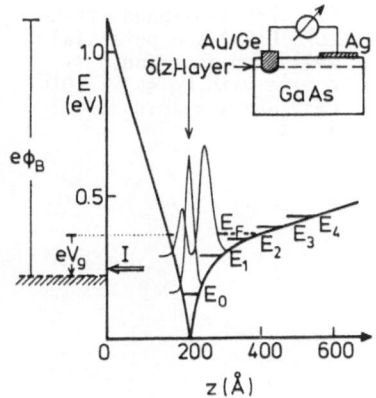

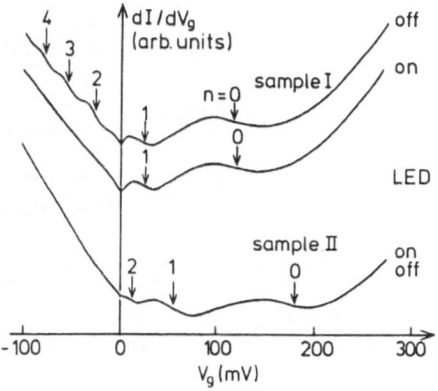

Fig. 5: Potential diagram and electronic levels of the doping layer for $N_D = 8 \times 10^{12}$ cm^{-2} and in the presence of an applied forward voltage eV_g. A barrier height of 900 meV and a p-type background doping have been assumed.

Fig. 6: Tunneling current derivative dI/dV_g vs. the applied gate voltage V_g. The arrows mark the energies of the subbands relative to the Fermi energy.

With positive V_g bias, electrons tunnel from the δ-layer to the Ag-gate giving rise to the dI/dV_g structures in Fig. 6. For the p-type Sample I structures are also found for negative V_g. They vanish when the sample is illuminated with a band gap radiation. The negative V_g signals are evidence for empty subbands above E_F in the depletion field potential of the p-sample. Sample II is known to be n-type and with a Se doped δ-layer. By fitting the observed dI/dV_g structures to a subband calculation the N_D is determined as 1.2×10^{13} cm^{-2}.

V. The Periodic δ-Layer

The tunneling experiments explicitly involve carrier motion through the self-consistent potential profile normal to the layer. It is a straightforward step from the tunneling through a single potential well, to consider the motion through a series of such δ-potentials. Following Döhler's nomenclature, the particular n-i-p-i structure that we examine here should be called a $\delta i \delta i$. It consists of a periodic n-type layer doping in otherwise intrinsic material.

We have calculated and dimensioned a series of such structures for future experiments. One that should prove suitable and that shows to advantage the typical effects to be expected is shown in Fig. 7. Choosing a spacing of 100 Å and $N_s = 3 \times 10^{12}$ cm^{-2} the band structure shown in the figure results. The n = 0 subband level broadens into a band whose width is $\sim$ 45 meV. There is a gap for motion in the $\hat{z}$-direction which is a little more than 20 meV. The Fermi energy lies in the middle of the next higher band.

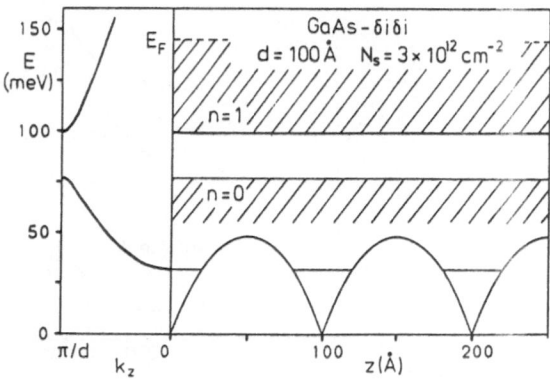

Fig. 7: Energy-band structure $E(k_z)$ and potential wells for a periodic δ-doping with $N_D = 3 \times 10^{12}$ cm^{-2} and a spacing of 100 Å

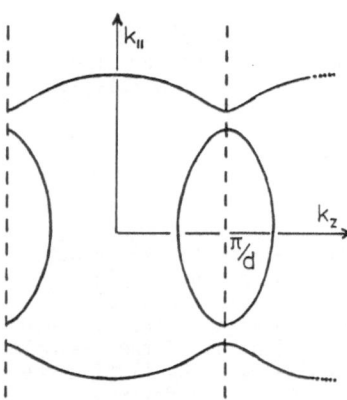

Fig. 8: Fermi surface cross-section in the k_z, $k_{\parallel}$-plane for a $\delta i \delta i$ with d = 100 Å and $N_D = 3 \times 10^{12}$ cm^{-2}

The construction that we have made here is one that serves as a "textbook case" in metal fermiology. In Fig. 8 we have gone on to construct a cross-section of the rotationally symmetric Fermi surface. There are two sheets, associated with the first and second subbands. The n = 0 band gives an open sheeted surface, the n = 1 band gives a lens-shaped electron surface centered about $k_z = \pi/d$. If extended along $\vec{k}_z$, we recognize the n = 0 surface as a cylinder with periodically modulated cross-section. It is a construction that should remind many a senior 2DEG-scientist of his early days in fermiology. Amazingly enough, however, this is degenerate GaAs with an average doping $n_D = 3 \times 10^{18}$ cm^{-3}! It may be necessary to choose a layer doping of more than 3×10^{12} cm^{-2} in order to clearly demonstrate the periodic character of the structure every 100 Å, but that is no great difficulty. Calculations using doping up to about 10^{13} cm^{-2} result in a qualitatively similar electronic structure. Enough of the speculations. Magnetotransport experiments on periodic δ-layers are in progress.

VI. Central Cell- and Valley-Orbit Effects. What's so Special about the δ-Layer?

After this excursion into band structure and Fermi surface construction, all of it based on the effective mass description of Γ-point electrons in the band scheme, we return to reality. The δ-layer is not a 2-d electron gas in the sense of a gated MOS-transistor or a remote-doped heterojunction. Electrons experience not only the electrostatic "physical" force of the

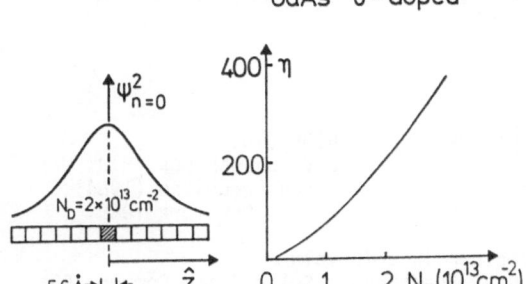

GaAs δ-doped

Fig. 9: The ground-state charge distribution on the scale of the 5.6 Å lattice constant for $N_D = 2 \times 10^{13}$ cm^{-2}. η is the ratio of electronic charge in the central cell for the n = 0 subband to that for the 1s hydrogenic state in GaAs.

charges, but they also sense the "chemical" nature of dopant atoms in the central cell. Also the density N_S is not limited by the breakdown strength of an insulator to about 10^{13} cm^{-2}, as it is for the best Si-SiO$_2$ interfaces. δ-layer samples have been prepared with up to 10^{14} Si or Se dopant atoms per cm^2 and there is no reason to stop there.

High densities N_S and related high values of E_F are a fact for doping layers. In Fig. 2, even for the low $N_D = 4 \times 10^{12}$ Si cm^{-2}, the subbands are filled a good part of the way from the Γ-minimum to the neighboring L- and X-points of the band structure of GaAs. The nonparabolicity of the Γ-conduction band minimum has to be properly included in order to correctly explain the subband energies.

Because it is a "chemical" layer, the δ-well subband electrons experience the central cell corrections in the way we are familiar from impurity-state physics in 3-d. In particular, the n = 0 level will experience a chemical shift depending on the central cell occupancy of a Si or Se ion. The interaction is expected to be much larger than the effect for bulk GaAs impurities. Whereas the hydrogenic 1s orbit diameter is ∿ 200 Å, the n = 0 subband for $N_D = 2 \times 10^{13}$ cm^{-2} in Fig. 9 extends only over ∿ 40 Å. The electron confinement to the central cell is hundreds of times more than for the hydrogenic states. These should be a chemical shift that will move E_0 up or down relative to E_1 or E_2 and thus give different ratios for different doping material, even if N_D were the same.

Moreover, for densities in the 1-3 x 10^{13} cm^{-2} range the L- and even the X-points need to be considered. These degenerate valleys would split into distinct levels in the presence of the chemical shift. In fact, it is sufficient to consider the strong Coulomb potential of the layer to get a significant valley splitting. This makes it quite difficult to exactly locate the next higher subbands, but they will be quite close when N_S is in the 10^{13} cm^{-2} range. Strong nonlinearities are expected when carriers get hot and transfer to other bands. These aspects will be considered in future work.

VII. Final Remarks

Perhaps the most important message of the paper is that in the brave new world of synthetic crystals, made by MBE, MOCVD, and other techniques, impurities don't just happen - they are caused. One particular way that they can be caused is the δ-layer, a novel 2-d system that we have described at some length.

We thank the Siemens Co. and the Deutsche Forschungsgemeinschaft for the financial support to carry out this work.

References:

1/ See for example, B. Vinter, Phys. Rev. B 26, 6808 (1982)
2/ See for example, N.C. Jarosik, B.D. McCombe, B.V. Shanabrook, R.J. Wagner, J. Comas, and G. Wicks, Proc. of the ICPS San Francisco (1984), Eds. J.P. Chadi and W.A. Harrison, Springer Verlag, New York 1985, p. 507
   and cf. B.D. McCombe, review lecture in this volume
3/ F. Crowne, T.L. Reinicke, and B.V. Shanabrook, Solid State Commun. 50, 875 (1984)
4/ A. Zrenner, H. Reisinger, and F. Koch, Proc. of the ICPS San Francisco (1984), Eds. J.P. Chadi and W.A. Harrison, Springer Verlag, New York 1985, p. 325
5/ E.F. Schubert and K. Ploog, Jap. J. of Appl. Phys. 24, L608 (1985)
6/ K. Ploog, A. Fischer, and E.F. Schubert, 2nd Int. Conf. on Modulated Semiconductor Structures, Kyoto (1985), Surface Sci. (in press)
   and cf. K. Ploog, lecture in this volume
7/ G. Döhler, cf. the lecture in this volume
8/ A. Zrenner, H. Reisinger, F. Koch, K. Ploog, and J.C. Maan, to be published
9/ M. Zachau, F. Koch, K. Ploog, P. Roentgen, and H. Beneking, to be published

Part V

Quantum Hall Effects and Density of States of Landau Levels

Quantum Hall Effect Experiments at Microwave Frequencies

R. Meisels[1], K.Y. Lim[1], F. Kuchar[1], G. Weimann[2], and W. Schlapp[2]

[1]Institut für Festkörperphysik, Universität, A-1090 Wien, and
Ludwig Boltzmann Institut für Festkörperphysik, A-1060 Wien, Austria
[2]Forschungsinstitut der Deutschen Bundespost,
D-6100 Darmstadt, Fed. Rep.of Germany

The observation of integer quantum Hall effect (IQHE)
behavior of σ_{xy} at microwave frequencies (~ 30 GHz)
is reported. This clears the discrepancy concerning
the so-called low-frequency breakdown of the IQHE. It
was observed in a.c. experiments in the MHz range
whereas 100 ns pulse experiment did not show a
deviation from the d.c. behavior. In this investigation, for the observation of σ_{xy} at microwave
frequencies, a crossed waveguide arrangement is
used. The common part of the cross-sections of
the two waveguides is much smaller than the sample
area. This is expected to minimize edge effects.
From the observed quantized behavior of σ_{xy} we
conclude that the mechanism of an effective
delocalization at low frequencies has to be ruled out.

1. Introduction

It is generally accepted that the integer quantum Hall effect
(IQHE) must be explained on the basis of the existence of localized and delocalized states in the broadened Landau levels[1].
Therefore the nature of the localization is of great interest for
the understanding of the IQHE. Dilute or dense short-range and
long-range scatterers have been considered as being responsible
for the localization. Experimentally, localization phenomena can
be investigated, e.g., by measuring the high-frequency conductivity of the two-dimensional electron gas (2DEG). In the past,
controversial results have been obtained for the components of
the high-frequency conductivity tensor in high magnetic fields.
Measurements were performed on Si-MOSFET's and on $GaAs/Al_xGa_{1-x}As$
heterostructures. KUCHAR et al.[2] and WOLTJER et al. [3] found
no difference between d.c. and 100-ns-pulse (comparable to 5 MHz)
experiments concerning the integer quantum Hall effect in GaAs/
$Al_xGa_{1-x}As$.

A strong frequency dependence in the Megahertz range was observed in $GaAs/Al_xGa_{1-x}As$ [4] and in Si MOSFET's [5,6] in measu-

ring the two-terminal resistance: Integer Hall plateaus were distorted or disappeared, the fractional plateaus were enhanced by
increasing the frequency. The frequency where the deviation from
the dc behavior sets in was lower in relatively low-mobility
samples [$\mu=(1.5-2)\times10^4$cm^2/Vs in MOSFETs, $\mu<10^5$cm^2/Vs in the hete-
rostructures]; it depended upon sample length and Landau-level
filling factor. In high-mobility GaAs/Al$_x$Ga$_{1-x}$As samples ($\mu=10^5$
cm^2/Vs) no deviation up to 50 MHz was observed. An explanation of
this behavior was suggested in the context of the localization
due to a disorder potential smooth on length scales comparable to
the magnetic length[7,8,9].The breakdown of the integer quantum
Hall effect at MHz frequencies would be caused by the effective
delocalization of semiclassical orbits in the sample. These orbits
are localized at d.c. even if they are quite extended spatially.
The delocalization of the states would cause a non-vanishing
diagonal component of the two-dimensional conductivity in the
previously quantized regime, and thereby destroy the integer plateaus.

Even lower frequencies were used by SMITH et al. [10]in experiments on GaAs/Al$_x$Ga$_{1-x}$ samples with capacitively coupled contacts.
They used frequencies below 100 KHz, their results agreed with
[4-6].While avoiding contact effects, they introduced the effect
of vanishing differential capacitance. It occurs, like the vanishing σ_{xx}, when there are no delocalized states at the Fermi level.
The Hall plateaus are thus replaced by minima. As recent measurements of the same group show, the frequency dependence appears
only in gated structures [11] . It is therefore attributed to the
coupling of the signal from the 2DEG to the gate and is no intrinsic effect of the 2DEG.

The main purpose of this investigation was to clarify the situation, i.e., to find out whether the integer plateaus disappear
because of delocalization at high frequencies. Microwave frequencies seemed to be particularly useful since they are 10^3-10^4 times
higher than those previously used.

In Chapter 2 we will calculate the response of a 2DEG to an
incident electromagnetic wave and show that the components of the
conductivity tensor can be determined. Chapter 3 describes the
experimental arrangement where crossed waveguides are used as a
polarizer-analyser system in order to determine σ_{xy}. From the experimental results on GaAs/Al$_x$Ga$_{1-x}$As presented in Chapter 4 we
will conclude that the model of semiclassical orbits in the treatment of Refs. 7 and 8 is not valid for a description of the localization in the 2DEG in high magnetic fields.

2. Response of a 2DEG to an Incident Electromagnetic Wave

We are now investigating the interaction of a 2DEG with an incident electromagnetic wave. We will use classical electromagnetic
theory. This approach is adequate as long as $\omega << \omega_c$,ω being the
circular frequency of the incident radiation, ω_c the circular
cyclotron frequency. For simplicity, we situate the 2DEG into

the infinite x-y plane at z=0. The 2DEG is embedded into a medium with dielectric constant $\kappa\kappa_o$ and permeability $\mu\mu_o$.

A static magnetic field is applied perpendicular to the sample leading to the conductivity tensor σ of the 2DEG:

$$\overleftrightarrow{\sigma} = \begin{pmatrix} \sigma_{xx} & \sigma_{xy} & 0 \\ -\sigma_{xy} & \sigma_{xx} & 0 \\ 0 & 0 & 0 \end{pmatrix} \tag{1}$$

The electromagnetic wave propagates parallel to the static magnetic field (Faraday configuration). Without loss of generality, we define the x-axis to be parallel to the electric vector E_{inc} of the incident wave. This electric field induces a current density j_{ind} in the sample. j_{ind}, in turn, generates a high-frequency field H_{ind}, an induction B_{ind}, and finally an induced electric field E_{ind}. We will now derive the relation between the incident E_{ind} and the induced E_{ind}.

With the above definitions we have

$$\vec{E}_{inc} = \begin{pmatrix} E_{inc} \\ 0 \\ 0 \end{pmatrix} e^{i(\omega t - kz)} \tag{2}$$

$$\vec{E}_{ind} = \begin{pmatrix} E_{ind,x} \\ E_{ind,y} \\ 0 \end{pmatrix} e^{i(\omega t - k|z|)} \tag{3}$$

$$\vec{E} = \vec{E}_{inc} + \vec{E}_{ind} \tag{4}$$

$$|\vec{E}_{inc}| \gg |\vec{E}_{ind}| \tag{5}$$

We will use the following Maxwell equations:

$$\operatorname{rot} \vec{H} = \vec{j} + \dot{\vec{D}} \tag{6a}$$

$$\operatorname{rot} \vec{E} = -\dot{\vec{B}} \tag{6b}$$

including the material equations

$$\vec{D} = \kappa\kappa_o \vec{E} \tag{6c}$$

$$\vec{B} = \mu\mu_o \vec{H} \tag{6d}$$

and Ohm´s law, modified for an 2DEG

$$\vec{j}_{ind} = \overleftrightarrow{\sigma} \vec{E}(z,t)\delta(z) \tag{7}$$

δ is Dirac´s Delta function.

Using Eq.(6b) for $\vec{E}_{ind}$ and $\vec{B}_{ind}$, and $\partial\vec{E}/\partial x = \partial\vec{E}/\partial y = 0$, $\vec{E}_{ind,z} = 0$ yields

186

$$-\vec{B}_{ind} = \begin{pmatrix} -E_{ind,y} \\ E_{ind,x} \\ 0 \end{pmatrix} e^{i(\omega t - k|z|)} (-ik)\ sgn(z). \tag{8}$$

Integration over time and using Eq.(6d) yields $\vec{H}_{ind}$ from which rot $\vec{H}_{ind}$ is obtained:

$$\text{rot } \vec{H}_{ind} = \frac{-1}{\mu\mu_o c_{med}} \begin{pmatrix} E_{ind,x} \\ E_{ind,y} \\ 0 \end{pmatrix} e^{i(\omega t - k|z|)} \{sgn^2(z)(-ik) + 2\delta(z)\} \tag{9}$$

$c_{med} = \omega/k$ is the velocity of light in the medium.

A second expression for rot $\vec{H}_{ind}$ can be obtained from Eq.(6a). The difference of rot $\vec{H}$ with and without the 2DEG for z=0 is:

$$\text{rot } \vec{H}_{ind} = \vec{j}_{ind} + \dot{\vec{D}}_{ind} = \vec{E}_{inc} \begin{pmatrix} \sigma_{xx} \\ \sigma_{xy} \\ 0 \end{pmatrix} e^{i\omega t} \delta(z) \tag{10}$$

According to Eq.(9) this is equal to

$$\text{rot } \vec{H}_{ind} = -\frac{1}{\mu\mu_o c_{med}} \begin{pmatrix} E_{ind,x} \\ E_{ind,y} \\ 0 \end{pmatrix} e^{i\omega t} 2\delta(z) \tag{11}$$

This yields

$$\begin{pmatrix} E_{ind,x} \\ E_{ind,y} \\ 0 \end{pmatrix} = - \begin{pmatrix} \sigma_{xx} \\ \sigma_{xy} \\ 0 \end{pmatrix} \frac{Z_{med}}{2} E_{inc}. \tag{12}$$

The impedance of the medium is $Z_{med} = \mu\mu_o c_{med}$. Eq.(12) shows that the induced electric field components are directly proportional to the components of the conductivity tensor. Of particular interest is $\vec{E}_{ind,y}$, which is perpendicular to the incident $\vec{E}_{inc}$ and therefore identical to the y-component of the transmitted wave. It is proportional to σ_{xy}.

The ratio $T_\perp$ of the intensities corresponding to $E_{ind,y}$ and E_{inc} is proportional to σ_{xy}^2 and given by:

$$T_\perp = \frac{P_{ind,y}}{P_{inc}} = \left(\frac{\sigma_{xy} \cdot Z_{med}}{2} \right)^2 \sim \sigma_{xy}^2. \tag{13}$$

Inserting $\mu=1$, $\kappa=13$, $\sigma_{xy} = (12.9\ k\Omega)^{-1}$ (i=2) we obtain for a single heterostructure $T_\perp = 0.9\times10^{-5}$. With multiple heterostructures the intensity is multiplied by the square of the number of the 2DEG layers.

3. Experimental arrangement

The controversy about the low-frequency breakdown of the IQHE
in the MHz range can be decided unequivocally by using measuring
frequencies several orders of magnitude above the proposed
thresholds of the frequency dependence. At these frequencies,
in the microwave region, the usual d.c. techniques with contacts
on the samples are no longer applicable, even if modified for
a.c. measurements [2,3,4,10]. Parasitic capacitances of the order
of pF, together with the impedances of the contacts (the Hall
resistance of typically 10 kΩ) form time constants of tens of
nanoseconds. They correspond to frequencies of about 10 MHz. So
these methods have come to their limits already in the previously used MHz ranges. Therefore, a contactless method using a
technique similar to an optical polarizer-analyser arrangement
was applied. As shown in Chapter 2 the component $P_{ind,y}$ of the
transmitted wave is proportional to the square of the Hall conductivity. To extract this component the waveguide arrangement
shown in Fig.1 was used. E_{inc} of Chapter 2 corresponds to the
polarization of the wave incident in the upper rectangular waveguide. The lower waveguide, at right angles to the upper one, acts
as an analyser and selects $E_{ind,y}$. In this way the 90° component
of the Faraday rotation is measured, which is equivalent to taking
the Hall conductivity. We would like to point out clearly that in
this experiment the Hall conductivity is directly measured. In
d.c. measurements it has to be computed from the values of ρ_{xx}
and ρ_{xy}.

The arrangement of the crossed waveguides is shown in Fig.1.
The intensity transmitted through the crossed waveguide (length
20 cm) was measured with a liquid helium-cooled bolometer. In
order to minimize interference effects in the transmitted intensity, an absorber was mounted behind the bolometer. A Hewlett-
Packard 8690B sweep oscillator with a 8697A rf unit was used as
the microwave source. Without a sample the transmitted intensity

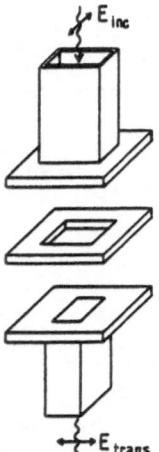

Fig. 1 Schematic drawing of the main part of the crossed
waveguide arrangement. The sample is positioned in the
middle flange (quadratic hole 8x8 mm^2). Over the length
of the cryostat the waveguide material is stainless
steel. Inner waveguide dimensions are 3.5x7 mm^2. E_{inc} is
the electric field vector of the incident wave. In the
transmitted wave, E_{trans} is the dominating field component after several wavelengths and is proportional to
σ_{xy} of the sample.

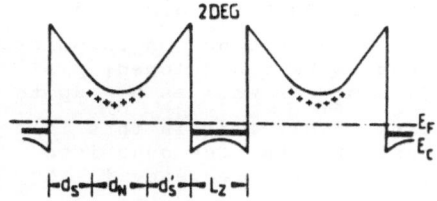

2DEG

$\vdash d_S \dashv d_N \vdash d_S' \dashv L_Z \dashv$

GROWTH DIRECTION ➔

Fig.2 Modulation doped multiple quantum-well structure.

Table 1. Characteristics of the samples. Meaning of symbols see Fig.2, n_s and μ at 2.2K. Si doping of the $Al_xGa_{1-x}As$ is $1.2 \times 10^{18} cm^{-3}$, x=0.35.

Sample	L_z (Å)	d_S (Å)	d_S (Å)	d_N (Å)	Number of GaAs wells	N_S (cm^{-2})	μ (cm^2/Vs)
1408	200	185	400	55	15	8×10^{12}	2×10^4
1325	500	55	450	570	10	13×10^{12}	1.7×10^5

was about 10^{-3} of the incident intensity. With a single hetero-structure the 2D-σ_{xy} effect was too weak to be observed with our experimental arrangement; multiple heterostructures had to be used instead. In principle, this measuring technique is a zero-background technique. Beside σ_{xy} contributions from the sample, mis-alignment of the waveguides (not ideally crossed) can add a background signal (without a sample) and a signal proportional to σ_{xx}^2 (with a sample.). These effects are observed to be fre-quency dependent, being weakest in a range of about 2 GHz. This range depended on the thickness of the sample flange (3 mm: 31.5-34 GHz).

Two different samples were used. Both were grown by molecular beam epitaxy [12]. They were multi-quantum-well structures, de-tails are given in Fig.2 and Table 1. One of the layers of sample 1408 was depleted as deduced from the QHE data, yielding an elec-tron density of $5.7 \times 10^{11} cm^{-2}$ per 2DEG layer. The mobility of this sample is comparable to that of the low-mobility samples of Refs. 4-6. In sample 1325 the doped AlGaAs regions were situated assyme-trically, so that 10 well-separated, single-sided heterostrucutres were produced. Even so, the evaluation of the d.c. quantum Hall effect data yielded a different number of 2DEG layers, viz.13. Sample areas were 7×7 mm^2 (1408) and 7×5 mm^2 (1325).

4. Experimental Results and Discussion

Fig. 3 shows the measured bolometer signal in the crossed wave-guide arrangement for sample 1408 together with the d.c. ρ_{xy} data and a calculated curve of σ_{xy}. The bolometer signal is proportional to σ_{xy}^2 as was shown in Chapter 2. Both the d.c. and the microwave data show a QHE behavior. No dependence on the microwave power was observed. The i=4 plateau is best developed; plateau-like structure can be recognized in ρ_{xy} as well as σ_{xy}^2 up to i=14. Although the i=4 plateau is not extremely flat in this 14-layer structure, the data clearly demonstrate that the quantized behavior of σ_{xy} is not destroyed at a measuring frequency in the 30 GHz range.

The classical Hall effect would yield a linear dependence of ρ_{xy} on B, with intersections at the "centers" of the plateaus. Therefore, we have fitted the classical formula for the high-frequency conductivity σ_{xy} to our experimental data of Fig.3. Generally the classical expression is [13] :

$$\sigma_{xy} = \frac{ne^2}{m^*} \frac{\omega_c \tau^2}{(1-j\omega\tau)^2 + \omega_c^2 \tau^2} \tag{14}$$

$1/\tau$ is the scattering rate at the Fermi energy, ω_c, the cyclotron frequency, and ω the measuring frequency. The other symbols have the usual meaning. For $\omega\tau \ll 1$, the essential dependence on magnetic field is given by

$$\sigma_{xy} \sim \omega_c / (1+\omega_c^2\tau^2) \tag{15}$$

Thus, the transmitted intensity in the crossed waveguide is proportional to $[\omega_c/(1+\omega_c^2\tau^2)]^2$ in the classical case. This can be fitted to the experimental σ_{xy}^2 curve in Fig.3. With $m^*=0.068m_o$, τ turns out to be 3.4×10^{-13} s. The overall shape of the experimental curve is nicely reproduced by the calculation. This proves that our bolometer signal corresponds to σ_{xy}^2 as predicted in Chapter 2. Moreover, most important is the agreement of the two σ_{xy}^2 curves at two particular positions: at the maximum and at the intersection at the center of the i=4 plateau. τ should

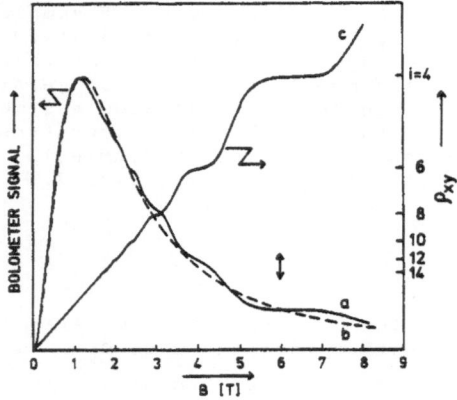

Fig.3 a: Bolometer-signal (proportional to σ_{xy}^2). b: calculated from the classical formula for σ_{xy}. c: d.c. results for ρ_{xy}. Sample 1408, T=2.2K, ν=33 GHz. The vertical arrow marks the intersection of the classical d.c.Hall resistance with the i=4 plateau of curve c.

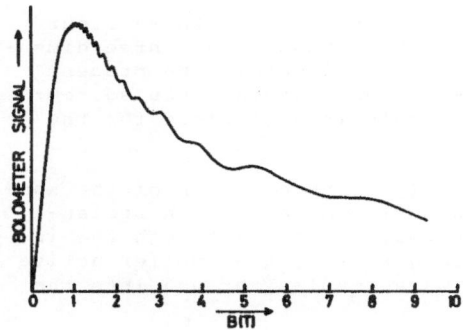

Fig.4 Bolometer signal
($\sim \sigma_{xy}^2$) of sample 1325,
T=2.2K, ν=32.7 GHz. In
terms of filling factors
structure in σ_{xy}^2 is seen
up to i=48.

be considered as a fitting parameter only, since its value obtained from the·mobility is much higher. The fitting procedure
using Eq.(2) is justified, since ω equals ω_c below 0.1 T and
$\omega_c \tau=1$ at about 1.1 T (maximum of the calculated curve).

We could also qualitatively reproduce the microwave data of σ_{xy}
of our samples from the d.c.values of ρ_{xy} and ρ_{xx}. Since for the
d.c. measurements only van-der-Pauw samples on very small pieces
of the same structures could be made, no quantitative comparison is
presented. Nevertheless,the σ_{xy} values calculated from the d.c. data
show the essential features that are also observed experimentally
in the microwave experiment: In the high mobility sample 1325 the
quantum oscillations of ρ_{xx} and the Hall plateaus of ρ_{xy} are visible to lower magnetic fields i.e. to higher filling factors
than in the low mobility sample 1408. Consequently structure in
the calculated σ_{xy} data also extends to lower fields. This can
also clearly be observed in Fig.4, which shows the σ_{xy}^2-proportional
bolometer signal of the high mobility sample 1325. Furthermore,
the more pronounced ρ_{xx} oscillations of sample 1325 have the
effect of producing minima in σ_{xy} at higher filling factors according to

$$\sigma_{xy} = \frac{\rho_{yx}}{\rho_{xx}^2 + \rho_{xy}^2} \qquad (16)$$

On the other hand, at high enough fields ρ_{xx}^2 is negligible
compared to ρ_{xy}^2 and a clear plateau behavior develops. The
σ_{xy} minima do not show up in sample 1408 because of the weaker
ρ_{xx} oscillations.

Another point should be also mentioned here: Although the low-
field d.c. mobilities of the two samples studied are quite different,the overall maxima of the two σ_{xy} curves occur roughly at
the same magnetic field strength. This allows to apply the fitting
procedure of Eq.(15) to both samples. The discrepancy with the
d.c. mobility data, however, cannot be explained at present.

Our microwave measurements are contactless. So, there occur no transitions of the electrons from the 2DEG to the three-dimensional degenerate electron gas of a contact metal. The process which underlies this transition is not understood. Even so, our microwave experiments show that its role is negligible for the occurrence of the QHE.

Also, edge effects are minimized. The active region of the sample is the small square $(3.56 \times 3.56 \text{ mm}^2)$ common to both rectangular cross-sections of the crossed waveguides. Although the incident wave hits the sample on a larger area, the smaller active area is the one where the induced wave couples through into the crossed waveguide.

5. Conclusions

From the direct observation of the quantized behavior of the Hall conductivity σ_{xy} at microwave frequencies several conclusions can be drawn:

a) The low-frequency ($\sim$MHz) breakdown of the IQHE does not occur. This confirms the results of Refs.2,3, and 11. It shows that the results of Refs.4-7 are incorrect; in gated structures this is possibly caused by the coupling of signals from the 2DEG to the gate, as argued in Ref.10.

b) The conclusions of Refs. 7 and 8 regarding very long semiclassical orbits cannot be confirmed. If semiclassical orbits exist, a proper theoretical treatment will have to consider that delocalization of the 2D electrons can only be expected at frequencies much higher than our measuring frequencies of about 30 GHz. Most likely, a frequency close to the cyclotron resonance frequency is neccessary (in the far infrared at magnetic fields of several Tesla).

c) Contact effects are irrelevant for the occurrence of the QHE since the σ_{xy} quantization was observed with the contactless microwave technique. In particular, the high-field regions [14] in the corners of usual Hall bars for d.c. measurements are avoided.

d) Edge current effects are minimized, can possibly be excluded, since the active sample areas in our crossed waveguide experiments were much smaller than the total sample areas.

Acknowledgement

The authors thank Professor K. Seeger for his continuous interest in this work. This work was partly supported by Fonds zur Förderung der wissenschaftlichen Forschung, Austria, -Project No. P5247.

References

(1) See e.g. the review by J. Hajdu: Advances in Solid State Sciences (Festkörperprobleme) XXV, 395 (1985)

(2) F. Kuchar, R. Meisels, G. Weimann, and H. Burkhard: Proc. 17th Int.Conf.Phys.Semicond., San Francisco, eds.J.D.Chadi and W.A. Harrison, Springer, 1985, p. 275.

(3) R. Woltjer, M. Mooren, J. Wolter, and J.P. André: Solid State Commun. 53, 331 (1985).

(4) C. McFadden, A.P. Long, H.W. Myron, M. Pepper, D. Andrews, and G.J. Davies: J. Phys. C 17, L 439 (1984).

(5) A.P. Long, H.W. Myron, and M. Pepper: Ref.2, p.279.

(6) T.G. Powell, R. Newbury, C. McFadden, H.W. Myron, and M. Pepper:J. Phys. C 18, L 497 (1985).

(7) M. Pepper and J. Wakabayashi: J. Phys. C 16, L 113 (1983).

(8) R. Joynt: J. Phys. C 18, L 331 (1985).

(9) S.M. Apenko and Yu.E. Lozovik: J. Phys. C 18, 1197 (1985).

(10) T.P. Smith, M. Heiblum and P.J. Stiles: Ref.2, p.393.

(11) B.B. Goldberg, T.P. Smith, M. Heiblum and P.J. Stiles: Proc. Yamada Conference XIII on Electronic Properties of Two-Dimensional Systems (to be published).

(12) G. Weimann and W. Schlapp: Appl.Phys. A 37, 3057 (1985).

(13) See e.g. E.D. Palik and J.K. Furdyna: Rep.Prog.Physics 33, 1193 (1970).

(14) F. Kuchar, G.Bauer, G. Weimann, and H. Burkhard: Surface Sci. 142, 196 (1984).

The Fractional Quantum Hall Effect
in GaAs-GaAlAs Heterojunctions

R.J. Nicholas[1], *R.G. Clark*[1], *A. Usher*[1], *J.R. Mallett*[1], *A.M. Suckling*[1],
J.J. Harris[2], *and C.T. Foxon*[2]

[1]Clarendon Laboratory, Parks Rd., Oxford, OX1 3PU, U.K.
[2]Philips Research Laboratories, Redhill, Surrey, U.K.

A description is given of the occurrence of fractional quantum effects
in GaAs-GaAlAs heterojunctions, covering the observation of an extended
set of fractional resistivity minima in both spin states of the lowest
Landau level. Features are observed at occupancies p/q with odd
denominator values up to q = 9. A brief description is given of the
theoretical approaches to the problem. The measurement of the energy
gaps by thermal activation, and the destruction of the effect by disorder
and non-exact occupation is described.

1. INTRODUCTION

The observation of the fractional quantum Hall effect is thought to be due to
the formation of a new state of matter, the quantum fluid, in which a two
dimensional gas of interacting electrons has condensed into a many body
ground state |1-6|. To date, this phenomenon has only been seen clearly in
GaAs-GaAlAs heterojunctions, where the disorder present due to interface and
doping fluctuations is sufficiently small in order to allow the electron-
electron interactions to dominate the properties. The main experimental
observations are that minima are observed in the electrical resistivity
component ρ_{xx}, at fractional Landau level occupancies $\nu = nh/eB$ of the form
p/q, where p is an integer and q is an odd integer |7-13|; while corresponding
Hall plateaus are seen at quantized resistivity values of $h/\nu e^2$. The theoretical treatments interpret the quantization at rational fractions $\nu = 1/q$,
and their electron/hole symmetric analogues at $\nu = 1 - 1/q$, as condensation
of the interacting 2D electrons into a ground state corresponding to an
incompressible quantum fluid. To date, fractional states have been reported
at ν = 1/3, 1/5, 2/5, 2/7, 3/7 and 4/9, and their symmetric analogues at
ν = 2/3, 3/5, 4/5, 4/7, and 5/9. Recently it has been shown that all of these
states can also exist in the second spin state of the N = 0 Landau level |13|,
at occupancies of the form ν = 1 + (p/q). The existence of minima in the

resistivity and quantized Hall plateaus may be shown, by using the gauge
invariance arguements of LAUGHLIN |14|, to be a consequence of the formation
of a mobility gap in the density of states. In other words, the degeneracy of
the individual Landau levels for isolated electrons has been lifted by the
residual Coulomb interactions, leading to the formation of an energy gap
between the ground and excited states of the quantum fluid. The energy of
formation of the collective ground states is relatively small |2-5|, being of
order $0.1\ e^2/4\pi\epsilon l_B$ for the 1/3 state ($l_B = \sqrt{\hbar/eB}$, the cyclotron radius), and
decreasing with increasing q as $q^{-2.5}$ |3|. This is of order 17 K for the 1/3
state at a magnetic field of 10 T, thus necessitating the use of very low
temperatures in order to see the higher order states. There is also evidence
|11, 12| to suggest that disorder reduces the magnitude of the energy gap.
Experimental values for the gap energy can be obtained by observing thermally
activated conductivity in the minima, although the values are likely to be
underestimates, since conduction will involve excitation only across the
mobility gap in the tails of the levels.

2. FRACTIONAL STATES

Experimental results are described here which were taken using Hall bridge
specimens with channel widths of 50 - 150 μm. These were modulation doped
GaAs-Ga$_{0.68}$Al$_{0.32}$As heterojunctions grown by M.B.E. |15|, using spacer layers
of 400 and 800 Å, and with resulting electron concentrations in the range
$0.6 - 4 \times 10^{11}$ cm^{-2}. The sample mobilities ranged from 0.5 to 2.1×10^6cm^2/Vs,
depending upon the sample and electron concentration. For any one sample it
was possible to change the electron concentration by factors of 2 - 3 by
excitation of persistent photoconductivity using a red L.E.D. The samples
were mounted on a laminated copper cold finger in a dilution refrigerator,
and were cooled to temperatures as low as 20 mK. The temperature was measured
by the use of a calibrated carbon resistor and were verified at the lowest
temperatures by the use of a ^{54}MnNi nuclear orientation thermometer. All
electrical connections were carefully screened and filtering was employed
both inside and outside the cryostat. A.C. currents were used at densities in
the range 10^{-2} to 10^{-6} A/m, in order to determine the onset of electron heating
effects. The magnetic field was produced by a 16T Nb$_3$Sn superconducting
solenoid.

Fig. 1 shows a typical recording of the resistivity ρ_{xx} and Hall component
ρ_{xy}, for the highest mobility sample G63 at an electron concentration of
1.9×10^{11} cm^{-2}, following the photo excitation of the majority of the carriers.

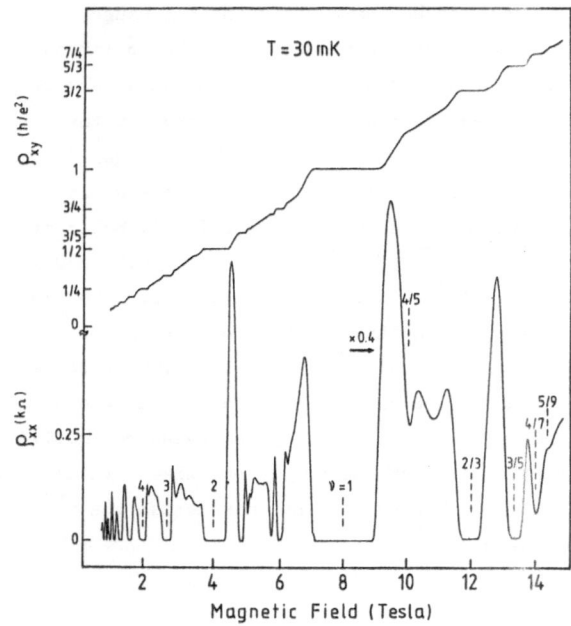

Fig. 1: The resistivity and Hall voltage (upper trace) for sample G63 at 30 mK. The fractional occupancies for $\nu < 1$ are indicated on the trace. The Hall resistivity is given in units of (h/e^2). The current density is 6×10^{-4} A/m.

This shows what is probably the most comprehensive set of fractional states observed to date. The region above 8T corresponds to the incomplete filling of the lower spin state of the N=0 Landau level. The most striking feature is the regular series of minima at the occupancies of 2/3, 3/5, 4/7 and a weak feature at 5/9. At the same time, corresponding plateaus appear in the Hall voltage, of progressively decreasing width. This behaviour is typical of that observed in the best samples, where the degree of disorder is relatively low. It originates from the series of collective ground states with odd symmetry, and progressively decreasing binding energy. In contrast to the strong features at 3/5 and 4/7, only a weak minimum is observed at 4/5, and no structure is evident from 5/7 or 6/7. This is not predicted directly by the theoretical calculations |1-6|, however it is reasonable to assume that it is the result of the increasing effects of disorder as the Fermi level moves into the tail of the broadened Landau level. Similarly the 1/5 and 2/7 states have only been observed as very weak features |8,9| in the resistivity. It is also clear that there are a similarly large set of features in the region $1 < \nu < 2$, when the upper spin state of N=0 is partially filled. This is illustrated in more detail in fig. 2, after a further 10% has been added to the carrier concentration. The minima form a similarly regular set of q = 3, 5, 7 fractions, which is obviously symmetric about the half-occupied level. Such behaviour is again

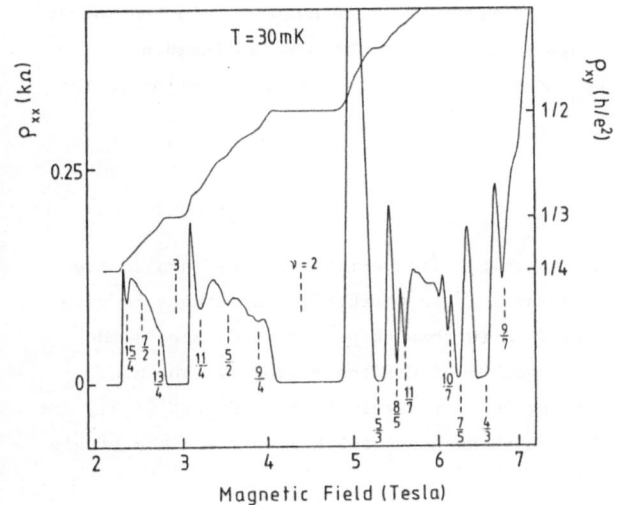

Fig. 2: Low field resis-
tivity and Hall traces in
in sample G63. Conditions
as in fig. 1.

entirely consistent with the picture of a series of states becoming progressively more bound as the denominator decreases or as they move towards the
centre of each level, where the influence of disorder is least. The presence
of a fully occupied level of opposite spin does not seem to have influenced
the formation of the states. The only slight anomalies are that there is a
relatively pronounced feature at 1 + 4/9 at 6T, compared with the weaker
minimum seen for 5/9 at 14T, and that the 5/3 minimum appears slightly more
strongly than the 4/3. Theoretically one should expect the converse, since
the theoretical ground state energies scale as $\sqrt{B}$, through the cyclotron radius.

Also visible in fig. 2 are further resistivity minima observed for the N=1
Landau level (2 < ν < 4), together with features in the Hall resistance. These
minima, reported recently by CLARK ET AL |13|, are considerably different from
the behaviour seen in the N=0 level. There are three regular minima for each
spin state, and these occur very close to even fractional occupancies of 1/4,
1/2 and 3/4. A further weak minimum is also seen at an occupancy of 2 + 1/3.
The Hall voltage, however, does not show any plateaus to be present, although
oscillatory features are occurring. The existence of some new ground state
at even fractional filling would be a very major obstacle to the majority of
theoretical descriptions of the collective ground state |1-5| but must remain
as only a possibility until Hall plateaus can be seen and a mobility gap
observed. In more recent measurements we have found that this effect is also
present in other samples, but the exact position of the outer ('1/4' and '3/4')
minima is dependent upon both temperature and current density. Under certain

conditions sharp 1/3 and 2/3 minima reappear, overtaking the outer structures
in strength, but the 1/2 minimum is a constant and pronounced feature at low
temperatures. This work will be the subject of a further publication in the
near future.

3.THEORY

The theoretical treatments of the fractionally quantized states involve the
use of rather complex formalism, however it is worthwhile to attempt to give
an intuitive picture of the problem. A further complicating factor is the
ability to describe the system in a number of different gauges. The most
widely accepted and successful theory is that developed by LAUGHLIN |1,2|, who
works in the symmetric gauge. The starting point for his work are the single
particle wavefunctions of the form

$$|n> \quad = \quad \frac{1}{\sqrt{2^{n+1}\pi n!}} \; z^n \exp(-\tfrac{1}{4}|z|^2), \tag{1}$$

where $z_j = (x_j - iy_j)/l_B$, and n is the angular momentum quantum number which
gives the degeneracy of the state. The next step is to introduce a variational
many body wavefunction of the form

$$\Psi_m(z_1,\ldots,z_N) \quad = \quad \prod_{j<k}^{N} (z_j - z_k)^m \exp(-\tfrac{1}{4}\sum_{1}^{N}|z_1|^2), \tag{2}$$

where m is an odd integer. LAUGHLIN |1,2| was led to the use of this form by
requiring it to be an eigenstate of angular momentum and by the need to make
the polynomial part antisymmetric, which leads to the requirement that m
should be odd. He was then able to show that the quasiparticle excitations of
this state could be created by piercing the ground state with a single flux
quantum. The resulting particle has an effective charge of -1/m, due to partial
screening of the induced charge. The excitation of a quasiparticle across the
energy gap separating the ground and excited states then requires the creation
of a quasi-electron - quasi-hole pair. He has calculated that this will require
energies of 0.056 and 0.014 $e^2/4\pi\epsilon l_B$ for the 1/3 and 1/5 states respectively.

In order to deduce the existence of quantized Hall plateaus from this calculation it is then necessary to introduce the idea of some disorder into the
system, which then creates a mobility gap separating the ground and excited
states. It is then possible to use the arguments of gauge invariance, as first
introduced by LAUGHLIN |14| to describe the integer quantum Hall effect, to
predict the existence of plateaus. This involves the introduction of the rather
unphysical geometry of an annulus or ring, which contains a solenoid in order
to introduce flux changes. This causes the Hall current to flow around the

ring, while at the same time transfering electrons from one side to the other through the Hall potential V. The fact that the conductivity and resistivity are zero at particular filling factors requires the wavefunction to be periodic around the ring, leading to flux quantization. The quantized Hall resistivity is then the ratio of the flux quantum to the electron charge. In the case of the fractional effect the reasoning is not quite so clear since it is not possible to introduce single quasi-particles externally. Instead one is led to the conclusion that motion of a single electron results in the introduction of m flux quanta. This seems a natural consequence of the ground state wavefunction (2), which is invariant upon a $2\pi/m$ rotation of the system. The periodic boundary condition will then introduce a $2m\pi$ phase change upon a complete rotation of the system, corresponding to motion of the single electron across the ring. The conclusion is then that the size of the flux quantum has changed, and the ground state is triply degenerate.

An important feature of the Laughlin approach to the ground state is that it is a quantum liquid, which is formed through the Coulomb interactions between the electrons, and therefore correlates their motion. Since the system remains a liquid there is only short range order, in contrast to the Wigner solid, and he has calculated |2| that the correlation lengths are of order $3 - 4$ l_B:

An alternative approach has been given by TAO and THOULESS |6|, who have worked in the linear gauge where

$$\Psi_{k,n} = e^{ikx} \phi_n(y-y_0) \tag{3}$$

with k as the degenerate quantum number. The single particle wavefunctions may be thought of as a set of parallel tracks of width l_B and centred at

$$y_0 = \hbar k/eB - E_0. \tag{4}$$

For a system of length L the track separation is

$$\delta y_0 = \hbar \delta k/eB = 2\pi l_B^2/L, \tag{5}$$

which is obviously much less than the width of any individual track. The argument of Tao and Thouless is that in the fractional state the system can lower its energy by an ordered filling of tracks such that every mth single particle state is filled. To show the correspondence with the Laughlin approach we take the linear geometry and bend it around to form a ring or annulus of diameter L. In this case the ground state is a set of concentric rings. If we now introduce a flux into the ring then we produce an azimuthal change in magnetic vector potential $\Delta A = \Delta\Phi/L$. This introduces a progressive phase change

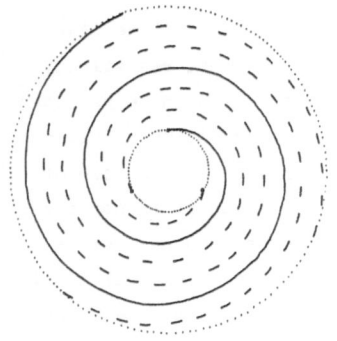

Fig. 3: A schematic view of the
application of a gauge transformation
for the m=3 ground state, based on the
Tao and Thouless approach. The occupied state is the solid line, and the
spiral corresponds to the progressive
application of the gauge potential
around the ring, such that one occupied
state is mapped into the next.

around the ring of $\exp(ie\Delta Ax/\hbar)$, which gives a shift in the track centre of

$$\Delta y_0 = \Delta A/B = \hbar\Delta k/eB. \qquad (6)$$

The occupied tracks have now been shifted outward by Δy_0, and in order for the
state to transform into the next occupied track we must have a shift of

$$\Delta y_0 = m \, \delta y_0, \qquad (7)$$

leading to a flux quantum of mh/e. The process of transformation is shown
schematically in fig. 3 as a spiral motion, for the m = 3 state. This picture
also has the $2\pi/m$ symmetry, corresponding to the m different starting points
for the spiral.

4. THE INFLUENCE OF TEMPERATURE AND DISORDER

The theoretical descriptions above are founded upon the assumption that the
system is at T = 0 and at exact fractional occupancy. In this section we shall
examine some of the influences of temperature and disorder, due both to non-
exact occupancy and scattering. Fig. 4 shows the temperature dependence of the
fractional structure in sample G63. The strongest feature at 2/3 is apparent
even at 1K, and rapidly develops into a deep flat conductivity minimum, covering
a range of occupancies. The weaker features, such as those for $1 < \nu < 2$, only
appear around 300 mK, not forming clear minima until very low temperatures.
Activation plots of the conductivity show a linear region at higher temperatures,
from which it is possible to deduce the activation energy for excitation across
the mobility gap. Typical values are 2.3K for 2/3 and 1.1K for 3/5, which are
substantially less than the calculated values for the ideal case, but comparable
with values found by other authors |11, 12|. At lower temperatures the conduc-
tivity decreases less quickly, as hopping conduction takes over.

When lower mobility, and hence more disordered, samples are studied, then a
number of changes can be seen. Some of these are illustrated in fig. 5, where

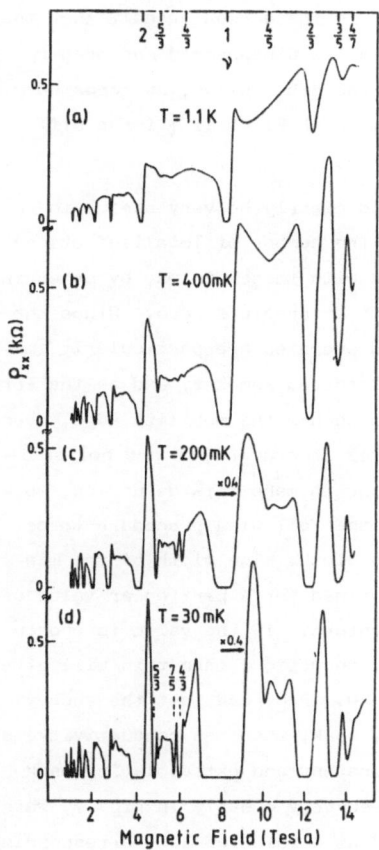

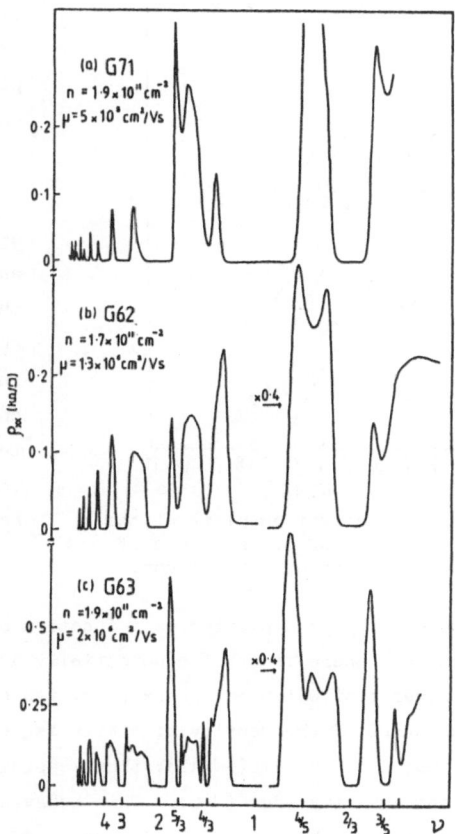

Fig. 4: The temperature dependence of the resistivity for sample G63.

Fig. 5: The influence of disorder on the fractional minima at 30 mK.

three different samples are shown with differing mobilities, but very similar electron concentrations. Looking at the 2/3 minimum it is clear that the increased disorder has had the effect of increasing its width, causing it to dominate the trace. In contrast the temperature dependence shows that the activation energy has been substantially reduced, from 2.4K to 1.0K in the lowest mobility sample. In the other spin state the majority of the minima have been suppressed by the disorder. After photoexcitation of further carriers the mobility increases and the higher order minima move to higher field, becoming stronger. This can be seen from the activation plot shown in fig. 6, for G71. With 1.9×10^{11} cm^{-2} carriers the activation energy is 1.0K for the 2/3 minimum (at 12T) and at 3.7×10^{11} it is 0.9K for the 5/3 minimum (at 9T). At the same

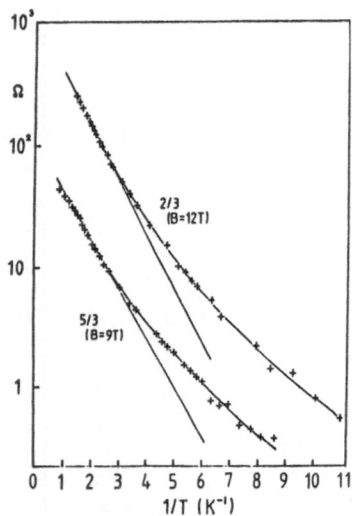

Fig. 6: Activation plots of the resistivity for G71 at two different electron concentrations.

time there is the unusual result that the 4/3 minimum has disappeared and been replaced by the 7/5 minimum, as reported earlier by CLARK ET AL |16| for a different sample.

It would clearly be very useful to determine the number of localised states associated with these minima, by measuring their width at absolute zero. Since the activation energies are particularly small in the disordered samples, and as the Fermi energy approaches the mobility edge, even temperatures of order 30 mK are not sufficiently low to make this deduction, so we have used the following procedure to determine the width. The width of the minimum is defined for a particular value of conductivity, and plotted as a function of temperature. If the value is chosen such that there is still exponentially activated conduction, then this will give a linear plot which may be extrapolated back to T=0. Provided that the energy dependence of the density of states is relatively weak, then the conductivity is dominated by the exp(-W/kT), thus keeping W/kT constant and extrapolating back to the mobility edge (W=0). This behaviour is seen very clearly in Fig. 7, which shows the width of the 5/3 minimum in G71 plotted at a value of 10Ω corresponding to the break point in fig. 6. A similar procedure for the 2/3 minimum also gives a full width of 0.08ν. For G63 the 2/3, 4/3 and 5/3 minima are all around 0.035ν wide. This implies that there is a critical level of 'occupational disorder'

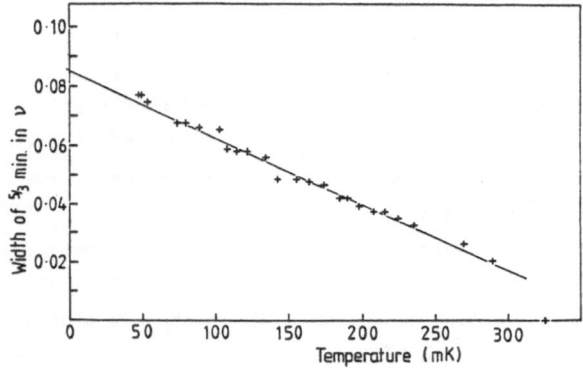

Fig 7: A plot of the temperature dependent width of the 5/3 minimum for sample G71 at 9T, using the procedure described in the text.

which destroys the fractional state and that purer samples with less random
disorder from potential fluctuations can tolerate less of this before becoming
de-localised. This behaviour is consistent with the idea that the fractional
state can only exist when the order length is greater than the correlation length
of the quantum fluid. For a full width of $\Delta\nu$ the ordering length (the distance
over which the system may still be considered to have the exact occupancy) will
be $l_B/\sqrt{\Delta\nu/2}$, which is $7\,l_B$ for G63, consistent with the calculations of LAUGHLIN
$|2|$ who estimates $\pm\,4\,l_B$. Random disorder reduces the correlation length of the
quantum fluid, and also reduces its binding energy. Thus the broader minima in
the lower mobility sample G71 exist to shorter order lengths ($5\,l_B$), but show a
smaller activation energy.

REFERENCES

1. R.B. Laughlin, Phys. Rev. Lett. 50, 1395 (1983); Surf. Sci. 142 163
    (1984)

2. R.B. Laughlin, in Two-Dimensional Systems, Heterostructures, and
    Superlattices, Ed. G. Bauer, F. Kuchar and H. Heinrich, (Springer
    Verlag) Solid State Sciences 53 p.279 (1984)

3. B.I. Halperin, Phys. Rev. Lett. 52, 1583 (1984); ibid 2390 (1984)

4. S.M. Girvin, Phys. Rev. B30, 558 (1984)

5. A.H. MacDonald, G.C. Aers and M.W.C. Dharma-wardana, Phys. Rev.
    B31, 5529 (1985)

6. R. Tao and D.J. Thouless, Phys. Rev. B28, 1142 (1983); R. Tao,
    Phys. Rev. B29, 635 (1984)

7. D.C. Tsui, H.L. Stormer and A.C. Gossard, Phys.Rev.Lett.48, 1559 (1982)

8. A.M. Chang, P. Berglund, D.C. Tsui, H.L. Stormer and J.C.M. Hwang,
    Phys. Rev. Lett. 53, 997 (1984)

9. E.E. Mendez, L.L. Chang, M. Heiblum, L. Esaki, M. Naughton, K. Martin
    and J. Brooks, Phys. Rev. B30, 7310 (1984)

10. G. Ebert, K. von Klitzing, J.C. Maan, G. Remenyi, C. Probst,
    G. Weimann and W. Schlapp, J. Phys. C17, L775 (1984)

11. G.S. Boebinger, A.M. Chang, H.L. Stormer and D.C. Tsui, EP2DSVI
    Proceedings, Kyoto, Japan (1985) to be published

12. J. Wakabayashi, S. Kawajii, J. Yoshino and H. Sakaki, EP2DSVI
    Proceedings, Kyoto, Japan (1985) to be published

13. R.G. Clark, R.J. Nicholas, A. Usher, C.T. Foxon and J.J. Harris (1985)
    EP2DSVI Proceedings, Kyoto, Japan (1985) to be published

14. R.B. Laughlin, Phys. Rev. B23, 5632 (1981)

15. C.T. Foxon, J.J. Harris, R.G. Wheeler and D.E. Lacklison (1985) to be
    published

16. R.G. Clark, R.J. Nicholas, M.A. Brummell, A. Usher, S.J. Collocott,
    J.C. Portal and F. Alexandre, Sol. Stat. Comm. 56, 173 (1985)

Density of States of Landau Levels from Activated Transport and Capacitance Experiments

D. Weiss [1*] , *K.v. Klitzing* [1], *and V. Mosser* [2+]

[1]MPI für Festkörperforschung, D-7000 Stuttgart, Fed. Rep. of Germany
[2]Physik-Department, Technische Universität München,
  D-8046 Garching, Fed. Rep. of Germany

Abstract. In this publication we demonstrate that a combination of
capacitance measurements with an analysis of thermally activated
conductivity seems to be useful for the determination of the density
of states (DOS) of Landau levels in two-dimensional systems. The
experimental results indicate that no real energy gap exists, but that
a nearly energy-independent background density of states (which decreases with increasing electron mobility) is present. Close to the
center of the Landau levels the DOS can be described by a Gaussian
lineshape with a line width which increases proportional to the
square root of the magnetic field. This result agrees with the prediction within the self-consistent Born approximation, but the expec-
ted variation $\Gamma \sim \sqrt{1/\mu}$ of the line width Γ with the mobility μ of the
electrons could not be confirmed.

1. Introduction

A microscopic theory of the quantum Hall effect should give a correct
description not only of the quantized resistivity values $\rho_{xy}=h/ie^2$ but
also of the transitions between the plateaus and the values of the finite
resistivity ρ_{xx}. Such transport calculations are extremely complicated
since the theory itself is complicated, and in addition not enough information is available about the scattering centers. The published theories
are based on certain approximations and assumptions about the distribution, the strength and the range of the scattering potential. A first test
whether such assumptions are realistic should be available from a comparison between the calculated and the measured density of states $D(E)$, since
calculations of $D(E)$ are much easier than a transport theory for $\rho_{xx}(B)$
which includes complicated phenomena like localization and correlation.
One of the first theories of the density of states (DOS) assumed short-
range scatterers, which leads, within the self-consistent Born approximation,
(SCBA) to a broadening of the discrete energy spectrum (expected for an
ideal two-dimensional electron gas without scattering) into an elliptic
lineshape for the DOS [1]. Higher order approximations show that an exponentially decaying DOS is expected for energies $E-E_n$ larger than the linewidth of the Landau levels E_n [2], so that a real energy gap with vanishing DOS may not be present, but the DOS at midpoint between two Landau
levels should decrease drastically if the magnetic field (energy separa-

---

*   Present address: Physik-Department, D-8046 Garching
+   New address: GIERS-Schlumberger, F-92124 Montrouge, France

tion between adjacent Landau levels) is increased. Experimental information about the DOS can be obtained from measurements of the specific heat
[3], from magnetization measurements [4], from temperature-dependent resistivity measurements in the regime of the Hall plateaus [5] or from
capacitance measurements [6,7]. In this article we will discuss in more
detail the derivation of the DOS from an analysis of the temperature-
dependent resistivity (chapter 2) and from temperature-dependent
capacitance measurements (chapter 3). The following discussion is based on
a picture which does not include many-body effects. The notation "density
of states (DOS)" in this paper is used to characterize the electronic
properties within a single particle picture.

2. Activated resistivity

The measurements of the resistivity ρ_{xx} in a strong magnetic field
(Shubnikov-de Haas oscillations) were carried out on GaAs-AlGaAs heterostructures with mobilities $14,000 < \mu < 550,000$ cm^{2}V^{-1}s^{-1} and carrier densities $1.4 \cdot 10^{11} < n_s < 4.2 \cdot 10^{11}$ cm^{-2}. The devices have Hall geometry with a
typical length of about 3 mm, a width of about 0.4 mm and a distance be-
tween potential probes of 0.5 mm. The device current was kept below 1μA
where electron heating is negligibly small.

The temperature dependence of ρ_{xx}^{min} (where ρ_{xx}^{min} means the minimum in
the resistivity which corresponds to a Fermi level position very close
to the midpoint between two Landau levels) in the temperature range
2K<T<20K is usually dominated by an exponential term corresponding to

$$\rho_{xx}^{min} \sim \exp\{-\frac{E_{a,\,max}}{kT}\} \tag{1}$$

where $E_{a,max}$ denotes the measured activation energy. Measured activation
energies $E_{a,max}$ for different samples at different magnetic field values
are shown in Fig.1. The filling factor i, defined as $i = n_s \cdot \frac{h}{eB}$ corresponds
always to a fully occupied lowest Landau level (i=4 for (100) silicon
MOSFETs and i=2 for GaAs-AlGaAs heterostructures). Since the measured
activation energy $E_{a,max}$ agrees fairly well with half of the cyclotron
energy $\hbar\omega_c$, this activation energy is interpreted as the energy difference
between the Fermi energy E_F and the center of the Landau level E_n. For the
sake of simplicity we assume that the mobility edge of the Landau level is

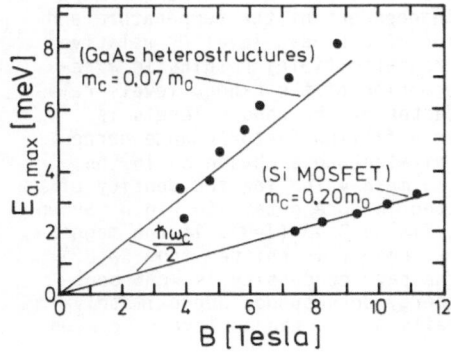

Fig. 1:

Measured activation energies
$E_{a,max}$ in the resistivity at
a filling factor corresponding to a fully occupied
lowest Landau level as a
function of the magnetic
field B. The solid lines
correspond to half of the
cyclotron energy.

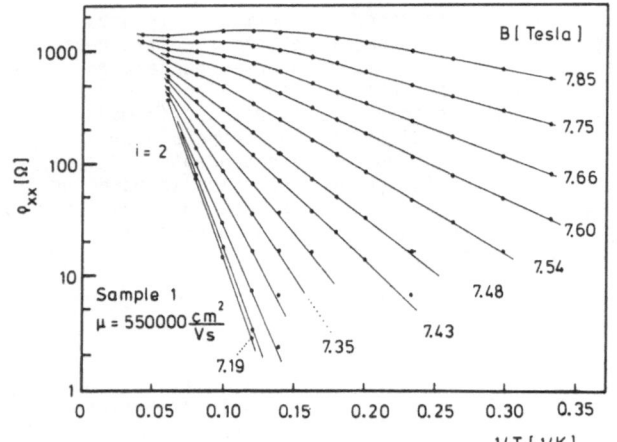

Fig. 2:

Temperature dependence of the resistivity ρ_{xx} at different magnetic fields close to a filling factor i=2.

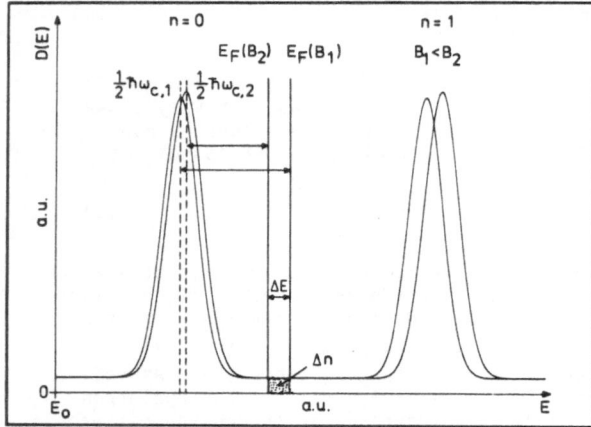

Fig. 3:

Model density of states at two different magnetic fields used to explain the reconstruction of the DOS from an analysis of the activation energies E_a (B_i).

located at the center of the Landau level, in agreement with calculations of the localization length [8] and percolation theories [9]. Furthermore the mobility edge should remain fixed, independent of the temperature and the carrier density. Changing the position of a Landau level E_n relative to the Fermi energy E_F (by changing the magnetic field) results in a reduced activation energy $E_a = |E_n - E_F|$. This motion of the Landau levels relative to the Fermi level if the filling factor of the Landau levels is varied is clearly visible in Fig.2. Since a filling factor change corresponds to a shift of the Fermi level, equivalent to a change Δn in the carrier density at fixed magnetic field, a mean value for the density of states can be deduced. This is demonstrated in more detail in Fig.3. Shown is a model DOS at two different magnetic fields $B_2 > B_1$ (T=0). If the magnetic field is raised from B_1 to B_2 the Fermi energy is shifted from $E_F(B_1)$ to the lower energy position $E_F(B_2)$ if the carrier density is kept constant. The variation of ΔE of the Fermi energy corresponds approximately (if the Fermi energy is located in the tails of the Landau levels between n=0 and n=1) to a change in the carrier density of

$$\Delta n \approx \frac{e}{\pi\hbar} (B_2 - B_1) \tag{2}$$

Since the energy difference between the two Fermi level positions is given by

$$\Delta E \approx E_a(B_1) - E_a(B_2) - \tfrac{1}{2}\hbar(\omega_{c,2} - \omega_{c,1}) \tag{3}$$

the density of states can be deduced:

$$D(E) \approx \frac{\Delta n}{\Delta E} \tag{4}$$

It should be noted that the "point by point" construction of the DOS discussed in this chapter becomes incorrect in the energy region where the DOS changes drastically with energy. From the above it is clear that the change of the activation energy with magnetic field contains information about the density of states in the localized region between Landau levels. The change of the activation energy E_a with magnetic field obtained from the ρ_{xx}-data shown in Fig.2 (sample 1, μ=550,000 cm^{2}V^{-1}s^{-1}, n_s=3.5·10^{11}cm^{-2}) is plotted in Fig.4a. Using (1) to (3) we can deduce the density of states in the high-energy tail of the Landau level n=0 (Fig.4b). Figure 4c shows the DOS of the lower mobility sample 3 (μ=180,000 cm^{2}V^{-1}s^{-1}, n_s=1.8·10^{11}cm^{-2}) over the whole Hall plateau region i=2. The density of states D(E) in Fig.4b,c is shown as a function of the energy relative to the center of the plateau region. The maximum of D(E) close to E=0 is an artifact, since for the Fermi energy at E=0 two Landau levels contribute to the thermally activated conductivity, which complicates the analysis of the experimental data. If the Fermi energy is shifted out of the midpoint between two Landau levels by more

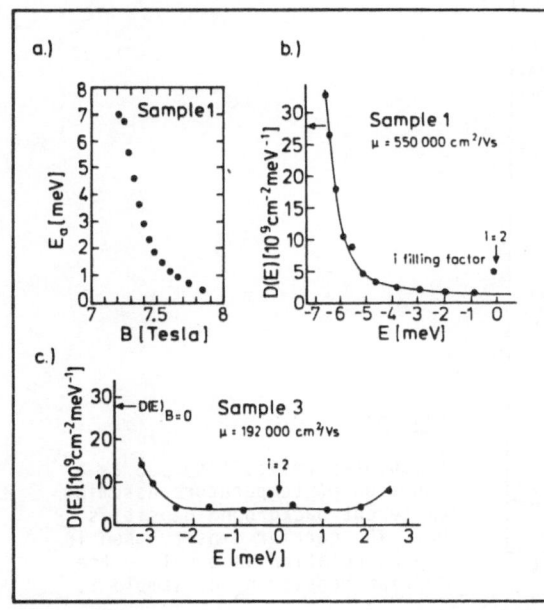

Fig. 4:

Change of the measured activation energy E_a as a function of the magnetic field B(a), and reconstructed DOS as a function of energy for sample 1(b) and 3(c).

than the thermal energy kT, the contribution of the Landau level with the higher activation energy becomes unimportant. The experimental result that the density of states is constant within 50% of the energy between Landau levels is typical for all samples investigated. Measurements of the DOS at filling factors i=2 and i=4 for one and the same sample show that the minimal density of states D_{UG} is nearly independent of the filling factor and therefore of the energy separation between Landau levels.

The ideas above are checked by computer simulations based on a model in which the temperature dependence of the resistivity in the plateau region is dominated by a term

$$\rho \sim \rho_0(T) \sum_n \exp\left(-\left|\frac{E_n - E_F}{kT}\right|\right) \tag{5}$$

where $\rho_0(T)$ is a temperature-dependent prefactor. The position of the Fermi level is determined by solving numerically the equation

$$n_s = \int_0^\infty D(E) f(E-E_F) dE \tag{6}$$

where $f(E-E_F)$ is the Fermi distribution function. The carrier density n_s is assumed to be constant in the investigated temperature range. The model DOS $D(E)$ in (6) was chosen as a Gaussian-like density of states superimposed on a constant background density of states D_{UG}

$$D(E) = A \cdot \frac{B}{\Gamma} \cdot \sum_n \exp\left\{-\frac{(E-(n+\frac{1}{2})\hbar\omega_c)^2}{2\Gamma^2}\right\} + D_{UG} \tag{7}$$

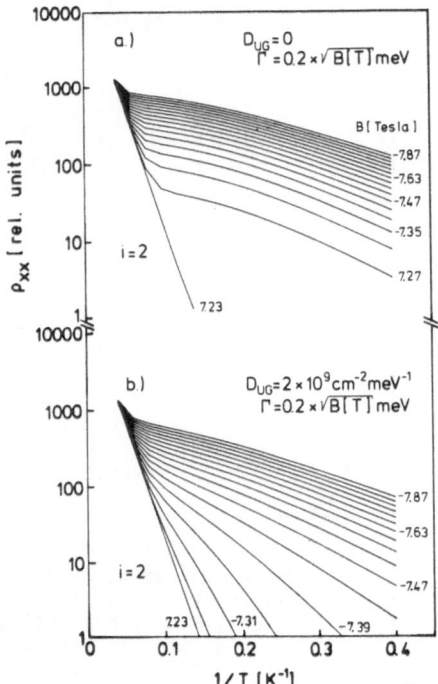

Fig. 5:

Calculated resistivity ρ_{xx} as a function of temperature assuming different background densities D_{UG}. The carrier density used in the calculation is equal to the carrier density n_s of sample 1.

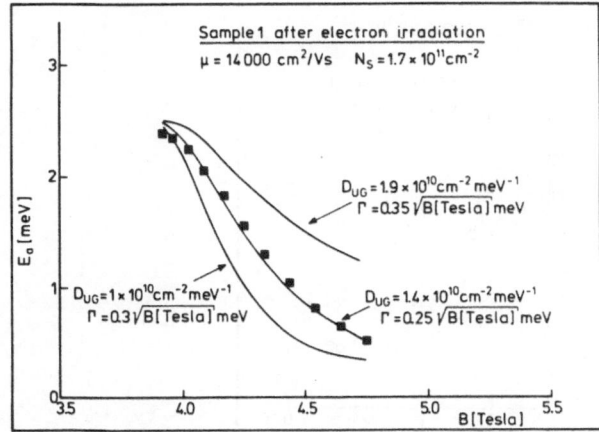

Measured activation
energies of sample 1* as a
function of the magnetic
field compared with
activation energies deduced
from calculated ρ_{xx}-curves
using different
combinations of background
density of states D_{UG} and
Gaussian linewidth Γ.

where the constant A is determined by the number of electrons in one
Landau level and Γ is the broadening parameter of the Gaussian distribution. Using (5) to (7) one can calculate temperature and magnetic field-
dependent ρ_{xx}-values and deduce activation energies in analogy to Fig.2.
Calculations based on (5) to (7) demonstrate that the finite density of
states D_{UG} in the mobility gap influences strongly the result. Fig.5 shows
numerical calculations with and without using a background DOS
$D_{UG}=2 \cdot 10^9 cm^{-2} meV^{-1}$. The linewidth of the Gaussian in this calculation was
$\Gamma=0.2 \cdot \sqrt{B[T]}$ meV. A reconstruction of the DOS by deducing activation energies from Fig.5b (the data used in the simulation correspond to sample 1
in Fig.2) in the temperature range of the experiments demonstrates that
the constant background DOS is reproduced within 20%.

Figure 6 summarizes the data obtained for a heterostructure (sample 1*)
with a mobility of only 14,000 $cm^2 V^{-1} s^{-1}$ at 4.2K. This low mobility is
achieved by irradiating sample 1 with 1 MeV electrons. The experimentally
deduced activation energies are compared to "calculated" ones - using
different combinations of Gaussian linewidth Γ and background DOS D_{UG} - as
a function of the magnetic field. The temperature-dependent prefactor was
chosen as $\rho_0(T)=const/T$. The experimental data are best described by a
Gaussian linewidth $\Gamma=0.25 \cdot \sqrt{B[T]}$ meV and a constant background
$D_{UG}=1.4 \cdot 10^{10} cm^{-2} meV^{-1}$.

Annealing of this electron-irradiated sample at 220°C leads to an increased mobility $\mu=28,000 cm^2 V^{-1} s^{-1}$ (sample 1**). The measured activation
energy versus magnetic field is shown in Fig.7a. The best fit is obtained
using a prefactor $\rho_0(T)=const$, a Gaussian linewidth $\Gamma=0.25 \cdot \sqrt{B[T]}$ meV and a
constant background DOS $D_{UG}= 9 \cdot 10^9 cm^{-2} meV^{-1}$. The use of the prefactor
$\rho_0(T)$ looks somewhat arbitrary, but in fact the prefactor does not change
remarkably the slope of $E_a(B)$, and therefore the value of the constant
background density of states D_{UG}. This has been checked for prefactors
const/T, const and const·T. This is shown in Fig.7b where the constant
background DOS is reproduced within 20% independent of the prefactor
$\rho_0(T)$. The determination of the linewidth Γ however depends on the knowledge of the energy difference between mobility edge and Fermi level, and
therefore on the absolute value of the activation energy. For this reason
there remains some uncertainty in the determination of the linewidth Γ by

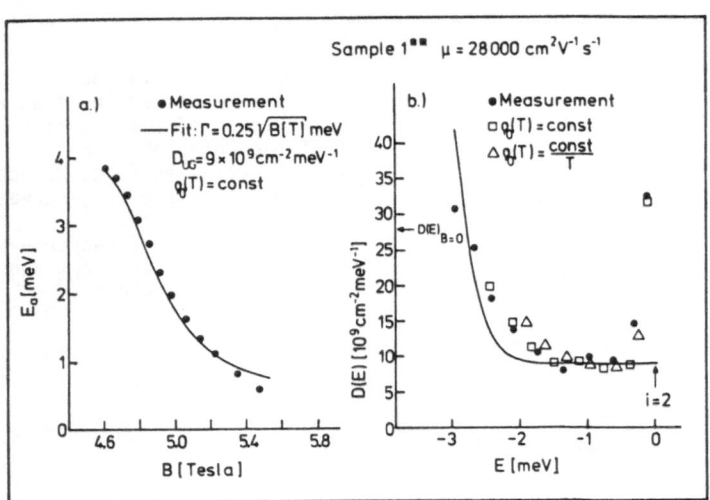

Sample 1** $\mu = 28000 \text{ cm}^2\text{V}^{-1}\text{s}^{-1}$

a.)
• Measurement
—— Fit: $\Gamma = 0.25 \sqrt{B[T]}$ meV
$D_{UG} = 9 \times 10^9 \text{cm}^{-2}\text{meV}^{-1}$
$\varrho(T) = \text{const}$

E_a [meV]

B [Tesla]

b.)
• Measurement
□ $\varrho(T) = \text{const}$
△ $\varrho(T) = \dfrac{\text{const}}{T}$

D(E) $[10^9 \text{cm}^{-2}\text{meV}^{-1}]$

$D(E)_{B=0}$

i=2

E [meV]

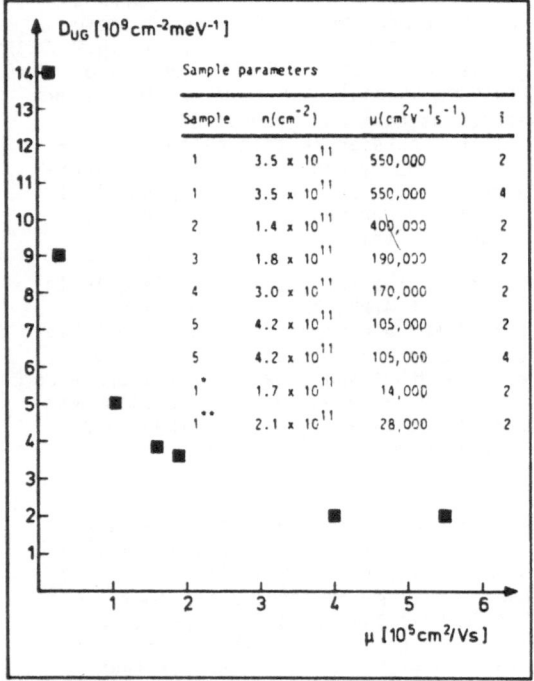

D_{UG} [$10^9 \text{cm}^{-2}\text{meV}^{-1}$]

Sample parameters

Sample	$n(\text{cm}^{-2})$	$\mu(\text{cm}^2\text{V}^{-1}\text{s}^{-1})$	i
1	3.5×10^{11}	550,000	2
1	3.5×10^{11}	550,000	4
2	1.4×10^{11}	400,000	2
3	1.8×10^{11}	190,000	2
4	3.0×10^{11}	170,000	2
5	4.2×10^{11}	105,000	2
5	4.2×10^{11}	105,000	4
1*	1.7×10^{11}	14,000	2
1**	2.1×10^{11}	28,000	2

μ [$10^5 \text{cm}^2/\text{Vs}$]

Fig. 7:

Measured activation
energies of sample 1**
compared with activation
energies deduced from
calculated curves (a). In
(b) the reconstructed DOS
both for measured $E_a(B)$
values and "calculated"
ones is plotted as a
function of energy. The
solid line shows the model
DOS (at B = 4.6 T) used in
the fit.

Fig. 8:

Background density
of states D_{UG} as a function of the mobility of
the devices. 1* denotes
sample 1 after electron
irradiation (1MeV elec-
trons) and 1** is sample
1* after annealing at
220 C⁰.

this method. The solid line in Fig.7b corresponds to the model DOS (filling factor i=2) used in these model calculations. The results of activated
resistivity measurements are summarized in Fig.8 where the constant background D_{UG} is plotted as a function of the mobility. Also included is a
table containing the parameters of the investigated samples. It should be

noted that the reconstruction of the density of states from activated resistivity measurements is restricted to the tails of the Landau levels. Furthermore the $\sqrt{B}$ dependence of the linewidth Γ assumed above cannot be proved with these measurements. Temperature-dependent measurements of the magnetocapacitance can overcome this problem and are discussed in the next section.

3. Magnetocapacitance

The capacitance experiments were carried out on gated GaAs-AlGaAs hetero-structures with a Hall geometry. The mobilities of the samples were 220,000 $cm^2V^{-1}s^{-1}$ and 800,000 $cm^2V^{-1}s^{-1}$. The corresponding carrier densi-ties were $2.25 \cdot 10^{11} cm^{-2}$ and $2.50 \cdot 10^{11} cm^{-2}$. For capacitance measurements all the Hall contacts were short-circuited and acted as a channel contact.

The signal was obtained by measuring the voltage drops at the sample and at a high-precision Boonton capacitance decade. This arrangement al-lows both a precise determination of the phase and the absolute value of the signal (see Fig.9b). The frequency chosen for the measurements was 223 Hz. Measurements between 22.3 Hz and 446 Hz showed no change in the signal. The modulation amplitude was 5 mV, which corresponds to a modula-tion of $\Delta n_s \approx 4 \times 10^9 cm^{-2}$. Further reduction of this amplitude showed no change in the signal. At each temperature the real part of the signal was monitored, and we checked that the signal was always purely capacitive for $C > C(B=0)$, even in the case of very low temperatures ($T \approx 1.64K$) and high magnetic fields. Warming up and cooling down of the sample introduced no change in the signal.

The capacitance of a system, consisting of a metal-insulator-(with ion-ized impurities) semiconductor-sandwich (e.g. Au-AlGaAs-GaAs-heterostruc-ture), depends not only on the thickness of the insulator but also on the DOS at the semiconductor side and on parameters of the material. Fig.9a

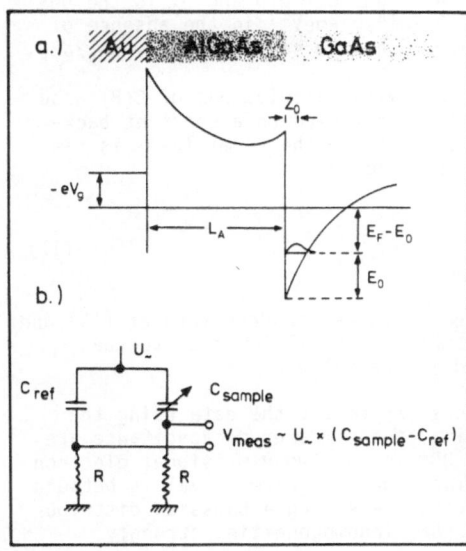

Fig. 9:

Schematic diagram of the conduction band edge (a) for a gated GaAs-AlGaAs heterostructure showing the quantities used in the derivations and schematic experimental set up (b). $U_\sim$ is the AC component of the applied voltage with an amplitude of 5 mV.

211

shows the band diagram of a heterostructure including a Schottky gate in contact with the AlGaAs. If the two depletion layers interpenetrate each other the gate voltage V_g is connected to the carrier density n_s by [10]

$$V_g = \frac{eL_A}{\varepsilon_i} n_s + \frac{1}{e} \left[E_0 + (E_F - E_0)\right] + K \tag{8}$$

where L_A is the thickness of the AlGaAs layer, ε_i is the dielectric constant of the "insulator" and K takes into account fixed charges in the AlGaAs and barrier heights at both interfaces. At low temperatures, carriers in the bulk of both materials are frozen out, so that K is a constant. Differentiating (8) with respect to n_s within the variational approximations of Stern [11], which take into account the n_s dependence of the subband edge E_0, one obtains for the capacitance:

$$\frac{1}{C} = \frac{1}{C_A} + \frac{\gamma z_0}{\varepsilon_s} + \frac{1}{e^2 \dfrac{dn_s}{d(E_F - E_0)}} \tag{9}$$

where C is the measured differential capacitance at a given magnetic field, C_A is the capacitance of the insulating AlGaAs layer, ε_s is the dielectric constant of GaAs, z_0 is the average position of the electrons in the channel, γ is a constant numerical factor between 0.5 and 0.7, and $dn_s/d(E_F-E_0)$ is the thermodynamic DOS at the Fermi level, in the following denoted as dn_s/dE_F. The first two terms on the right-hand side of (9) are assumed to be constant in a magnetic field, and thus changes of the capacitance are directly related to changes in the thermodynamic DOS of the 2DEG. At T=0 the total inverse capacitance in a magnetic field can be expressed as

$$\frac{1}{C} = \frac{1}{C_0} - \frac{1}{e^2 D_0} + \frac{1}{e^2 D} \tag{10}$$

where C_0 denotes the value of the total capacitance at B=0, D is the DOS at the Fermi level in the presence of a magnetic field and D_0 is the DOS within the lowest subband, equal to $2.9 \times 10^{10} \mathrm{cm}^{-2}\mathrm{meV}^{-1}$ in the absence of a magnetic field. At finite temperatures D has to be replaced by dn_s/dE_F.

The experimental results were compared with calculations of C(B) assuming a Gaussian-like density of states superimposed on a constant background DOS D_{UG} according to (7). The position of the Fermi level is determined again by solving numerically (6) and then

$$\frac{dn_s}{dE_F} = \int_0^\infty D(E) \frac{df(E-E_F)}{dE_F} dE \tag{11}$$

is calculated numerically. With the temperature-dependent form of (10) and (11) one obtains C(B). Spin splitting,which is small compared to the cyclotron energy for GaAs,is neglected in the calculations.

Some further considerations are necessary to fit the data using the expressions above. The minima and maxima of the measured capacitance are connected to minima and maxima in the DOS in the two-dimensional electron gas. A minimum in capacitance is obtained when the Fermi level is between two Landau levels. Additional calculations, assuming a Gaussian distribution of the electron density n_s, show that inhomogeneities strongly

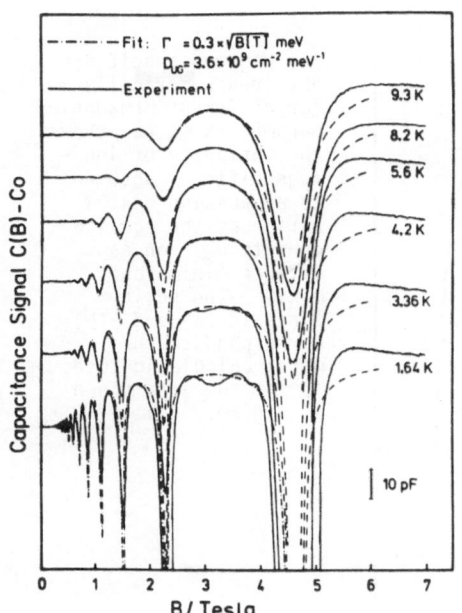

Fig.10:

Measured magnetocapacitance of a
sample with a mobility of
220,000 cm^{2}V^{-1}s^{-1} (same material
as sample 3) and corresponding
fit using our model DOS with the
parameters given in the plot.
For the sake of clarity the
curves are shifted vertically.

influence the minima but not the maxima of the capacitance at sufficient
high magnetic fields. Furthermore, at low temperatures and high magnetic
fields the capacitance signal is no longer purely capacitive, if the Fermi
level position is between two Landau levels ($\sigma_{xx} \approx 0$). For this reason, it
is advis-able to concentrate on the maxima of the measured capacitance to
fit the data.

Figure 10 shows the capacitance data at different temperatures. Also
shown is the fit to these data assuming a linewidth $\Gamma=0.3\sqrt{B[T]}$[meV] and a
background DOS of 3.6×10^9cm^{-2}meV^{-1}. The value of the constant background
is obtained from the temperature-dependent resistivity measurements in the
regime of the Hall plateaus carried out on the same material (sample 3,
see Fig.4c). At all investigated temperatures the calculated maxima of the
magnetocapacitance are in excellent agreement with the experimental ones
in the magnetic field range up to 5 Tesla. In Fig.11a calculated magnetocapacitance data using a $\sqrt{B}$ dependent Gaussian linewidth in the model DOS
are compared to those assuming a magnetic field independent linewidth.
Assuming a constant linewidth $\Gamma=0.54$ meV the magnetocapacitance for the
Landau level n=1 is fitted correctly, but for higher Landau levels the
resulting capacitance maxima are too small. On the other hand a linewidth
$\Gamma=0.34$meV describes correctly the capacitance for the Landau level n=3 but
results in larger magnetocapacitance maxima for lower Landau levels.

The solid lines in Fig.11 correspond to the model calculation used to
fit the data shown in Fig.10. Therefore a fit of the experimental data
with a magnetic field independent linewidth as well as a linewidth which
differs from the assumed value by more than 10% was not possible. It
should be mentioned that the assumption of a vanishing background D_{UG}
broadens Γ only by about 10%. The depths of the measured capacitance minima at low temperatures are smaller than the calculated ones as long as

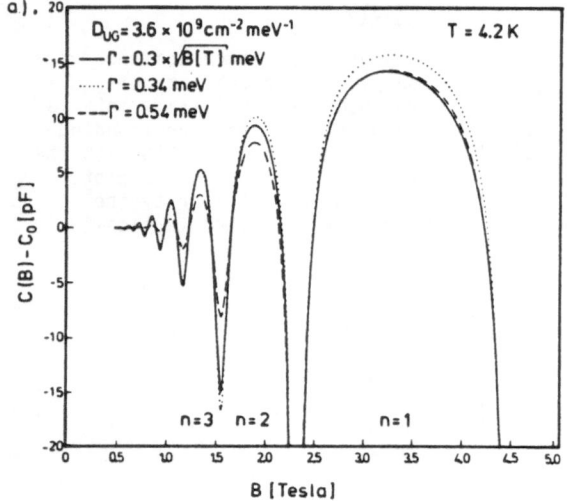

a).

$D_{UG} = 3.6 \times 10^9 \, cm^{-2} \, meV^{-1}$

—— $\Gamma = 0.3 \times \sqrt{B[T]} \, meV$

······ $\Gamma = 0.34 \, meV$

– – – $\Gamma = 0.54 \, meV$

$T = 4.2 \, K$

$C(B) - C_0 \, [pF]$

$n=3$ $n=2$ $n=1$

B [Tesla]

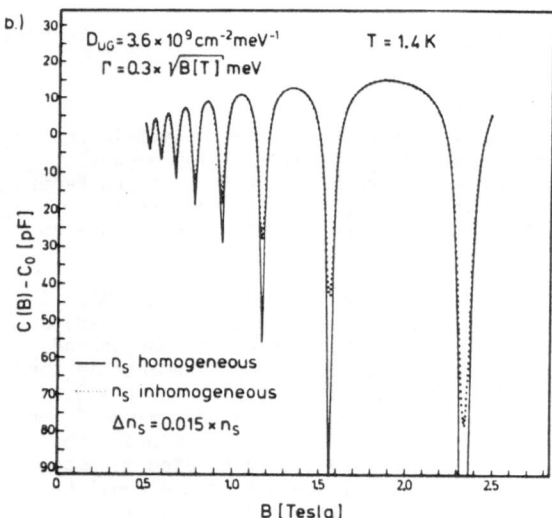

b.)

$D_{UG} = 3.6 \times 10^9 \, cm^{-2} \, meV^{-1}$

$\Gamma = 0.3 \times \sqrt{B[T]} \, meV$

$T = 1.4 \, K$

$C(B) - C_0 \, [pF]$

—— n_s homogeneous

······ n_s inhomogeneous

$\Delta n_s = 0.015 \times n_s$

B [Tesla]

Fig. 11:

(a) Magnetocapacitance obtained numerically for different broadening parameters Γ. (b) shows the influence of inhomogeneities – assumed as a Gaussian distribution of the carrier density n_s – on capacitance minima and maxima. The solid lines in (a) and (b) correspond to the model calculation used to fit the data shown in Fig.10.

resistivity effects in the channel are negligible (low B-field). This is attributed to inhomogeneities. Their influence is demonstrated in Fig.11b. The assumption of a Gaussian distribution of the carrier density n_s with a broadening parameter $\Delta n_s = 0.015 \cdot n_s$ leads at 1.4K to a remarkable reduction of the depth of the capacitance minima compared to the homogeneous case, but the maxima in the magnetocapacitance remain unchanged. The influence of the inhomogeneities decreases with increasing temperature. At higher magnetic fields the capacitance signal at the minima is governed by the small conductivity σ_{xx} which becomes less important at higher temperatures. Therefore the fit works well for minima and maxima of magnetocapacitance at higher temperatures. The difference between experiment and cal-

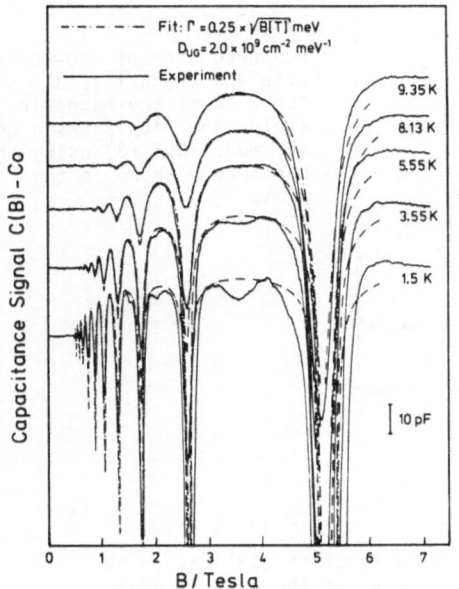

Fig.12:

Measured magnetocapacitance of a sample with a mobility of 800,000 $cm^2V^{-1}s^{-1}$ and corresponding fit using our model DOS with the parameters given in the plot. For the sake of clarity the curves are shifted vertically.

culation if only one Landau level is filled cannot be explained yet. In Fig.12 the temperature-dependent magnetocapacitance data for a high-mobility sample (μ=800,000$cm^2v^{-1}s^{-1}$, n_s=2.50·$10^{11}cm^{-2}$) are shown. The data are fitted with our model DOS using a Gaussian linewidth Γ=0.25·$\sqrt{B[T]}$ meV and a constant background DOS D_{UG}=2·$10^9cm^{-2}meV^{-1}$ according to Fig.8.

It should be emphasized that the constant background D_{UG} is not necessary to fit the capacitance data,since the reduction of the measured capacitance minima compared to the calculated ones can be explained by inhomogeneities. On the other hand it is not possible to exclude the existence of a constant background density of states from our capacitance experiments.

Our model used to calculate numerically the magnetocapacitance requires some supplementary remarks. Actually not the carrier density n_s but the Fermi level is kept constant during capacitance experiments. Using the notation of Fig.9a this means that the gate voltage V_g is kept constant. Varying the magnetic field B then leads to oscillations of the surface potential (bottom of the potential well) and to a charge transfer between gate and channel of the heterostructure. Since the amount of transferred charge is small compared to the two-dimensional carrier density n_s the subband edge (taken relative to the bottom of the potential well) is assumed to be constant. Equation (8) can then be rewritten as

$$n_s + \frac{C_A}{e^2} (E_F-E_0) = const \tag{12}$$

Combining (12) with (6) and taking all energies relative to the subband edge E_0 leads to

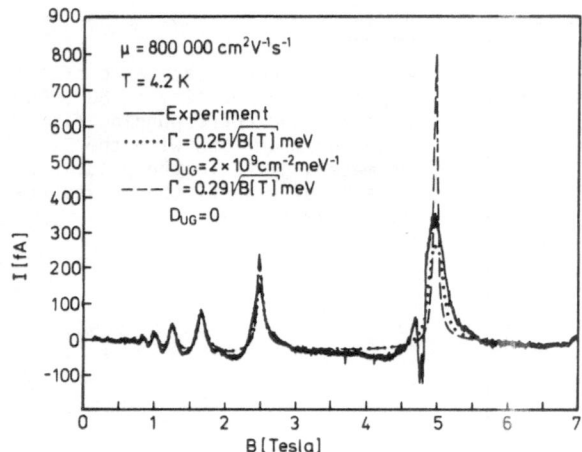

Fig.13:

Measured current flow between
gate and channel as a
function of the magnetic
field. The fit is based on
our model DOS (6) using the
parameters shown in the
plot.

$$\int_0^\infty D(E)f(E-E_F)\, dE + \frac{C_A}{e^2} E_F = \text{const} \tag{13}$$

where the constant can be determined at B=0. Equation (13) has to be
solved numerically to give the correct position of the subband edge
relative to Fermi level, and then the magnetocapacitance can be calculated
using (11) and the temperature-dependent form of (10). Calculating the
magnetocapacitance in the way described above results in a broadening of
the width of the capacitance minima compared to calculations assuming a
constant carrier density n_s. The difference, however, is small, and cannot be
resolved in Fig.10 or Fig.12. The charge flow mentioned above can be
determined by measuring the current between gate and channel as a function
of the magnetic field B. The current flow is given by

$$I(B) = A \cdot e \cdot \frac{dn_s}{dt} = A \cdot e \frac{dn_s}{dB} \cdot \frac{dB}{dt} \tag{14}$$

where A is the area of the two-dimensional electron gas and dB/dt the
sweep rate of the magnetic field. dn_s/dB can be determined by solving
(13) at different magnetic fields since the first term on the left-hand
side is equal to the carrier density n_s. Therefore measurements of the
current flow between gate and channel contain information about the density of states and may be a method to obtain new information about the density of states. Preliminary results are shown in Fig.13 where the current
flow between gate and channel is plotted as a function of the magnetic
field. This measurement was carried out on the high-mobility sample
(μ=800,000cm^{2}v^{-1}s^1). The experimental data are compared with numerical
calculations using our model DOS with the same parameters as shown in
Fig.12. An additional calculation assuming a vanishing background D_{UG}=0
but a somewhat broadened linewidth Γ=0.29·$\sqrt{B[T]}$ meV is shown, too. It
should be mentioned again that both sets of parameters fit the capacitance
data shown in Fig.12.

4. Summary

The discussion above has shown that temperature-dependent resistivity
measurements and capacitance measurements are complementary methods for
the experimental determination of the DOS since the analysis of the acti-

vated resistivity is restricted to the tails of the Landau levels,whereas magnetocapacitance experiments give information about the DOS close to the center of Landau levels. The analysis of the activated resistivity leads to the result of a constant or weak magnetic field-dependent background density of states whose magnitude increases if the mobility of the sample decreases. Measurements of the magnetocapacitance are compatible with the assumption of a background DOS but give no evidence for its existence, since the differences between experiment and calculation are well described by the influence of inhomogeneities. Inhomogeneities may also influence the results of heat capacity [3] and magnetization measurements [4]. Furthermore magnetocapacitance measurements demonstrate the $\sqrt{B}$ dependence of the Gaussian linewidth assumed. The linewidth Γ however is larger than expected from the selfconsistent Born approximation (SCBA) assuming short-range scatterers,and does not follow the mobility dependence $\Gamma \sim \sqrt{1/\mu}$ [1].

Further information about the density of states,especially about the background D_{UG} may be obtained by analyzing carefully the current flow between gate and channel as a function of the magnetic field and temperature.

References

1. T. Ando, Y. Uemura, J.Phys.Soc.Jap. 36, 959 (1974)
2. R.R. Gerhardts, Surf.Sci. 58, 227 (1976)
3. E. Gornik, R. Lassnig, G. Strasser, H.L. Störmer, A.C. Gossard, W. Wiegmann, Phys.Rev.Lett. 54, 1820 (1985)
4. J.P. Eisenstein, H.L. Störmer, V. Narayanamurti, A.Y. Cho, A.C. Gossard, Phys.Rev.Lett. 55, 875 (1985)
5. E. Stahl, D. Weiss, G. Weimann, K. v.Klitzing, K. Ploog, J.Phys. C18, L783 (1985)
6. T.P. Smith, B.B. Goldberg, P.J. Stiles, M. Heiblum, Phys.Rev. B32, 2696 (1985)
7. V. Mosser, D. Weiss, K. v.Klitzing, K. Ploog, G. Weimann, Solid State Commun., to be published
8. T. Ando, J.Phys.Soc.Jap. 53, 3101 (1984)
9. S.A. Trugman, Phys.Rev. B27, 7539 (1983)
10. D. Delagebeaudeuf, N.T. Linh, IEEE Trans. ED-29, 955 (1982)
11. F. Stern, Phys.Rev. B5, 4891 (1972)

Density of States of Landau Levels from Specific Heat and Magnetization Experiments

R. Lassnig and E. Gornik

Institut für Experimentalphysik, Universität Innsbruck,
A-6020 Innsbruck, Austria

The density of states of two-dimensional electrons in GaAs-GaAlAs structures is determined from thermodynamic experiments in high magnetic fields. From specific heat and magnetisation measurements a Gaussian form at the center of the Landau levels and a substantial density of states between the levels is found. The results can be described with a $\Gamma = \Gamma_0 \sqrt{B}$ relation for the Gaussian half-width on top of a flat background consisting of about 10 to 20% of the states. From theoretical considerations, the level width can be explained due to scattering on ionized impurities. The flat background has a different origin and can be interpreted as resulting from short range scattering and point defects.

1. Introduction

In a 2-dimensional electron system (2DES) a magnetic field perpendicular to the plane of electrical confinement leads to full quantization of the electron motion. The energy spectrum consists of Landau levels (LL) separated by the cyclotron energy $\hbar\omega_c$. In an unperturbed system the LL are discrete and highly degenerate, with $1/2\pi l^2$ (with $l = \sqrt{\hbar/eB}$ the cyclotron radius) possible states in each level. For an electron concentration n_s the filling factor is defined as $\nu = (n_s/\pi l^2)$, neglecting spin splitting.

In a real system the LL are broadened due to scattering by impurities, phonons and other scattering mechanisms. In the simplest approximation the levels are described by a level width Γ. In the case of a high magnetic field, where $\hbar\omega_c \ll \Gamma$, real gaps appear between the LL. This leads to an oscillatory structure of practically all physical quantities as a function of the magnetic field.

The most fundamental quantity underlying all these physical properties of the system is the form of the density of states (DOS). The most pronounced effects are the Quantum Hall effect [1] and the fractional Quantum Hall effect [2]. Current theoretical understanding of the Quantum Hall effect (QHE) [3] and the fractional QHE [4] relies on assumptions about the DOS; the distinction between localized and extended electronic states being essential. The form of the density of states can be obtained directly by measurement of thermodynamic quantities such as specific heat or magnetization, where localized and extended states contribute equally in equilibrium. Measurements of transport and optical properties yield also information on the DOS, however, the extraction of the pure DOS is more complicated.

In this paper we will describe the experimental determination of the DOS by thermodynamic techniques including specific heat and magnetization measurements. A brief discussion of the current theoretical understanding of the DOS will be given in the light of the experimental findings.

2. Experimental Techniques to Determine the DOS

Up to now four different methods have been applied to determine the DOS of 2DES in high magnetic fields. Beside specific heat and magnetization, a capacitance measurement and the determination of activation energies in the Hall plateau range were applied. In principle the capacitance of the system yields direct information of the DOS. However, the capacitance technique is applicable only for systems with high conductivity. The 2DES has a strongly oscillatory and even vanishing diagonal conductivity. As a consequence reasonable results are only achieved at low magnetic fields [5] and at high fields in the narrow highly conducting regions, when the Fermi level is within a LL [6]. In the localization region the method is not useful.

The measurement of activation energies is a transport experiment measuring the activation energy of carriers at the Fermi energy with respect to the next higher LL. Electrical conduction is assumed to take place only in the delocalized states in the LL center. The DOS for localized states between LL can be extracted from the data by shifting the Fermi level with the magnetic field. This method allows to determine the DOS to high accuracy [6, 7].

a) Specific Heat

The most direct method to determine the DOS is the measurement of the electronic specific heat given by $C_v = dU/dt$, where U is the internal energy. An externally induced temperature change leads to a reordering of the electrons at the Fermi energy. The heat capacity of the electron system is proportional to the DOS at the Fermi energy.

A heat pulse technique was applied to determine the electronic specific heat. In this technique a short heat pulse heats the sample adiabatically. The temperature change of the total sample is measured with a temperature detector, which is not sensitive to the magnetic field. Details of the experimental technique are given in Ref. [8,9]. As the specific heat of the 2D electrons is extremely small, samples with a large number of 2D layers have to be used to make the electronic specific heat comparable with the lattice specific heat.

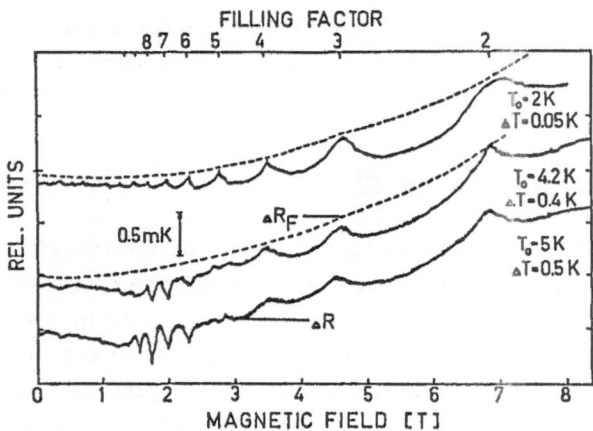

Figure 1: Temperature change of multilayer sample with 94 layers measured with a Au-Ge film as a function of magnetic field (curves ΔR). A heat pulse raised the sample temperature by ΔT. The d.c. dependence of the Au-Ge film resistance is shown by the curves ΔR$_F$ on an extended scale.

Figure 1 shows the observed temperature change of a sample consisting of 94 layers of 220 Å GaAs and 500 Å GaAlAs as a function of magnetic field The sample has a density of $7.7 \times 10^{11} cm^{-2}$ per layer and a mobility of about 80 000 cm^2/Vs at 4.2 K. The experimental curves show the resistance variation of the temperature sensor (Au:Ge film) for three different temperatures. The curves ΔR$_F$ represent the background d.c. resistance variation of the detector film on an extended scale. Oscillations of the sample temperature are clearly observed with a spike-like behaviour for integer filling factors at higher fields. Additional spikes are observed for 4.2 K and 5.0 K at lower fields.

The interpretation of the structures at low and high magnetic fields which have different signs can be made by comparison with theoretical calculations performed by ZAWADZKI and LASSNIG for the same system [10]. They found that in the case of a Gaussian density of states two contributions to the specific heat are present: intra-LL contributions due to excitations within one LL (dominant when the LL splitting $\hbar\omega_c \gg k_B T$) and inter-LL contributions due to excitations between adjacent LL (dominant when $\hbar\omega_c \sim k_B T$). Consistent with the prediction, intra-LL contributions are found at low temperatures and high magnetic fields [10]. Inter-LL contributions are observed at higher temperatures and low magnetic fields.

From the experimental data we can determine first the specific heat and by comparison with calculations the form of the density of states over a rather wide range of magnetic fields. A constant heat input ΔQ is applied to the sample at a given T. The resulting temperature change ΔT is directly evident from Fig. 1 as difference to the curve ΔR_F. The measured specific heat is given by

$$C_{el}(T,B,\Gamma) = \frac{\Delta Q}{\Delta T} \div \frac{1}{4} \frac{\alpha}{\Delta T} \left[(T + \Delta T)^4 - T^4 \right] \tag{1}$$

with $C_{lat} = \alpha T^3$ and the assumption $C_{el}(T + \Delta T) \sim C_{el}(T)$.

The experimental results are compared with calculations of C_{el} assuming a Gaussian density of state

$$\rho_G(E) = (\pi l^2)^{-1} \sum_n (2/\pi)^{1/2} \Gamma^{-1} \exp\{ \frac{(E - \lambda_n)^2}{2\Gamma^2} \} \tag{2}$$

and a Gaussian density with a constant background

$$\rho_{GB}(E) = \rho_G(E) (1 - x) + (\pi l^2)^{-1} (x/\hbar\omega_c)\cdot\Theta(E) \tag{3}$$

where $\lambda_n = \hbar\omega_c(n + 1/2)$, x is the percentage of background states and Γ is the RMS half-width of the levels.

A comparison of calculated C_{el} values with the experiment is shown in Fig. 2. A fit to the data (curve 1) was tried with a magnetic field independent width with $\Gamma = 0.75$ meV and $\Gamma = 1.5$ meV (curve 3 and 4). It is evident that inter-LL peaks are very sensitive to Γ, while intra-LL peaks at high fields are less sensitive to Γ. In previous papers only a magnetic field independent linewidth was used [8,9] where Γ was defined a factor of 2 larger. The best linewidth fit to the data was always achieved in the range of 2 T where inter-LL contributions were dominant given a width of $\Gamma = 0.75$ meV at

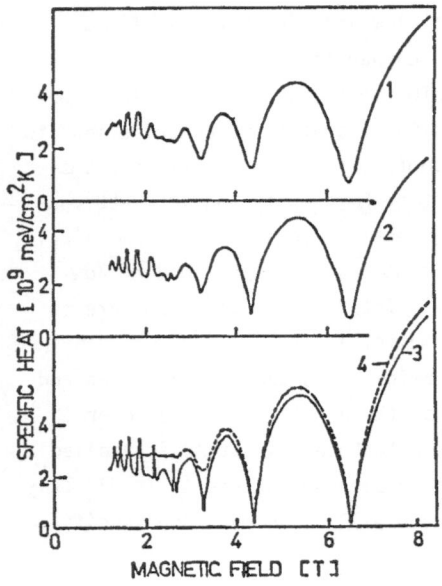

Figure 2: Comparison of calculated C_{el} over magnetic field
for sample 2 at 4.2 K with experimentally determined values.
Curve 1: experiment; curve 2:
$\Gamma = 0.6$ meV·$\sqrt{B}$, background
$x = 0.25$; curve 3: $\Gamma = 0.75$meV,
$x = 0$; curve 4: $\Gamma = 1.5$ meV,
$x = 0$.

2 T. Thus we get the best fit to the data if we assume a magnetic field dependent linewidth with $\Gamma = 0.6$ meV·$\sqrt{B}$ (B in T) and a background of $x = 0.25$.

The assumption of a background was necessary to explain the flat part of
the data at high fields for integer filling factors and low temperatures as
was demonstrated in Ref. [10].

If we try to fit the sample with 172 double layers and a low mobility
of 40 000 cm^2/Vs with a magnetic field dependent width we obtain $\Gamma =$
$= 0.8$ meV·$\sqrt{B}$ and a background $x = 0.3$. We have to increase the background
quite considerably to make the inter-LL vanish.

The presence of a flat and constant background density of states which
is even present in high mobility single layer heterostructures was very recently also demonstrated by WEISS et al [6,7]. From temperature dependent
resistivity measurements in the Hall plateau range they found a flat background density of states for all samples between the LL peaks and a $\sqrt{B}$
linewidth dependence. The background decreases with increasing mobility from
a value of $x = 0.2$ for a sample with a mobility of 10^5cm^2/Vs to x-values of
0.07 for a sample with a mobility of 550 000 cm^2/Vs, while the value for the
linewidth depends only weakly on the sample. An extrapolation of their data
to our mobilities of 80 000 cm^2/Vs give x-values of $\sim$0,25 in agreement with
our observed values.

b) Magnetization (de Haas-van Alphen effect)

The magnetization measurements were performed by EISENSTEIN et al. [11]
with a recently developed torsional technique [12]. The samples are mounted
on a thin fiber held perpendicular to the applied magnetic field and are
oriented so that the normal of the 2D plane, along which the orbital magnetic moments must lie, is tilted away from the field direction by a small
angle; this geometry is seen in the insert of Fig. 3. In the experiment the
magnetic field is swept very slowly; the torque is registered by a capacitive technique.

The samples used were a GaAs/GaAlAs multilayer (50 periods) and a single
heterostructure. Fig. 3 shows normalized magnetization data from a multilayer sample with a mobility of $8.0 \times 10^4 cm^2/Vs$ (50 layer) at 4.2 K and a
single heterostructure with a mobility of $2.85 \times 10^5 cm^2/Vs$. We will only
discuss the multilayer data since the specific heat experiments were performed on similar samples. The magnetization varies smoothly over the entire field range. The lack of discontinuities suggests the absence of gaps
in the DOS.

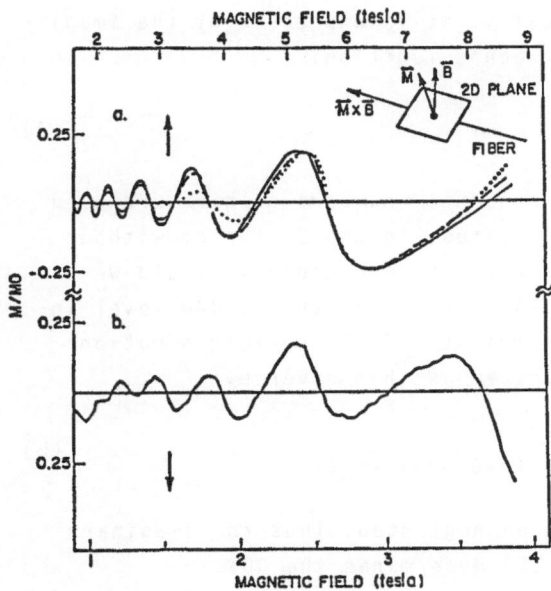

Figure 3: Normalized magnetization for (a) a sample with 50 periods (n_s =
$= 5.4 \times 10^{11} cm^{-2}$, $\mu(4.2 K) = 8.0 \times 10^4 cm^2/Vs$ and (b) single heterolayer
($n_s = 3.7 \times 10^{11} cm^2/Vs$, $\mu(4.2K) = 2.85 \times 10^5 cm^2/Vs$) after Ref. [11]. Dotted
and dashed lines are fits; $\Gamma = 2.4$ meV and $\Gamma = 1$ meV$\cdot\sqrt{B}$ respectively. The
basic geometry of the magnetization measurements is shown in the insert.

In Fig. 3 a comparison with calculations using a Gaussian density of states for the multilayer sample is given. The fit is tried with a magnetic field independent width of Γ = 2.4 meV and a field dependent width of Γ = = 1 meV$\cdot\sqrt{B}$. It is evident that the narrower and field dependent width fits the data better. However, this width is somewhat larger than the results from the specific heat. We have tried to fit the data by including a flat background of x = 0.25. This reduces the width to a value of Γ = 0.80 meV$\cdot\sqrt{B}$ which is quite close to the specific heat value of Γ = 0.6 meV$\cdot\sqrt{B}$ determined for a comparable sample.

However, it should be mentioned that the magnetization experiment measures the contributions of all electrons. The information about the DOS has to be extracted from the shape of the oscillation. The analysis is therefore not extremely linewidth dependent.

3. Theoretical Considerations of the DOS

The density of states of an electron system denotes the number of possible states within an energy interval, divided by the interval width. Mathematically, it is expressed by the imaginary part of the retarded Green's function:

$$D(E) = \frac{1}{\pi} \sum_\alpha \text{Im } G_{ret}(\alpha, E) \tag{4}$$

The summation goes over the quantum numbers α which are used to classify the individual states. In the 2D system without magnetic field, $\alpha = (k_x, k_y)$. For a strong magnetic field we use the Landau gauge and $\alpha = (n, X)$, where n is the Landau level index and X is the center coordinate of the cyclotron motion. The DOS of the perturbed system is then given by:

$$D(E) = \frac{1}{\pi l^2} \frac{1}{\pi} \sum_{n=0}^{\infty} \text{Im } \left(\frac{1}{E - (n+0.5)\hbar\omega_c + \Sigma_n(E)} \right) \tag{5}$$

where spin splitting has been neglected. Thus the imaginary part of the self-energy $\Sigma_n(E)$ determines the DOS.

Theoretically, diagrammatic [13,14] as well as path integral [15-17] techniques are used to calculate the explicit form of the self-energy. In the self-consistent Born approximation [13,14], one obtains a rather simple semi-elliptic

form for the DOS (see Fig. 4a). Using a path integral technique within lowest order cumulant expansion GERHARDTS [15]
obtained a Gaussian DOS (see Fig. 4b) for long-range potentials. Exact results have been obtained by WEGNER [16] and
BREZIN et al. [17] for some restricted types of short-range
scattering distributions (expressed by different correlation
functions). For a white-noise potential one obtains a DOS
decaying as a Gaussian function. A Poisson distribution of
scatterers (with nonzero higher order correlations) yields
a strong peak at the center of the LL and a weakly decaying
DOS between the levels.

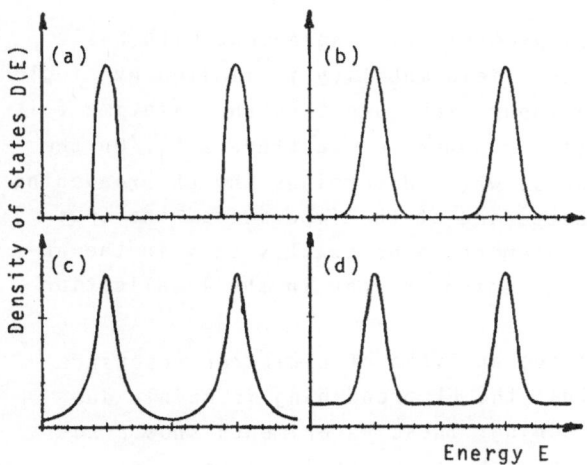

Figure 4: Comparison of model density of states.
(a) elliptic   (b) pure Gaussian   (c) Lorentzian   (d) Gaussian with a
constant background.

In a real system, the effective potential depends strongly
on the screening strength of the electron gas [13], which in
term is a function of the Fermi energy and the filling factor.
Thus, for an exact description of a real experiment, a self-
consistent screening calculation must be performed [14]. On
the other hand, for model calculations the effective (screened) potential can be interpreted as a pseudopotential.
    For the analysis of experiments, quite often a Lorentzian
DOS and a Gaussian DOS with a constant background are used.
These two types are depicted in Fig. 4c and Fig.4d.

4. Discussion

The analysis of the experimental data reveals that the largest amount of the DOS is concentrated at the LL centers. These exponentially decaying peaks sit on a flat background (Fig. 4c). The level width can be expressed by $\Gamma \cong \Gamma_o \sqrt{B}$. Summarizing the experimental results of several groups we find that the widths of the Gaussian peaks Γ_o vary by a factor of three while the mobility at B=0 varies over an order of magnitude for all samples investigated [6-9,11]. The amount of background necessary to fit the data has to be increased from about 7% for the highest mobility sample to about 30% for the lowest mobility sample.

The results for the level widths are in agreement with basic scattering theory: The zero field mobility is limited by single scattering events at the impurities, and thus the lifetime τ is inversely proportional to the number of scatterers N_i. On the other hand, the main process which determines the LL broadening is the virtual double scattering at the impurities, and thus $\Gamma \sim \sqrt{N_i}$ [13]. The latter statement holds exactly only in the absence of screening, but screening is weak in the localisation regime.

It is well known from the analysis of cyclotron resonance experiments [18,14,19] that the LL broadening is mainly due to ionized impurity scattering. These experiments show a Lorentzian linewidth; in contrast, a Lorentzian level width (Fig. 4c) can be excluded from the above experiments.

Therefore we conclude that the experimentally detected form of the DOS can only be explained by an interplay of two different scattering mechanisms. First of all, the LL are broadened by ionized impurities, but the states remain still close to the LL center. Due to screening, this part oscillates with the filling factor.[14]. Second, some sort of neutral, short-range scatterers smear out a part of the states continuously over the spectrum. This kind of scattering was investigated by BREZIN et al.[17]. The origin can be either dislocations or alloy scattering. However, to make conclusive theoretical statements about the form of the DOS a realistic analysis for experimentally determined distributions of scatterers is necessary, which has not been done up to now.

Acknowledgements: This work is partly supported by the Stiftung Volkswagenwerk. We thank Prof. Janos Hajdu and Prof. Klaus v. Klitzing for valuable discussions.

References

1. K.v.Klitzing, G.Dorda, M.Pepper; Phys.Rev.Lett. 45, 494 (1980)
2. D.C.Tsui, H.L.Störmer, A.C.Gossard; Phys.Rev.Lett. 48, 1559 (1982)
3. R.B.Laughlin; Phys.Rev.B 23, 5632 (1981)
   Phys.Rev.Lett. 50, 1395 (1983)
4. H.Levine, S.B.Libby, A.M.M.Pruisken; Phys.Rev.Lett. 51, 1915 (1983)
5. T.P.Smith, B.B.Goldberg, M.Heilblum, P.J.Stiles; Surf.Science (1986)
6. D.Weiss, E.Stahl, G.Weimann, K.Ploog, K.v.Klitzing; Surf.Science (1986)
7. D.Weiss, K.v.Klitzing; this issue, preceding paper
8. E.Gornik, R.Lassnig, G.Strasser, H.L.Störmer, A.C.Gossard; Phys.Rev. Lett. 54, 1820 (1985)
9. E.Gornik, R.Lassnig, G.Strasser, H.L.Störmer, A.C.Gossard; Surf.Science (1986)
10. W.Zawadzki, R.Lassnig; Solid State Comm. 56, 537 (1984)
11. J.P.Eisenstein, H.L.Störmer, A.Y.Cho, A.C.Gossard, C.W.Tu; Phys.Rev. Lett. 55, 875 (1986)
12. J.P.Eisenstein; Appl.Phys.Lett. 46, 695 (1985)
13. T.Ando, Y.Uemura; J.Phys.Soc. Japan 36, 959 (1974)
14. R.Lassnig, E.Gornik; Solid State Comm. 47, 959 (1983)
15. R.R.Gerhardts; Z.Phys.B 21, 275 (1975)
16. F.Wegner; Z.Phys.B 51, 279 (1983)
17. E.Brezin, D.J.Gross, C.Itzykson; Nuclear Phys.B 235, 24 (1984)
18. T.Englert, J.C.Maan, Ch.Uihlein, D.C.Tsui, A.C.Gossard; Solid State Comm. 46, 545 (1983)
19. W.Seidenbusch, G.Lindemann, R.Lassnig, J.Edlinger, E.Gornik; Surf. Science 142, 375 (1984)

The Integer Quantum Hall Effect:
An Introduction to the Present State of the Theory

J. Hajdu

Institut für Theoretische Physik, Universität zu Köln,
D-5000 Köln 41, Fed. Rep. of Germany

1. Introduction

Let me briefly summarize the basic experimental facts: The Hall conductivity σ_H and the longitudinal conductivity σ were analysed as functions of the magnetic field B perpendicular to the 2d system, or of the filling factor $n = n/(eB/h)$, where $n = N/Ar$ is the carrier concentration. It was found /1/ that at high magnetic fields and low temperatures

$$\sigma = 0 \tag{1}$$

$$\sigma_H = i\frac{e^2}{h} \tag{2}$$

within symmetric intervals around $n = i$, $= 0,1,2,\ldots$. In the transition regime between two adjacent plateaux (given by (2)) $\sigma \neq 0$ and σ_H is continuous and monotonically increasing (sometimes steplike). Although, for the time being no predictive theory of the integer quantum Hall effect (IQHE) is available - in the sense of explicit calculations of σ and σ_H as functions of n,B,T (temperature), ω (frequency of the external electric field)- the effect can be explained /2/ on the basis of (strong) localization of the carriers by a random potential in the presence of a high magnetic field /3/. This explanation is strongly supported by numerical experiments /4/ and (classical) percolation arguments /5/. It is generally believed that once localization is assumed, the QHE results as a consequence of gauge invariance /6/. It has been demonstrated by self-consistent perturbation theory /7/ and rather elaborate field theory /8/ that for certain model systems all states are localized at $T = 0$ except for some states at the center of each Landau band (corresponding to $n = i/2$) which are always delocalized. Under quite general conditions which seem to be fulfilled by localization the QHE has been shown to occur for topological reasons /9/. This may explain the high accuracy of the observed plateau values (2) ($< 10^{-7}$).

At integer fillings the measured Hall conductivity (2) coincides with that of free electrons,

$$\sigma_H^0 = \frac{en}{B} = n\frac{e^2}{h} \tag{3}$$

This remarkable fact can be explained by applying a suitable form of the Kubo formula to a non-interacting electron system at $T = 0$ and assuming that there exist (quasi) gaps in the energy spectrum at energies corresponding to $n = i$ /10/. Whereas the existence of such gaps is theoretically well established for high mobility systems at sufficiently high magnetic fields /11/, experiments indicate a rather high density of states at $n = i$ /12/. (This contradiction can possibly be resolved by taking into account the spatial fluctuation of the Fermi energy in the random potential /13/. The coincidence of the observed Hall conductivity at η = i with the corresponding free electron values has been explained by scattering theory /14/: the loss of the current caused by electrons occupying bound (localized) states is exactly compensated for by an additional phase shift (acceleration) of the scattering (delocalized) states. (For an explicit demonstration of this compensation in the case of an arbitary number of point impurities cf. /15/). The corresponding explanation in the percolation picture is: the total Hall voltage drops across the channel of percolating states (each finite perimeter cluster being equipotential).

The distribution of the current and the electric field in the QHE regime is essentially inhomogeneous and asymmetric /16/. The possible role of the Coulomb interaction between the electrons in the IQHE has not yet been investigated with appropriate care. The role of the edge states (edge currents) - see below - is obscure and controversial (even more so than in the case of the conduction electron diamagnetism /17/).

The QHE has also been related to the chiral anomaly (field theory) and the zero eigenspace of the Dirac wave operator, but unfortunately I am not able to report on these relativistic attempts to explain the QHE.

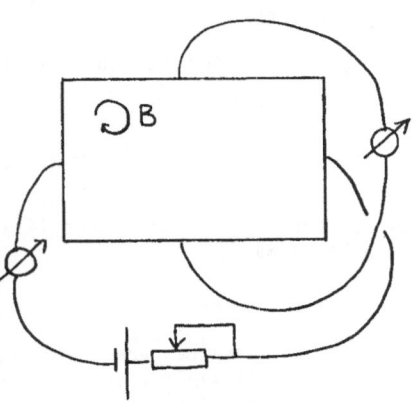

The usual 4 contact and Corbino disc Hall measurements consist of 2 current loops. This is possibly of great topological importance /9/. Such a system , however, is rather unwieldy to treat theoretically. A possible way out of this dilemma is to consider a model with appropriate boundary conditions (b.c) for the wave functions, allowing for current propagation in both directions /9/. The Landau model - periodic b.c. in one direction and infinite extension in the other - has this property. But the infinite extension gives rise to mathematical problems (in particular if a homogeneous electric field is required in this direction). If the extension of the system is limited in one direction by an infinite confinement potential (Teller model), a homogeneous electric field can be imposed but no current flow in this direction is possible. (The "pumping" of electrons from one edge to the other /6/ by an omnipotent agent acting outside of the system has been invented to overcome this difficulty). This model is possibly suitable to describe the Hall effect in a Corbino disc as long as σ = 0 (QHE regime). It is rather embarrassing that one has to worry about such things as boundary conditions - but certainly less embarrassing than the QHE.

In the following two chapters I will recall some basic facts about the 2d free electron and electron-impurity system in high magnetic fields.

2. Free Electrons

Landau model:

$\underline{B}$ = curl $\underline{A}$
$\overline{\underline{A}}$ = (0,Bx,0)

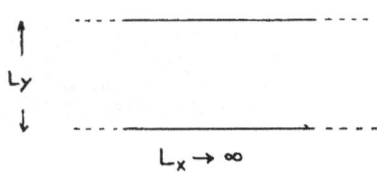

L_y

$L_x \rightarrow \infty$

b.c.:

$$\psi_\alpha(x,y) \to 0 \quad \text{for} \quad x \to \pm\infty \tag{4a}$$

$$\psi_\alpha(x,y+L_y) = \psi_\alpha(x) \tag{4b}$$

Due to the periodic b.c. (4b) the system is - with respect to connectivity - equivalent to an infinitely long cylinder with radial magnetic field and a Corbino disc with infinite radius. Since the momentum in the y direction is conserved the 2d energy eigenvalue problem separates. The energy eigenvalues are

$$\varepsilon_\alpha = \varepsilon_\nu = \hbar\omega_c(\nu+\tfrac{1}{2}), \quad \nu = 0,1,2,\ldots \tag{5}$$

ω_c = eB/m and the corresponding eigenfunctions ψ_α are products of plane waves in y direction with momentum p_y = $\hbar$k and oscillator wave functions centered around X = -1^2k, 1^2 = $\hbar$/eB. (1 is the radius of the classical cyclotron orbit at the ground state energy $\varepsilon_0 = \hbar\omega_c/2$). The quantum numbers are $\alpha = (\nu,k)$. For each fixed ν there are $2\pi 1^2$ states per cm^2 (degeneracy of the Landau levels (5)). In the presence of a homogeneous electric field $\underline{E}$ = (E,0,0) the degeneracy is lifted,

$$\bar\varepsilon_\alpha = \varepsilon_\nu + eEX \tag{6}$$

The expectation value of the velocity in the state $\bar\psi_\alpha$ is

$$<\bar\alpha|v_x|\bar\alpha> = 0 \tag{7a}$$

$$<\bar\alpha|v_y|\bar\alpha> = \frac{\partial\bar\varepsilon_\alpha}{\partial p_y} = -\frac{1}{eB}\frac{\partial\bar\varepsilon_\alpha}{\partial X} = -\frac{E}{B} \equiv v_D = \text{const.} \tag{7b}$$

Teller model: L_x is finite and (4a) is replaced by

$$\psi(x,y) = 0 \quad \text{for} \quad x = \pm L_x/2$$

The energy eigenvalues are in the bulk essentially the same as in the Landau model. Near X = $\pm L_x/2$ the energy levels bend up. Thus, the degeneracy is lifted, $\varepsilon_\alpha = \varepsilon_\nu(X)$. In the presence of the electric field $\bar\varepsilon_\alpha = \varepsilon_\nu(X)+eEX$. Consequently

$$<\bar\alpha|v_y|\bar\alpha> = -\frac{1}{eB}\frac{\partial\varepsilon_\nu(X)}{\partial X} + v_D \tag{8}$$

Here the first term is the contribution of the edge states.

Hall conductivity:

X is conserved. Thus, $f_\alpha(X) = f(\bar\varepsilon_\alpha - \xi(X))$ (where $f(\varepsilon)$ is the Fermi function and $\xi(X)$ the chemical potential) is a stationary distribution. The homogeneity of the distribution in the bulk requires $\xi(X) = \xi + eEX$. Hence $f_\alpha(X) = f(\varepsilon_\alpha - \xi) = f_\alpha$. With this distribution function the average current is

$$J_y = \frac{-e}{Ar} \sum_\alpha f_\alpha <\bar{\alpha}|v_y|\bar{\alpha}> = \frac{-e}{Ar} \sum_\alpha f_\alpha \left(-\frac{1}{eB}\frac{\partial \varepsilon_\alpha}{\partial X}\right) + \frac{e}{B}\frac{1}{Ar}\sum_\alpha f_\alpha E \tag{9}$$

The first contribution vanishes by symmetry. In the second

$$\frac{1}{Ar}\sum_\alpha f_\alpha = n \tag{10}$$

Thus, with $J_y = \sigma_H E$ we get (3). Linear response theory (Kubo formula) leads to the same result. For free electrons there is no QHE (neither in the Landau nor in the Teller model).

Let us consider σ_H^0 at $T = 0$ as a function of the Fermi energy ε_F. This is obviously a step function. Writing

$$\frac{d\sigma_H}{dn} = \frac{d\sigma_H}{d\varepsilon_F}\frac{d\varepsilon_F}{dn} \tag{11}$$

we recognize that $\sigma_H(n)$ shows the ideal QHE (steps) if $\sigma_H(\varepsilon_F)$ behaves in the same way as in the case of free electrons (step function) and $\varepsilon_F(n)$ is smooth (i.e. radically different from the free electron case).

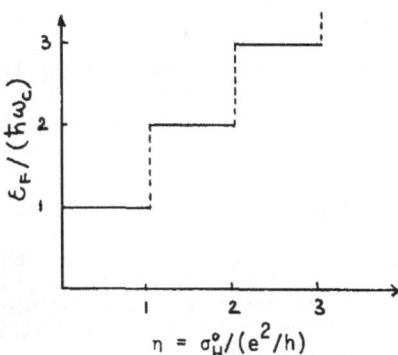

3. Non-Interacting Electrons in a Random Potential

Both requirements formulated above are fulfilled by a non-interacting system of electrons in a random potential (impurites) at zero temperature: the first by localization and the second by level broadening. The ideal QHE for such a system has been proved by Pruisken et al. /8/. A considerably simpler (but less systematic) proof can be given by combining earlier particular results achieved by Tsukada /3/, Aoki and Ando /2/, Streda /10/ and Ono /7/. In the following we retrict ourselves to the limit of very high magnetic fields and slowly varying random potentials. In this limit the sudden change of $\sigma_H(\varepsilon_F)$ is due to a sharp percolation transition at $\varepsilon_F = \varepsilon_\nu$. The quantized Hall current appears as topologically stable edge effect.

Let us consider a single electron in a random potential $U(x,y)$ in the presence of a magnetic field,

$$H = \frac{m}{2} v^2 + U(x,y) \tag{12}$$

For high B it is advantageous to introduce "center and relative" coordinates defined by

$$x = X + \xi, \quad y = Y + \eta \tag{13}$$

$$\xi = v_x/\omega_c, \quad \eta = -v_y/\omega_c \tag{14}$$

(η,ξ) and (X,Y) are canonically conjugate variables in the sense

$$\left[\eta,\xi\right] = i1^2, \quad \left[X,Y\right] = i1^2 \tag{15a,b}$$

For U = 0 $|\xi(t)|$, $|\eta(t)| < 1$. Assuming that U(x,y) is slowly varying and $|\text{grad } U|1 << |U|$

$$H = \frac{m\omega_c^2}{2} (\xi^2 + \eta^2) + U(X,Y) \tag{16}$$

The eigenvalues of the first term in (16) are determined by (15a) to be $\varepsilon_v = \hbar\omega_c(v+1/2)$. For closed equipotential lines U(X,Y) = const (15b) can approximately be satisfied by the semi-classical quantization role

$$\oint Y dX = 2\pi 1^2(\mu+\gamma) \tag{17}$$

$\mu = 0,1,2,...$; $0 \leqslant \gamma < 1/2$. In the limit B → 0 $[X,Y] = ih/eB \to 0$ and, therefore, X and Y can be treated as classical variables. The classical description with the Hamiltonian H = ε_v + U(X,Y) and the resulting equations of motion

$$\dot{X} = \frac{1}{eB} \frac{\partial U}{\partial Y} , \quad \dot{Y} = - \frac{1}{eB} \frac{\partial U}{\partial X} \tag{18}$$

represent a reliable zeroth order approximation as long as $|\partial^2 U/\partial X^2|$, $|\partial^2 U/\partial Y^2| < |U|/1^2$. The equations (18) describe motion along equipotential lines. In the presence of a homogeneous electric field $\underline{E}$ = (E,0,0), U has to be replaced by U + eEX. The distribution function $\overline{f}$ of non-interacting electrons obeys the equation

$$\frac{\partial\overline{f}}{\partial t} + \frac{\partial\overline{f}}{\partial X} \dot{X} + \frac{\partial\overline{f}}{\partial Y} \dot{Y} - \frac{E}{B} \frac{\partial\overline{f}}{\partial Y} = 0 \tag{19}$$

and the normalisation condition

$$\sum_v \int \frac{dXdY}{2\pi 1^2} \overline{f} = n \tag{20}$$

Putting E(t) = Ee^{nt}, the linear response solution $\overline{f}$ = f(H) + g, g ~ E of this equation is at t = 0

$$\overline{f} = f + eE\frac{\partial f}{\partial U} (X-\overline{X}) \tag{21}$$

where $\overline{X}$ is the infinite time average of X(-t)

$$\overline{X} = \lim_{n\to 0} \int_0^\infty e^{-nt} X(-t)dt \tag{22}$$

X(t) ≡ X(t,X,Y) being the solution of (18) with initial condition X(0) = X, Y(0) = Y . Calculating the average of the current

$$J_y = - \frac{e}{Ar} \dot{Y} + \frac{1}{Ar} \frac{E}{B} \tag{23}$$

we assume that the system is confined in the X direction by sufficiently high walls at X = ±L_X/2 and obtain

$$\sigma_H = - \frac{e^2}{h} \sum_v \frac{1}{Ar} \iint dXdY \frac{\partial f}{\partial X} \overline{X} \tag{24}$$

$\partial f/\partial X = (\partial f/\partial U)(\partial U/\partial X)$. In the limit T = 0

$$\frac{\partial f}{\partial U} = -\delta(\varepsilon_\nu + U(X,Y) - \varepsilon_F) \tag{25}$$

As long as the equipotential lines defined by $\varepsilon_F = \varepsilon_\nu + U(X,Y)$ are well separated we can perform the integration with respect to X and Y by integrating in tangential and normal directions with respect to the lines

$$\iint dXdY = \int ds \int \frac{dU}{|\text{grad } U|} \, , \tag{26}$$

$$\sigma_H = \frac{e^2}{h} \sum_\nu \frac{1}{Ar} \int ds \; n_x(s) \; \overline{X}(s,\nu) \tag{27}$$

and similarly

$$n = \frac{eB}{h} \sum_\nu \frac{1}{Ar} \int ds \; n_x(s) \; X(s,\nu) \tag{28}$$

Here ds is the line element and $\underline{n} = \text{grad } U/|\text{grad } U|$ the unit vector normal to the line. Eq. (24) is the high field classical limit of the corresponding Kubo formula /18/. For an alternative treatment of the high field classical limit (considering a non-confined ergodic system) cf. Ono /5b/. In the present model $\sigma = 0$.

Discussion:

1. Closed equipotential lines do not contribute to σ_H since for such lines $\overline{X}$ is independent of X and $\oint ds\underline{n}(s) = 0$. But they do contribute to n (i.e. they can be occupied by electrons). This is exactly what we required.

2. According to classical percolation theory open lines (percolating from edge to edge) in the bulk can only exist at $U(X,Y) = 0$ (large Ar and $<U(X,Y)> = 0$ is assumed). Since $\varepsilon_F = \varepsilon_\nu + U(X,Y)$, bulk percolation occurs only at the centers of the energy bands (the width of which is $U_{max} - |U_{min}|$).

3. Due to the confinement potential, open equipotential lines exist at the edges $X = \pm L_x/2$ at any value $\varepsilon_F \geqslant \varepsilon_0$ of the Fermi energy. These lines correspond to the skipping orbits (edge currents).

4. Since for the open equipotential lines at the edges $\overline{X} = X = \pm L_x/2$, $n_x = \pm 1$ and $\int ds = L_y$, (28) yields

$$\sigma_H = i\frac{e^2}{h} \, , \quad i = \sum_\nu \Theta(\varepsilon_F - \varepsilon_\nu) \tag{29}$$

i.e. the ideal quantization.

5. The open equipotential lines at the edges are topologically different from the closed ones in the bulk and cannot be converted into closed lines by any local perturbation (topological stability of the quantization).

6. Finite temperatures lead to finite transition regimes between the plateaux.

7. For free electrons X is a conserved quantity and, therefore, $\overline{X} = X$. Combining (27) and (28) we get again (3).

The high field classical limit analysis presented here should only be considered as a first step of a satisfactory theory. The task is to take into account higher order effects in 1, finite temperatures and processes leading to finite σ in the transition regimes. It is very well possible that the linear response theory (Kubo formula) is not the appropriate approach to the QHE and, instead, quantum percolation in the presence of a local electric field has to be investigated /5a/.

4. Some Remarks on Other Approaches

4.1. A convincing approach to the IQHE from the theoretical point of view is the field theory developed by Pruisken et al. /8/. They consider non-interacting electrons in a white noise random potential. The average with respect to the potential fluctuation is worked out by converting the problem into an equivalent field theory. They prove that (in the bulk) all states are localized except those at the center of the Landau bands and σ_H takes the quantized values (2). This quantization is due to topological structures in the field theory which are intimately related to persistent edge currents. The treatment of the transition regimes is on less solid ground.

4.2. Topological quantisation of σ_H in units of e^2/h has also been demonstrated by Niu et al. and by Avron and Seiler /9/. Referring to the actual two current loop Hall measurement, the last mentioned authors consider an arbitrary Hamiltonian $H(\phi_1, \phi_2)$ depending periodically on two fluxes. Assuming only that the ground state of H is non-degenerate and separated from the rest of the spectrum by an arbitrarily small gap (finite size effect) they were able to prove that at $T = 0$ the Kubo Hall conductivity averaged over one period of the fluxes is quantized in units of e^2/h and, in these units, $\Delta\sigma_H = 1$. However, Avron and Seiler's theory does not provide a complete understanding of the observed QHE since it leaves the widths and the succession of the plateaux (on the n or B scale) completely open. On the other hand, in this theory the Coulomb interaction between the electrons is not excluded. For non-interacting electrons in a random potential the assumptions concerning the spectrum are satisfied if ε_F lies in a mobility gap.

4.3. Scattering theory can hardly explain localization. But it allows one to demonstrate that at integer fillings the Hall conductivity of a disturbed system exactly coincides with that of free electrons /14/. This can be proved elegantly by making use of Levinson's theorem. The physical reason for this phenomenon is that in the limit of a weak electric field and a strong magnetic field, $eEL_x \ll \hbar\omega_c$, the momentum p_y perpendicular to the electric field cannot be changed in an elastic scattering process:

$$\overline{\varepsilon}_\alpha = \hbar\omega_c(\nu + \tfrac{1}{2}) - \frac{E}{B}p_y \rightarrow \overline{\varepsilon}_\alpha' = \varepsilon_\alpha \text{ implies } p_y' = p_y \tag{30}$$

As long as ε_F lies in an energy gap, no diffusion in the direction of the electric field is possible. Consequently J_y is conserved. Starting with a free electron initial state outside the scattering regime $J_y = i(e^2/h)E$, the final state (again outside that regime) has to carry the same amount of current - even if a part of the electrons are adiabatically transferred into bound states.

5. Resumé

From several indications one may conclude that the IQHE is a topological phenomenon connected with localization at $\varepsilon_F \neq \varepsilon_\nu$ brought about by a random potential and delocalization at $\varepsilon_F = \varepsilon_\nu$ caused by the high magnetic field. I am confident that the various current approaches which at first glance seem different will, within the next few years, merge and form a reliable and practical theory.

References

1. K. von Klitzing, G. Dorda and M. Pepper: Phys. Rev. Lett. 45, 494 (1980)
2. H. Aoki and T. Ando: Solid State Comm. 38, 1079 (1981)
3. M. Tsukada: J. Phys. Soc. Japan 41, 1466 (1976)
4. T. Ando: J. Phys. Soc. Japan 52, 1740 (1983), 53 , 3101, 3126 (1984);
   H. Aoki and T. Ando: Phys. Rev. Lett. 54, 831 (1984); L. Schweizer,
   B. Kramer and A. MacKinnon: J. Phys. C17, 4111 (1984), Z. Phys.
   B59, 379 (1985); Proc. LITPIM Suppl. (PTB-PG-1 Braunschweig 1984).
5a. D. Tsui and S.J. Allen: Phys. Rev. B24, 4082 (1981), S.V. Iordansky:
   Solid State Comm. 48, 1 (1982), R.F. Kazarinov and S. Luryi: Phys.
   Rev. B25, 7626 (1982), S. Luryi and R.F. Kazarinov: Phys. Rev. B27,
   1386 (1983), S.A. Trugman: Phys. Rev. B27, 7539 (1983), R. Joynt
   and R.E. Prange: Phys. Rev. B29, 3303 (1984)
5b. Y. Ono, in: Anderson Localization, edited by A. Nagaoka and H. Fukuyama,
   Solid State Sciences 39 (Springer, Berlin, Heidelberg, New York
   1982) p. 207, S.M. Apenko, Yu. E. Lozovik: J. Phys. C18, 1197
   (1985), JETP (russ.) 89, 573 (1985)
6. R.B. Laughlin: Phys. Rev. B23, 5632 (1981), in: Two-Dimensional Systems,
   Heterostructures, and Superlattices, edited by G. Bauer, F. Kuchar,
   and H. Heinrich, Solid-State Sciences 53 (Springer, Berlin, Heidelberg, New York, Tokyo 1984)
7. Y. Ono: J. Phys. Soc. Japan (Suppl.) 52, 247 (1983), 53, 2342 (1984)
8. A.M.M. Pruisken: Nuclear Phys. B235, 271 (1984), H.Levine, S.B. Libby,
   and A.M.M. Pruisken: Phys. Rev. Lett 51, 1915 (1983), Nuclear Phys.
   B240, 30,49,71 (1984), A.M.M. Pruisken: LITPIM (Braunschweig 1984)
9. Q. Niu and D.J. Thouless: J. Phys. A17, 2453 (1984), Q. Niu, D.J. Thouless and Y.-S. Wu: Phys. Rev. B31, 3372 (1985), J. Avron and R. Seiler, Phys. Rev. Lett. 54, 259 (1985), JGP 1, 13 (1984)
10. P. Streda: J. Phys. C15, L717 (1982), Cf. also L. Smrčka, J. Phys. C17,
   L63 (1984)
11. T. Ando: J. Phys. Soc. Japan 37, 622, 1233 (1974), R. Gerhardts: Z. Phys.
   B21, 275, 285 (1975), Surf. Sci. 58, 227 (1976), F. Wegner: Z. Phys.
   B51, 279 (1983)
12. Cf. E. Gornik and D. Weiss, K. von Klitzing and V. Mosser in this volume.
13. R. Gerhardts, private communication
14. R.E. Prange: Phys. Rev. B23, 4802 (1981), R.E. Prange and R. Joynt: Phys.
   Rev. B29, 3303 (1984), J. Chalker: J. Phys. C16, 4297 (1983), Surf.
   Sci. 142, 182 (1984), W. Brenig: Z. Phys. B50, 305 (1983), W. Brenig
   and K. Wysokiński, Preprint. Cf. also E.B. Hansen: Physica 123B,
   183 (1984)
15. J. Kosch, U. Gummich and J. Hajdu: Z. Phys. 62, 295 (1980)
16. A.H. MacDonald, T.M. Rice and W.F. Brinkman: Phys. Rev. B28, 3648
   (1983), O. Heinonen and P.L. Taylor: Phys. Rev. B28, 6119 (1983), J.
   Reiss, Phys. Rev. B31, 8265 (1985), W. Maass and U. Krey: Europhys.
   Lett. in press
17. S. Luryi: Preprint
18. J. Hajdu, U. Gummich: Acta Phys. Pol. in press

The Fractional Quantum Hall Effect

M. Jonson

Institute of Theoretical Physics, Chalmers University of Technology,
S-41296 Göteborg, Sweden

There is by now a rather well established *standard picture* of the *fractional quantum Hall effect* . The theoretical background of this picture is reviewed here. The basis is Laughlin's conjecture for the many-body wavefunction describing the "1/m-state" and the hierarchy of states obtained from these by adding "quasi-hole" and "quasi-electron" excitations. We shall in particular describe how the standard theory relates to three experimental facts: (1) certain electron densities corresponding to a fractional filling of the lowest Landau level are "special", (2) the excitations from the corresponding ground states have a *finite* energy, hence dissipationless flow is possible at these densities, (3) the Hall steps have a "finite width".

The low-temperature Hall resistance for two-dimensional electron systems in strong magnetic fields is quantized in units of (h/e^2). This extraordinary fact, known as the quantum Hall effect (QHE), was first discovered by Klaus von Klitzing and won him the Nobel Prize for physics in 1985.

The QHE appears as plateaus or "steps" when the Hall resistance is traced as a function of electron density or magnetic field. In the original experiment the measured Hall resistance at the steps was found to be very close to (h/ie^2), with (i) an integer. This is the *integer quantum Hall effect* (IQHE). Later, Hall steps were found corresponding to (i) being a rational number, a fraction. This is the *fractional quantum Hall effect* (FQHE). The steps in the trace of the Hall resistance are associated with minima in the trace of the ordinary resistance.

Although similar in appearance, the physics behind the FQHE is different from the physics of the IQHE. The IQHE is essentially a single - particle phenomenon caused by quantization of the electron orbits in a magnetic field (and hence quantization of the kinetic energy into Landau levels) and by electron localization due to disorder. The main features of the IQHE ar quite well understood [1] and the quantization of the macroscopic Hall resistance can be elegantly expressed in terms of topology. The remaining problems in the IQHE, such as those related to the precise width and shape of the Hall steps,probably have to wait for a better understanding of the localization problem in the presence of a magnetic field. This is a hard problem.

In contrast to the integer effect, the FQHE is a subtle many-particle effect which is caused by electron-electron interactions.

The scope of this lecture is firstly to describe in rather qualitative terms how the FQHE might come about. Secondly we will review the current "standard picture" of the FQHE. This picture is based on Laughlin's ingenious conjecture for the many-body wavefunction describing the "1/m ground state" (i=1/m) [2] and the formation of a hierarchy of fractional states by adding "quasi-electron" and "quasi-hole" excitations to these states as first suggested by Haldane [3].

Our discussion of the FQHE will necessarily be rather incomplete. For a recent extensive review of the QHE see for instance the proceedings of a series of lectures given at the University of Maryland in the fall of 1985 [4]. In passing we note that, although rather well established, the standard theory is not unchallenged [5]. The implications of the alternative theories and their relation to the standard picture seem, however, unclear at present and we shall not discuss them here.

1 Brief Account of Experimental Results

Before we turn to the theory, let us review som experimental facts. In the first experiment [6] steps in the Hall resistance with $R_H = h/(e^2 i)$, i=1/3 and 2/3 was reported (actually in the 2/3 case only a dip in the longitudinal resistance and a slight kink in the Hall resistance was seen). This result may be interpreted using the free electron formula

$$R_H = B/nec \quad ; \qquad n = \nu \, (eB/hc) \tag{1}$$

where n is the electron density, (eB/hc) is the degeneracy factor of a Landau level and hence (1) defines the *filling factor* ν . One concludes that the plateaus correspond to a *fractional filling* of the lowest Landau level, ν = 1/3 and 2/3.

Subsequently, quantization of the Hall resistance was reported for other fractional fillings, viz. ν=p/q with q=5, 7 and possibly 9 (only dips in the longitudinal resistance) (see for instance ref. [7]). Fractional quantization in higher Landau levels has also been reported, i.e ν=1+p/q with q=3,5,7 for the higher spin state of the lowest Landau level [8]. In all these cases q has been an odd integer. For the higher Landau level quantization for n=2+p/q with q even has been reported [8]. This result is not understood and cannot readily be explained by the theories reviewed in this lecture. In what follows we will only deal with the lowest spin state of the lowest Landau level, 0<ν<1.

The experimental results can be summarized as follows:

(i) Hall steps are seen at parameters corresponding to fractionally filled Landau levels, ν=p/q, where p is an even or odd integer but where q is always odd (at least in the lowest Landau level).

(ii) In the region of magnetic fields corresponding to the Hall steps the longitudinal resistance is close to zero, $R_{xx} \approx 0$. At low finite temperatures the resistance is activated $R_{xx} \propto \exp(-\Delta/2k_BT)$.

(iii) Small deviations in magnetic field do not affect the Hall resistance but large ones do (i.e. steps have a finite width)
(iv) The FQHE (in the Hall resistance) disappears at $T \approx 1$ K (at B=15 Tesla).

The FQHE occurs in high mobility samples. (Samples used in the early experiments had a mobility of 100 000 [cm^2/Vs] corresponding to a level broadening for GaAs of 0.1 Kelvin, in recent experiments [10] the mobility has been as high as 2 000 000 [cm^2/Vs])
It is interesting to compare the behaviour in high and low mobility samples:

High mobility samples:	Weak IQHE with narrow steps, FQHE readily seen
Low mobility samples:	Strong IQHE with wide steps, FQHE hard to see

From these observations it is not surprising that the physical mechanism behind the integer and the fractional QHE turns out *not* to be the same. The two effects are indeed similar in that the observed Hall steps depend on the formation of a mobility gap, as signalled by the activated behaviour of the ordinary resistance. As we stated before this is a single-particle effect in the IQHE. We shall now argue that it is a many-body effect in the FQHE.

2 Towards a Theoretical Understanding: A Qualitative Argument

For a start let us consider a qualitative argument for why the FQHE might occur.
Suppose the motion of two charged particles is described by classical physics. Due to their mutual Coulomb interaction they repel each other, hence in the absence of a magnetic field they would fly apart. If a magnetic field is present, however, the particles experience a Lorentz force that results in a motion that is perpendicular to both the Coulomb force and the magnetic field. It follows that the two charged particles will circle around each other with *the total angular momentum conserved* during the motion.

Now, let the motion of the two particles be governed not by classical physics but by *quantum mechanics* . In quantum mechanics courses we learn that angular momentum is quantized. In our two-particle system it is easy to see how angular momentum is related to density (distance between the particles). Large distance (low density) corresponds to large angular momentum. This relation holds also for many particles, and hence *there is a relation between electron density and angular momentum*, the latter being quantized. This is the key to understanding the FQHE where quantization of the macroscopic Hall resistance happpens at discrete, "special" values of density. Let us now turn to a somewhat more quantitative description and first study *one* electron in a magnetic field.

3 Quantum Mechanics for One Electron in a Magnetic Field

Consider the Schrödinger equation for one electron in a magnetic field

$$H_0 = (1/2m) [p + eA/c]^2 . \qquad (2)$$

We are free to chose the most convenient gauge for the vector potential, A, as long as the correct applied magnetic field B obtains from the relation $B = \nabla \times A$. For discussions of the IQHE one often uses the Landau gauge, $A = B(0,x,0)$, which is a natural choice considering the boundary conditions appropriate to a rectangular (Hall bar) geometry. For a discussion of the FQHE we shall find it more convenient to use the symmetric gauge, $A = (B/2)(-y,x,0)$.

When the Schrödinger equation is solved, a specific gauge corresponds to a particular wavefunction, but measurable quantities, such as electric currents and the magnetic field, do not depend on gauge. The reason for the choice of the symmetric gauge here is that the corresponding one-electron solutions to the Schrödinger equation are *eigenfunctions not only of the Hamiltonian but also of the angular momentum operator*, and we have argued above that the angular momentum plays a cruical role in the FQHE.

The solutions to the Schrödinger equation $H_0\Phi = E\Phi$, with H_0 given by (2) and the symmetric gauge can be labelled by quantum numbers n,m:

$$\Phi_{nm}(x,y) \propto \exp[(x^2+y^2)/4a_0^2] \, (\partial_x - i\partial_y)^m \, (\partial_x + i\partial_y)^n \, \exp[-(x^2+y^2)/2a_0^2] . \qquad (3)$$

The length scale is set by a_0, the magnetic length ($a_0^2 = hc/2\pi eB$). The angular momentum operator is $L = r \times p = -i(h/2\pi)r \times \nabla$ and hence $L_z = -i(h/2\pi)(x\partial_y - y\partial_x)$. It is easy to verify that the wavefunction Φ_{mn} is an eigenfunction of both H_0 and L_z :

$$H_0\Phi_{mn} = (h\omega_c/2\pi) (n+1/2) \Phi_{mn} \qquad (4)$$

$$L_z\Phi_{mn} = i(h/2\pi)(m-n) \Phi_{mn} \qquad (5)$$

We recognize n as the Landau level index, while m is related to the angular momentum. For convenience we restrict ourselves to the lowest Landau level (n=0) and describe the position of an electron by the complex variable z = x - iy.

The wavefunction (3) reduces to

$$\Phi_m(z) \propto z^m \exp(-|z|^2/4a_o^2) \tag{6}$$

To establish the connection mentioned above between "electron density" and angular momentum, we note that the expectation value of r^2, the area of the electron orbit, is related to m via

$$<m|r^2|m> = 2(m+1)a_o^2 \tag{7}$$

We are now ready to take the important qualitative step of considering *two* interacting electrons.

4 Two Interacting Electrons in a Magnetic Field

The Hamiltonian for a system of two interacting electrons is

$$H = (1/2m)[\ \mathbf{p}_1 + e\mathbf{A}_1/c]^2 + (1/2m)[\ \mathbf{p}_2 + e\mathbf{A}_2/c]^2 + e^2/|\mathbf{r}_1 - \mathbf{r}_2| \tag{8}$$

It is important to notice that the Hamiltonian conserves the total angular momentum (easy to check). Hence the total angular momentum of the interacting system is a good quantum number. It then makes sense to separate the internal part from the center of mass part of (8) and try the following approximate two-particle wavefunction for the internal portion of the interacting two-particle system.

$$\Phi(z_1, z_2) = (z_1 - z_2)^m \exp[(|z_1|^2 + |z_2|^2)/4a_o^2] \tag{9}$$

This wavefunction closely resembles the free electron wavefunction with the relative coordinate as argument ($z = z_1 - z_2$). The wavefunction has the necessary antisymmetric property of a Fermion system *if m is odd.* Note also that (9) contains correlations between the positions of the electrons, $|\Phi|^2 \rightarrow 0$ if $|z_1 - z_2| \rightarrow 0$.

The wavefunction (9), which is an eigenfunction of the total angular momentum as well as the Hamiltonian, was studied by Laughlin [9] who found that it was a good approximation to the exact wavefunction (which of course can be obtained for a two-particle system). The trial wavefunction gave an error in the energy of about 7% and in the wavefunction (using some measure) of roughly 20%. Clearly we are on the right track. Let's go on to three particles.

5 Three Interacting Electrons in a Magnetic Field.

In going from two to three particles there is again a qualitative step in the sense that three particles define a plane and enclose an area and a certain amount of magnetic flux. This system was also studied by Laughlin in his early work [9]. For the three-particle system we have to use two complex relative coordinates z_a and z_b to describe the internal motion. It is possible two construct a complete orthonormal set of antisymmetric two-particle wavefunctions of the form

$$|m,n> = f(m,n,z_a,z_b) \exp[(|z_a|^2 + |z_b|^2)/4a_0^2]$$ (10)

where f is a polynomial of order M (the total angular momentum$\cdot(2\pi/h)$) and $M=3m+n$ (in the particular set of wavefunctions used in (9)). Recall that we have restricted ourselves to working in the lowest Landau level (which is an approximation that becomes exact in the limit of infinite magnetic field strength), hence n is *not* a Landau level index here.

If we now calculate the matrix elements of the interaction part of the Hamiltonian, H_{int}, we get a set of values $V_{mn,m'n'}$ where

$$V_{mn,m'n'} = <m,n|H_{int}|m'n'>$$ (11)

We have already argued that H_{int} conserves the total angular momentum. Hence only matrix elements where $M=3m+2n=3m'+2n'$ are nonzero. To calculate the energy as a function of total angular momentum (related to density) we have to diagonalize the hamiltonian for each M. The gross features of the result are then that we get a smooth variation of energy with M (or density). However for certain values of the total angular momentum, M, the energy tends to be lower.

In our three-particle system we can take the specific example of M=9. As $M=3m+2n$ this value can be realized with two sets of (m,n), viz. (3,0) and (1,3). The M=9 state which in the absence of interactions is two-fold degenerate is split by the interaction, and it turns out that the lowest energy level corresponds to (m,n)=(3,0). The angular mometum of each electron is *not* a good quantum number, but from the explicit form of the wavefunction [9] one notes that (3,0) corresponds to the angular momentum, being "on the average" evenly distributed among the thre electrons. In general one finds that when M (the total angular momentum) is a multiple of the number of electrons (i.e. three in this case) the energy is particularily low.

The method of explicit diagonalization of the Hamiltonian can be extended to a somewhat larger (but still small) number of electrons (about 10). A particularily useful method for such calculations have been deviced by Haldane (see below).

6 Many Interacting Electrons: Laughlin's Wavefunction

By looking at a small number of interacting electrons we have tried to argue that electron-electron interactions are responsible for the FQHE. We have still to describe a theory for an infinite (or at least very large) number of interacting particles. Such a theory has been proposed by Laughlin [2].

Lauglin's theory for the FQHE was inspired by the theory of another quantum fluid, ^{4}He, where wavefunctions with explicitly built-in correlations are used (Jastrow functions). Guided by his studies of two and three particles described above he proposed a ground-state wavefunction of the form

$$\Psi = \prod_{i<j} f(z_i - z_j) \, \exp[- \sum_k |z_k|^2/4a_0^2] \tag{12}$$

to describe the system having the Hamiltonian

$$H = (1/2m) \sum_i \{ [p_i + eA_i/c]^2 + V(r_i) \} + \sum_{i<j} (e^2/|r_i - r_j|) \tag{13}$$

Several important observations restrict the form of the function $f(z)$ in (12)

(1) with all electrons in the lowest Landau level $f(z)$ must be analytic in z (i.e. $f(z)$ must be a polynomial in z and cannot contain for instance z^* (cf. the form of the single-particle wavefunctions which form a complete basis set)

(2) Ψ has to be antisymmetric, hence $f(z)$ must be an odd function of z

(3) Because the Coulomb interaction conserves angular momentum it is required that $\prod f(z_i - z_j)$ is a *homogeneous* polynomial.

The simplest form of $f(z)$ that conforms with these restrictions is $f(z) = z^m$ with m an odd integer. This form was indeed chosen by Laughlin and his Ansatz for the wavefunction can hence be written

$$\Psi_m = \prod_{i<j} (z_i - z_j)^m \, \exp[- \sum_k |z_k|^2/4a_0^2] \tag{14}$$

It is now necesssary to calculate the ground-state energy from this wavefunction. This is a non-trivial calculation. The energy should be lower than the energy associated with any competing wavefunction that might be considered (corresponding perhaps to a different state of matter like a Wigner lattice). Even so, to explain the FQHE it is also necessary to show that there is a gap in the excitation spectrum. To achieve these objectives Laughlin discovered a very useful and beautiful analogy between the quantum fluid we are discussing and a different system - the one-component *classical* plasma of charged particles interacting via a logarithmic potential. The latter system is well studied and a lot of numerical results and methods of calculation are available.

The trick involved in seeing the analogy is to write the square of the many-body wave-function in the form

$$|\Psi_m(z_1 ... z_N)|^2 = \exp[-\beta \Phi (z_1 ... z_N)] \tag{15}$$

The left-hand side of (15) is a measure of the probability that the N electrons of the quantum-fluid has a certain configuration corresponding to the wavefunction $\Psi(z_1 \ldots z_N)$. The right-hand side resembles a Boltzmann factor, which gives the probability for a configuration of N classical particles with potential energy $\Phi(z_1 \ldots z_N)$. The most probable configuration minimizes Φ and hence maximizes the Boltzmann factor *and* $|\Psi_m|^2$. If we know the configuration for the classical system, in particular if we know, say, the pair correlation function $g(|z_1-z_2|)$, then we also know the pair correlation function for the quantum system as it is exactly the same. We can then readily calculate the ground-state energy. But first we need to know Φ. It is convenient to use dimensionless units where the units of charge, length and energy is respectively e, a_0 and e^2/a_0. If we let $\beta=1/m$ then

$$\Phi = -\sum_{i<j} 2m^2 \ln|z_i-z_j| + (1/2)m^2 \cdot (1/2\pi m)\, 2\pi \sum_k |z_k|^2 + \text{constant terms} \tag{16}$$

The first term in (16) represents the repulsive (logarithmic) interaction between particles of charge (-m). The second term comes from the attractive (logarithmic) interaction between particles of charge (-m) and a uniform positive background of positive charge m of density $1/2\pi m$. This is nothing but a particular member of a family of one-component classical plasmas which have been studied extensively. In particular we conclude from (16) that Laughlin's wavefunction, Ψ_m, describes a *liquid*.

The most likely configuration of the classical system minimizes the potential energy. But clearly Φ is minimized when the positive background density $1/2\pi m$ equals the charge density that generates the interaction part of the Hamiltonian (13), i.e to the average density of the particles which is $\nu/2\pi$ in dimensionless units. It follows that *m is related to the filling factor ν* as m = $1/\nu$.

From a knowledge of the spatial distribution of classical particles of charge (-m) we hence also know the spatial distribution of quantum particles of charge (-1). By integrating over all particle coordinates except those of two electrons we get the pair correlation function, $g(|z_1-z_2|=r)$, of the quantum system,and so we can calculate the ground-state energy as (dimensionless units again):

$$E = (\nu/2\pi) \int dr\, [g(r)-1] . \tag{17}$$

To obtain the pair correlation function from the wavefunction is not trivial. Laughlin originally used the so-called hypernetted chain approximation. More reliable numerical calculations were done later [10]. Laughlin's ground state (14) indeed seems to have lower energy than competing states except for small filling factors. For $\nu < 1/7$ a Wigner lattice is thought to be favoured energetically [11].

7 Accuracy of the Laughlin Wavefunction

One believes that Laughlin's wavefunction, Ψ_m, describes the true $1/m$ ground state quite accurately, at least for small m. The most convincing evidence is Haldane's numerical comparison of ground state properties derived from the exact ground state wavefunction, Ψ, and Laughlin's wavefunction, Ψ_m, for few electron systems [12]. By mapping the circular, disc-like geometry implicitly assumed in Laughlin's analysis onto a sphere, he was able, as it were, to reduce the boundary "edge" of the disc to a boundary "point" on the sphere, hence minimizing the effects of boundaries. Calculated ground-state properties for few (<10) electron systems have a surprisingly weak dependence on the number of electrons and can be reliably extrapolated to a large system. The extrapolated ground state energies, for example, compare well with the energies obtained from Laughlin's Ψ_m. Perhaps more significant is the comparison between the exact and approximate ground state for a few electron system. Laughlin's Ψ_m (adopted to spherical geometry) gives a ground state energy with a relative error of $1:10^3$ in the $1/3$ state (m=3) for a system of six particles. Other quantities like the pair correlation function also come out in excellent agreement with the exact result for few particles (and small m).

Note that the interaction potential plays no *quantitative* role in the arguments leading to Ψ_m in (14). Obviously the exact wavefunction must depend on the form of the potential to some extent. One can prove that Ψ_m is exact in the limit of a shortrange potential [12]. Laughlin's wavefunction is a less good approximation when the range of the potential is comparable with or larger than the size of the "exchange-correlation hole" (the region where $g(r)<1$).

8 Why is the Laughlin Wavefunction so Accurate?

How can we understand that Laughlin's wavefunction, Ψ_m, is so accurate? The following argument is due to Halperin [13]: Fix the positions z_j of all electrons *except one* and consider the (exact) wavefunction, as function of the position z_1 of the remaining electron, $\Psi = \Psi(z_1)$. Then move this electron around in a closed loop enclosing an area Ω (avoiding the other electrons). From quite general arguments the phase of Ψ has to change by a factor $\Delta\Phi = 2\pi\Phi/\Phi_0$ to avoid having an enormous kinetic energy for the electron. In the expression for $\Delta\Phi$, $\Phi = B\cdot A$ is the enclosed magnetic flux and hence there must be (Φ/Φ_0) elementary flux lines or vortices in Ω, each carrying a flux quantum $\Phi_0 = hc/e$ (cf. the theory of type II superconductors!). At each vortex the wavefunction vanish, which costs kinetic energy (as $\nabla\Psi$ is nonzero).

Each vortex contributes a phase change 2π if encircled by our travelling electron. In the enclosed area there are $N_\Omega = v(eB/hc)\cdot\Omega = (1/m)(\Phi/\Phi_o)$ electrons. Hence the number of vortices equals m times the number of electrons, $\Phi/\Phi_o = mN_\Omega$.

Because of the Pauli principle $\Psi(z_1)$ vanish if z_1 coincides with the position of any other electron, hence at least one vortex (and a corresponding cost in energy) is associated with each electron. Laughlin's wavefunction corresponds to distributing the vortices in the most economical way, *putting m vortices at each of the N_Ω electron sites!*

9   An Incompressible fluid: The Excitation Spectrum has a Gap

In order to explain the FQHE it is also necessary to explain why the Hall current can flow without dissipation. This is presumably related to the existence of a gap in the excitation spectrum of the system. Now with Halperin's picture of Laughlin's wavefunction, Ψ_m, in mind a natural way of "kicking" the electron system in the gentlest possible manner, is to add two (well-separated) magnetic vortices of different sign (so as not to change the magnetic field). We know what adding a positive (negative) vortex at the origin would do to the single electron wavefunctions (6), viz. it would increase (decrease) the power of the factor z^m by 1, $z^m \rightarrow z^{m+1}$ (z^{m-1}). Bearing this in mind, Laughlin's trial wavefunctions

$$\Psi_m^+ = \prod_k \{z_k - z_o\}\, \Psi_m\, ; \qquad \Psi_m^- = \prod_k \{2\partial/\partial z_k - z_o\}\, \Psi_m \tag{18}$$

for "quasi-holes" and "quasi-electrons" respectively, seem reasonable.

To show that Ψ_m^+ corresponds to a quasihole, consider the absolute value of the square:

$$|\Psi_m^+|^2 = \exp[-\beta\Phi'] \tag{19}$$

where again (in dimensionless units) $\beta = 1/m$ and

$$\Phi' = -\sum_{i<j} 2m^2 \ln|z_i - z_j| + (m/2)[m\cdot(1/m)]\cdot \sum_k |z_k|^2 - \sum_i 2m\cdot 1 \ln|z_i - z_o| \tag{20}$$

This describes the same classical plasma as earlier, except for an additional phantom particle of "charge" -1 located at z_o. In screening this phantom particle the plasma accumulates a charge of +1. In the quantum fluid the particles that move in to screen out the "charge" at z_o have charge -1 rather than -m. Hence the screening charge is +1/m (in units of e) which is the charge then of the quasihole.

The electrostatic energy associated with the quasi-particle formation must roughly scale as the square of their charge, i.e. as $(1/m)^2$. Hence the energy $\Delta_+ + \Delta_-$ required to create a quasielectron-hole pair is finite. The energy gap is larger for smaller m, in accordance with the experimental fact that it is easier to detect the FQHE for $v = 1/m$ with m small.

Are there perhaps other type of excitations with lower energy? In an ordinary electron gas we have not only electron-hole excitations of the kind discussed above but also collective excitations - plasmons. What collective excitations are there in the present system?

In many-body physics a standard technique (linear response theory) is to express the excitation spectrum of a system in terms of its ground-state properties. A famous example is Feynman's theory of the excitation spectrum of ^{4}He. He argues that the low-energy excitations are density waves and constructs a wavefunction ϕ_k for the excited state by letting the density operator ρ_k act on the ground state $|0\rangle$: $\phi_k = \rho_k|0\rangle$. The excitation energy can be written as

$$\Delta(k) = \langle \phi_K|H-E_0| \phi_K\rangle / \langle \phi_K| \phi_K\rangle = f(k)/s(k) \qquad (21)$$

where $f(k) = h^2k^2/2m$ is the single-particle energy and $s(k)$ is the static form factor which can be obtained from neutron scattering experiments. As function of wavevector $\Delta(k)$ starts out from zero and has the famous roton (local) minimum at a finite wavevector k_0.

In a beautiful analogy with Feynman's theory for ^{4}He Girvin et al [14] have calculated the excitation spectrum of Laughlin's incompressible fluid. The essential technical difference is that they project all operators (like the density operator) onto the lowest Landau level to suppress inter-Landau level excitations. Also, instead of taking $s(k)$ from experiments, they have to obtain it directly from Ψ_m. The result is a spectrum $\Delta(k)$ that starts out at a *finite* energy at k=0 and has a a "magneto-roton" minimum at finite k_0. It is furthermore conjectured that $\Delta(\infty)$ corresponds to the quasiparticle-hole excitation discussed above, $\Delta(\infty) = \Delta_+ + \Delta_-$. The numerical value for $\Delta(\infty) \approx 0.1$ in dimensionless units for the 1/3 state. This is larger than the experimentally determined gap. The discrepancy is smaller, but still significant, if the calculation is redone with the Coulomb potential replaced by a more realistic potential. Impurity scattering is expected to close the gap and account for the remaining error.

10   A Hierarchy of Fractional States

We have so far only talked about the special densities $v = 1/m$. What about other fractions? To see what happens, suppose the electron density and magnetic field, B, correspond to $v=1/m$ and we have $1/m$ state. Then decrease the magnetic field. The energy of the state scales with B and is lowered, but fewer electrons can now be accomodated as the degenaracy also scales with B. The excess electrons must be put in excited states - quasielectron states. These quasielectrons may, however, interact among themselves and *can themselves form a higher order Laughlin state* . (A similar argument works for quasiholes). This mechanism was proposed by Haldane [3] and generates in principle *all possible fractions.*

The hierarchical picture leads to a prediction: the FQHE can be seen at a certain fractional filling only if it is seen for fractions higher up (earlier) in the hierarchy. This seems always to have been the case so far.

11   The Effects of Disorder on the FQHE

The picture of the FQHE outlined above is not quite good enough as it does not explain the finite width of the Hall steps. No argument has furthermore been given for the exact quantization of the macroscopic Hall resistance to (h/e^2v). The effects of disorder are certainly essential for these effects, but little is known about them at least in quantitative terms. One generally relies on analogies with the IQHE. For example it is assumed that with only few quasiparticles excited, disorder would localize them and produce a finite Hall step in much the same way as in the IQHE.

Perhaps it is encouraging for the participants of this winter school to find that there are still important, fun problems to work on.

1   J. Hajdu: this volume; M. Jonson: "The Quantum Hall Effect" in
        NORDITA Spring School,Tvärminne, 1984 (NORDITA, 1985)
2   R.B. Laughlin: Phys. Rev. Lett. **50**, 1395 (1983)
3   F.D.M. Haldane: Phys. Rev. Lett. **51**, 605 (1983)
4   S.M. Girvin and R. Prange eds.: in *Solid State Sciences* (Springer Verlag 1986)
5   S. Kivelson, C. Kallin, D.P. Arovas and J.R. Schrieffer:
        Phys. Rev. Lett. **56**, 873 (1986)
6   D.C. Tsui, H.L. Störmer and A.C. Gossard: Phys. Rev. Lett. **48**, 1559 (1982)
7   A.M.Chang, P.Berglund, D.C.Tsui, H.L.Störmer and J.C.M.Hwang:
        Phys. Rev. Lett. 53, 997 (1984)
8   R.J. Nicholas: this volume; R.G. Clark, R.J. Nicholas, A. Usher, C.T. Foxon
        and J.J. Harris: EP2DS VI Proceedings, Surface Science (1986)
9   R.B. Laughlin: Phys. Rev. **B27**, 3383 (1983)
10 D.Levesque, J.J. Weiss and A.H. MacDonald: Phys. Rev. **B30**, 1056 (1985)
11 P.K. Lam and S.M. Girvin: Phys. Rev. **B30**, 473 (1984); ibid. **B31**, 613E
12 F.D.M. Haldaneand E.H. Rezayi: Phys. Rev. Lett. 54, 237 (1985)
13 B.I. Halperin: Helv. Phys. Acta **56**, 75 (1983)
14 S.M. Girvin, A.H. MacDonald and P.M. Platzmann:
        Phys. Rev. Lett. 54, 81 (1985); se also ref. 4

Part VI

New Structures and Devices

Microwave Performances
of GaAlAs/GaAs Heterostructure Devices

J. Chevrier and D. Delagebeaudeuf

Thomson Semiconducteurs Departement Arseniure de Gallium,
Rd. 128/ BP 48, F-91401 Orsay, France

1) Introduction

GaAlAs/GaAs heterostructures have recently led to devices presenting high performances at microwave frequencies. The first microwave heterostructure device now commercially available is the two-dimensional electron gas field effect transistor (TEGFET, also referred to as HEMT or MODFET) presenting extremely low noise figures up to 40 GHz [1][2][3]. Multiple stage TEGFET amplifiers have already been realized to serve as input circuit in receivers for high frequencies communications. Most of them have been described at the last GaAs ICs Conference, at Monterey, CA, in November 85 [4].

The intrinsic characteristics of this heterostructure also bring optimism to digital circuits designers : ring oscillators have already been fabricated with switching delays of 10 picosecondes at room temperature and 5.8 ps at 77 K [5].

Elsewhere, multiple channel heterostructures are now capable of high sheet carrier concentration ($5 \ 10^{12} \ cm^{-2}$ at 77 K) -[6]. First microwave results [7] indicate that the output power (0.55 W/mm) at 20 GHz is quite comparable to that obtained with GaAs MESFET developed in several laboratories.

In parallel, research on heterostructure bipolar transistor (HBT) is making progress : the wider band gap (AlGa)As is used as an emitter to inject electrons into the thin heavily doped GaAs base, leading to a high current gain. Circuits simulations suggest that HBT delays as low as 10 ps may be achievable.

This paper summarizes the noise and speed properties of TEGFETs for analog and digital applications. Some aspects of the characteristics of other heterostructures are also given.

2) High Frequencies

The 1 to 100 GHz range is divided into 10 bands (D to M) which have been progressively attributed to civilian (4/6 GHz - G-band) or military (6/8 GHz - E-band) communications. Then, as it was not sufficient, the 11/14 GHz range (in the 10/20 GHz J-band) was reserved to satellite communications (TV-DBS at 12 GHz for example) and, in 1979, a 3.5 GHz bandwith was assigned to satellite communications in the 20/30 GHz K-band because the traffic had risen at a faster rate than original predictions.

However, it has to be understood that this rise in frequency is related to the progress in device technology which permits to fabricate GaAs Microwave Integrated Circuits (MICs): these high density circuits present higher performances (lower noise, higher gain) and smaller size, then smaller weight.

It can be expected that in the near future, the new performances reached with TEGFETs will be used in more efficient circuits working at higher frequencies,in the 50-70 GHz range.

3) TEGFET as a Low Noise Transistor

As far as it is the first high-performance heterostructure device now available, it is important to understand its noise properties and its potential applications.

3-1: Heterostructure.

2DEG AlGaAs/GaAs heterostructures are usually grown by molecular beam epitaxy (MBE). By this technique,high purity material, abrupt interface and sharp control of growth thickness down to 10 Angstrœms can be obtained; however, attempts to grow such structures by metal-organic deposition (MOCVD) are becoming more and more successful [8][9].

A cross-sectional view of the usual structure for microwave applications is presented in Fig.1: it is composed of a doped GaAs cap layer and an AlGaAs layer grown on an undoped GaAs buffer layer; electrons transfer from the higher band gap (AlGaAs) material to the lower band gap (GaAs), where they are confined by the energy barrier at the interface.

Since they are separated from their parent donors and flow in the high purity sub-layer, they experience only weak ionized impurity scattering.

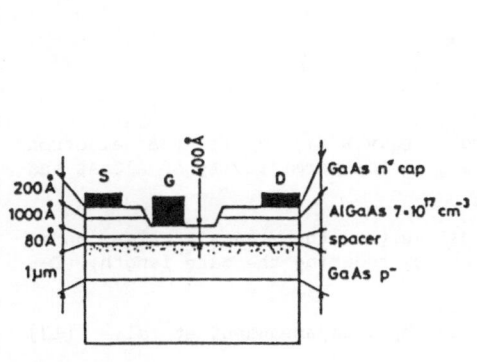

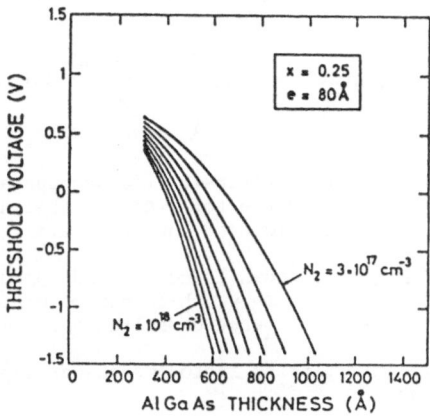

Fig.1 : Typical cross-sectional view of an heterostructure after processing for microwave operation.

Fig. 2 : Calculated threshold voltages of TEGFETs versus AlGaAs thickness for different doping concentrations N2 (x=Al content)

The carrier mobility in the channel (8000 cm^2/V.s) is higher than in a conventional GaAs MESFET (5000 cm^2/V.s), especially at low temperatures (enhanced mobility). Values ranging from 5 10^5 to 1 10^6 compared to 1.5 10^5 are measured at 4.2 K .

The performances of the transistor will depend on some growth parameters (like the doping level 'N$_2$'or the thickness 'd' of the AlGaAs layer,...) which can be selected according to the well-known analytical model first developed by Delagebeaudeuf and Linh [10],[11]. For example, the theshold voltage V$_T$ of the transistor which is one of the key parameter to be controlled can be "a priori" evaluated from the curves drawn in Fig.2.

3-2: TEGFET

It consists of two Au/Ge-Ni ohmic contacts deposited on the n$^+$ layer to form source and drain contacts .This n$^+$ doped GaAs layer permits to reduce the access resistance (R$_S$) between the source and the gate ; then, the gate zone is defined by electronic lithography which allows the definition of 0.25/0.3 microns line width (referred as gate length L$_G$) inside a 2.5/3 microns of source-drain spacing; let us note that after metallisation, the yield of such lines over a 50 or 75 mm diameter wafer can be poor: so, it is advisable to work with 0.5 µm gate length; moreover, this leads to a lower gate resistance R$_G$. Then , the ungated source-drain current I$_{DS}$ is adjusted to a pre-determined value by recessing the gate zone (chemical etching). A Ti-Pt-Au Schottky gate is then deposited and annealed at 380 C.

3-3: Noise parameters

For microwave evaluation,TEGFET chips are bonded onto carriers which are mounted in a test fixture.
The noise figure which is usually reported in the literature is the minimum value measured after correct bias and adaptation of the transistor; this minimum is found at an optimal source-drain current I$_{opt}$ (for example, 4-8 mA for TEGFET (fig.3), 7-12 mA for MESFET, for a same geometry 0.5μm x 150μm); at the operating frequency "f", this minimum value is usually well approximated by the FUKUI equation [12] :

$$F_{min}(dB) = 10\log \left\{ 1 + \frac{2\pi C_{GS}}{g_{mo}}.f.K_F.((R_S+R_G)g_{mo})^{1/2} \right\}$$

- g$_{mo}$ is the intrinsic transconductance (ε_s.v.W/d); v is the electron velocity, d the total AlGaAs thickness, ε_s the permittivity of AlGaAs and W the gate width of the transistor;

- C$_{GS}$/g$_{mo}$ is known to correspond approximatively to the transit time under the gate, or L$_G$/v, which can be minimized by reducing the gate length; (C$_{GS}$ is the gate-source capacitance);

- K$_F$ is a fitting factor which was expressed by Delagebeaudeuf et al. [13] as:

$$K_F = 2.[I_{opt}/g_{mo}.Ec.L_G]^{1/2}$$

where Ec is the critical field ($\sim$3.5 10^5 V/m) for which the electron velocity becomes saturated ($\sim$1.3 10^5 m/s). This original expression

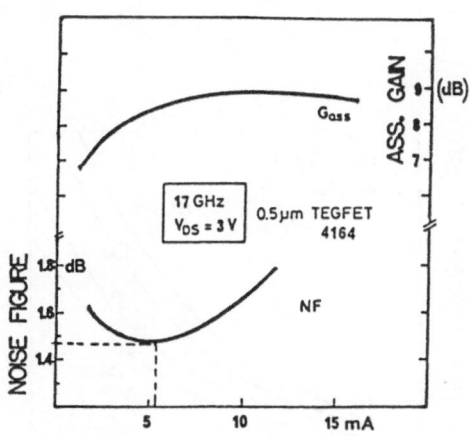

Fig. 3 : Typical noise figure variation of a 0.5μm TEGFET with the
drain current at high frequency.

has been well verified for TEGFETs and MESFETs [13]: K_F has been shown to be systematically smaller for a heterostructure transistor than for a normal MESFET (1.6 instead of 2.5). This can be essentially due to the higher quality of the layers grown by molecular beam epitaxy and to the general properties of the heterostructures.

We usually define the cut-off frequency (f_T) as $\quad g_{mo}/2\pi C_{GS}$ and the maximum available gain (to first order) as

$$MAG \sim (f_T/f)^2 \qquad \text{or} \qquad \sim (g_{mo}^2/4\pi^2 f^2 L_G^2).$$

For the viewpoint of microwave applications, short-gate lengths lead to higher maximum available gain. For TEGFETs, the linear variation of the I_{DS}-V_G characteristics can promote a higher value of the extrinsic transconductance, ($g_{me} = g_{mo}/(1+R_S \cdot g_{mo})$) in the low noise bias conditions (fig.4), and consequently results in a higher gain than in GaAs FETs, for which the I_{DS}-V_G variations remains quadratic.

Generally, the geometry of the transistor is adapted to the frequency range which has to be covered; the shorter the gate length , the shorter the transit time under the gate, the higher will be the maximum available frequency; the working range also depends on the gate width ,W : 250/300 microns gate width transistors are designed for the 2-15 GHz range; 30/75 microns transistors will more easily work up to 60 GHz : this can be seen in fig.5 which shows the variation of the noise figure (calculated according to the previous expression) as a function of the frequency for various gate-finger widths W_u (W/n, n being the number of unit fingers).

The principal advantage of the TEGFET is to give the same noise figure and associated gain for a 0.5 micron gate length (now currently realized in a production line) as conventional 0.25 micron GaAs MESFETs. Fig.6 shows the experimental noise figure measured on a TEGFET at room temperature in the 8-40 GHz range; the full line represents at the same time the results obtained with 0.5 μm TEGFETs and those obtained with 0.25 μm GaAs FETs.

This obvious superiority is explained through the higher electron mobility which tends to reduce the access resistance R_S , a possibly higher electron velocity (v) which reduces the C_{GS}/g_{mo} ratio, or a more favorable aspect ratio L_G/d.

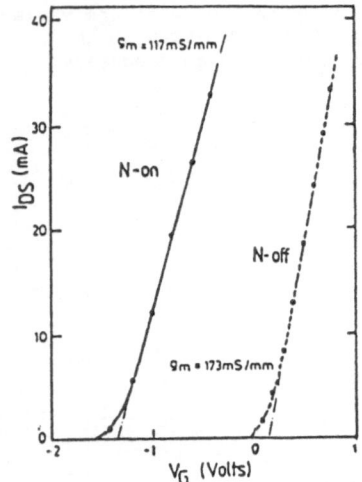

Fig. 4 : Linear variation of the source-drain current with the gate voltage for N-on and N-off TEGFETs.

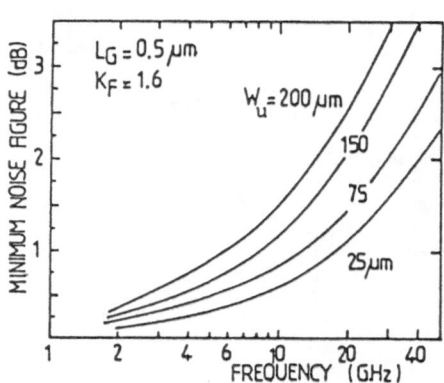

Fig.5 : Calculated values of the minimum noise figure as a function of the frequency for a 0.5μm gate length and for one gate finger.

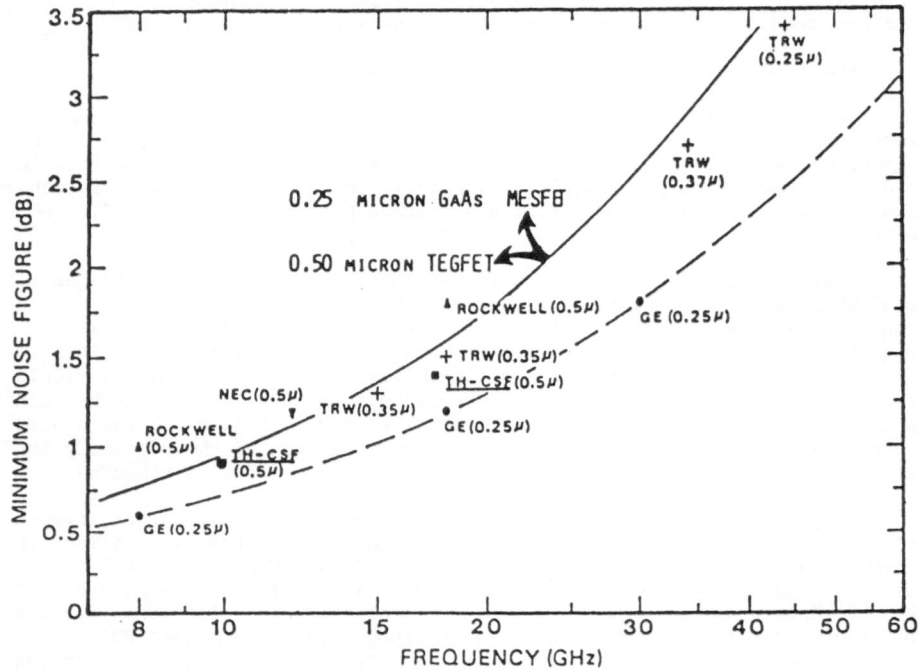

Fig. 6 : Best reported results of 2-DEG FETs at room temperature. (after J. Berentz, TRW, [22])

Finally, the two-dimensional gas may be profiting from the fact that the electrons only have two degrees of freedom in the TEGFET as compared to three in conventional devices.

3-4: Cryogenic operation

At low temperature, the TEGFET is more interesting than any other device because of its enhanced mobility; it can operate at considerably lower noise levels than GaAs FETs.

In cryogenic operations, it is usual to define a noise temperature, T_n, related to F_{min} through
$$F_{min}(db) = 10 \log (1 + Tn/T_0)$$
T_0 is the reference temperature normalized to 290 K; T_n varies almost linearly with frequency according to
$$T_n = \alpha(T) . f(GHz)$$
Delagebeaudeuf et al.[14] give:

$$\alpha(T) = 2\pi 10^9 \, K_{FO} . T . C_{GS} . [(RS+RG)/gmo]^{1/2}$$

K_{FO} still being the quality factor.

In Table 1 are shown the room temperature and 77 K measured values of R_S, R_G, g_{mo} and C_{GS}, as given by Weinreb [15] for a commercial GaAs MESFET. Also given are the measured values of noise temperature at 4.65 GHz and the deduced experimental value of α . For comparison, we show in Table 2 the various parameters measured for a TEGFET (Thomson) at 18 GHz: at 77K, R_S becomes immeasurable as a result of the mobility enhancement effect, which is not obtainable in normal structures. At the same time, between 300 K and 77 K, the ratio C_{GS}/G_{mo} decreases by a factor 1.55 instead of 1.3 for a MESFET. The marked improvement in transconductance (g_{mo}) is still a result of a larger saturation velocity ; as much as 2 times velocity enhancement has been observed for TEGFETs whereas such an effect is not seen on MESFET structures.

Table 1: GaAs FET characteristics at 4.65 GHz for 300 and 77 K

T[K]	$R_G[\Omega]$	$R_S[\Omega]$	g_{mo}[mS]	C_{GS}[pF]	T_n[K]	Fmin	αexp
300	2.66	2.3	43	0.5	102.4	1.32	22
77	1.4	2.1	56	0.5	30	0.43	6.4

Table 2: TEGFET characteristics at 17.5 GHz between RT and 77 K

T[K]	$R_G[\Omega]$	$R_S[\Omega]$	g_{mo}[mS]	C_{GS}[pF]	T_n[K]	Fmin	αexp
300	2.5	2.3	69	0.38	155	1.86	8.9
77	1.2	0	106	0.38	24	0.35	1.4

Some typical variations of T_n with the cooled temperature T are shown in fig.7.

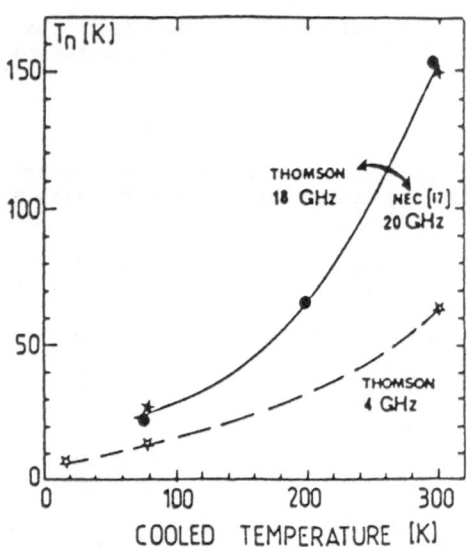

Fig.7 : Noise temperature given by cooled TEGFETs at different frequencies.

3-5: Applications.

Present-day satellite communication needs receivers with ultra-low noise devices in the input circuits; this is why multistage TEGFET amplifier circuits have been realized by several companies in the World. All of them have got excellent results never obtained before.

Toshiba [16] reports on the fabrication of a four-stage 0.4 μm TEGFET amplifier (7.2 dB NF and 19 dB gain in the 18 - 26 GHz range) still surpassing the 5-stage 0.25 μm GaAs FET amplifier (12 dB NF and 23 dB gain over the same range);

TRW [2] has developed a single and a three stage 0.25 μm TEGFET amplifier working at 36-40 GHz that surpasses that of GaAs FET technology (4 dB of Noise figure for 15 dB of associated gain). Tests with the single-stage amplifier cooled to 200 K (- 73 C) have yielded numbers even more impressive. When cooled to that temperature, the single-stage amplifier's noise dropped to 1.6 dB at 37 GHz with a corresponding device noise figure of 1.3 dB. So far, this is the lowest noise figure reported for a TEGFET amplifier circuit, regardless of temperature.

General Electric [3] has successfully produced a TEGFET device capable of very low noise figures at 40 GHz (2.1 dB NF for 7-db associated gain). At 12 K and 3.3 GHz, it has a noise figure of only 0.05 dB; so, Scientists of Jet Propulsion Laboratory (Pasadena,CA) plan to use this transistor as part of their radiotelescope detectors which must maintain extremely low noise figures in order to discriminate low-level deep-space signals from surrounding noise.

NEC [17] has developed for earth stations (in the 4 GHz band) the same type of circuit yielding a 48 K noise temperature at room temperature and 35 K (T_n) in a thermoelectrically cooled regime (- 45°C).

4) TEGFET for Digital ICs

With its inherent speed advantage, the TEGFET has naturally generated optimism among digital circuits designers.

4-1 Heterostructures

The heterostructures grown for logic circuits are similar to those used for analog devices. It is expected that in the future, the Direct-coupled FET logic family (DCFL), using enhancement-mode drivers and depletion-mode loads (which is the simplest configuration) will be used in commercially advanced circuits. The power consumption of elementary DCFL circuits is already the lowest possible. The key point will be more and more the control of the threshold voltage V_T (at 0.2 Volts with a dispersion of 20 mV for example) accross a single wafer and from wafer to wafer.

4-2 Ring Oscillators

The simplest integrated circuit which permits speed and power evaluation is the ring oscillator (RO). Fig.8 represents the propagation delay time t_{pd} between two stages, measured at room temperature versus the power dissipated per gate for different geometries of the driver (10, 20, microns gate width, 0.7 μm gate length); these results obtained at Thomson [18] are compared to general results in the literature. The best value ever published is given by Bell Labs.[5] with 10.2 ps at room temperature and 5.8 ps at 77 K for respectively 1.03 mW and 1.76 mW of power use. At 77 K, it has been shown that the TEGFET consumes 5 to 10 times less power than GaAs FETs; this is a great advantage for large scale integration circuits.

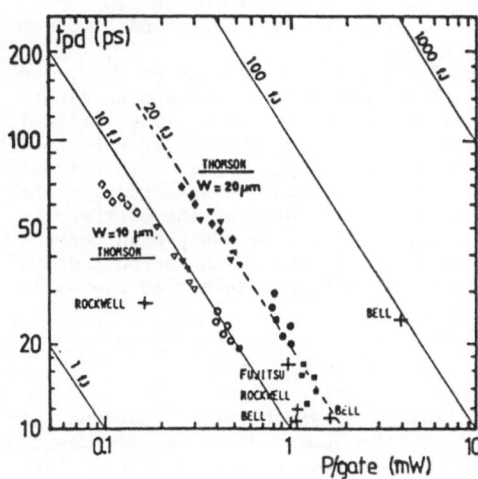

Fig.8: Propagation delay time versus power dissipated per gate at room temperature for different types of TEGFET ring oscillators. The diagonal lines are lines of constant power-delay product.

4-3 Integrated Circuits

Frequency dividers made from two-dimensional electron gas heterostructures can also be considered as valuable test circuits ; they already work at 8 GHz at room temperature [18] and 10 GHz at 77 [19]. Circuits of large scale complexity have emerged as the 4 Kb static random access memory of Fujitsu [20] which already integrates 27,000 transistors.

5) Other Heterostructures

5-1 Multiple Channel High Electron Mobility Transistors

The sheet carrier concentration is known to increase as the undoped AlGaAs spacer layer is decreased, and the maximum concentration of 2DEG obtained is about $1\ 10^{12}\ cm^{-2}$, which is limited by the doping concentration and the Al content in the n-AlGaAs layer.

By stacking up several double heterostructures, the number of conductive channel increases and so does the sheet carrier density. At 77 K, the sheet carrier density is linearly proportional to the number of channel (5 to 6 10^{12} cm^{-2} for 6 channels [6]. So, high performance depletion and enhancement mode devices (1 μm x 50 μm) have been fabricated using a conventional process : 600 to 800 mA/mm drain current have been measured (to be compared to 150 mA/mm in a normal single channel TEGFET and to 300 mA/mm for the typical channel current of GaAs power MESFETs).

At the same time, Fujitsu [7] fabricated power double-channel HEMT (1um x1200 um) ; after optimisation of the doping density of the n-AlGaAs layer and that of the different thicknesses, they obtained 0.55 mW/mm output power at 20 GHz, with 3.2 dB gain at a bias voltage of 9 Volts. This output power is comparable to that of the best GaAs MESFETs (0.5-0.6 W/mm) developed in several laboratories.

Finally, the multiple channel TEGFET has high power capability and is a potential rival to GaAs MESFET for higher frequency operation.

5-2 H.B.T.

Increasing interest in the AlGaAs/GaAs heterostructure bipolar transistor results from the high-speed potential, the performance dependence mainly on the grown buried layers and graded junctions (high transconductance, high current per chip area).

The usual structure consists of an n^+ sub-collector region of GaAs, an n^- GaAs collection-depletion region , a p^+ GaAs base region and an AlGaAs emitter region; in this structure, the electron flow is essentially vertical across regions whose dimensions are established during growth, and therefore is controllably fabricated in the 0.05-0.1 μm range. The small distance between emitter and collector in the vertical devices reduces the electron transit time. At the same time, the entire area of the emitter is effective in contributing to current flow, so that relatively high current can be maintened in small devices. As a result, HBT have high output drive capability. Circuits simulations suggest that HBT gate delays as low as 10 ps may be achievable [21].

6) Conclusion

The market of analog and digital circuits which will continue to progress in the next decade will need the best discrete devices or integrated circuits based on GaAs and AlGaAs heterostructures. These last ones will supplant the previous ones in specific applications where expense and complexity are secondary and where the highest performances are required.

Acknowledgments : It is a great pleasure for the authors to thank their co-workers and colleagues for their help or advice , especially N.T.Linh, J.Magarshack, M.Briere, C.Rumelhard, M.Gloanec, P.Delescluse, J.F.Rochette and M. Derewenko.

REFERENCES

1) P.R.Jay, D.Adam,D.Delagebeaudeuf, H.Derewenko, and L.Chusseau : " high gain low noise TEGFET devices for 18-40 GHz use ", Cornell Conf., July, 27-31,

2) M.Scholley, J.Berenz, A.Nichols, K.Nakano, R.Sawircs and J.Abell : " 36-40 GHz HEMT low noise amplifier" in IEEE-MTT-S Digest, 555

3) P.C.Chao, S.C.Palmeeter, P.M.Smith, U.K.Mishra, K.H.G.Duh, J.C.M.Hwang : "Millimeter-wave low noise HEMT" in IEEE Trans. Elect. Dev.Letters, 10, 531,

4) GaAs ICs Symposium, Monterey, CA, (1985).

5) ATT-Bell Laboratories, Electronics, Dec 16, (1985).

6) N.H.Sheng, C.P.Lee, R.T.Chen, D.L.Miller, and S.J.Lee: IEEE Elect. Dev. Letters, 6, 307,

7) W.R.Wisseman :"Advances in GaAs monolithic power amplifier", in 1985 GaAs ICs symposium, pp

8) Y.Takanashi, N.Kobayashi : "AlGaAs/GaAs 2-DEG FET's fabricated from MOCVD wafers" in IEEE Elect. dev. Letters ,3, 154

9) J.P.Andre, A.Briere, M.Rocchi and M.Riet : "Growth of AlGaAs/GaAs heterostructure for HEMT devices" in J.Cryst. Growth, 68, 445,

10) D.Delagebeaudeuf and N.T.Linh : "Charge control of the heterojunction two-dimensional gas for MESFET applications" in IEEE Trans. Elect. Dev.

11) D.Delagebeaudeuf and N.T.Linh : "Metal-AlGaAs-GaAs Two-dimensional electron gas FET" in IEEE Trans. Elect. Dev. 29, 955

12) H.Fukui : Optimal noise figure of microwave GaAs MESFET's" in IEEE Trans. Elect. Dev. 26, 1032,

13) D.Delagebeaudeuf, J.Chevrier, M. Laviron and P.Delescluse :"A new relationship between the Fukui coefficient 'and optimal current value for operation of FET's" in IEEE Elect. Dev. Letters, 9, 444

14) D.Delagebeaudeuf, P.Delescluse and P.Jay : "Extremely low noise and low temperature TEGFET operation " in Eur. Microw. Conf.,Paris, 15,

15) S.Weinreb, "Low noise cooled GaAs FET Amplifiers" in IEEE Trans.MTT 28,10,

16) K.Shibata, B.Abe, H.Kawasaki,S.Hori and K.Kamei : "Broadband HEMT and GaAs amplifiers for 18-26.5 GHz" in IEEE -MTT-S Digest, U-3,547,

17) T.Mochizuki, K.Honma, K.Handa, W.Akinaga and K.Ohata : "Low noise amplifiers using two-dimensional electron gas FETs" in IEEE-MTT-S Digest,

18) P.N.Tung , Thomson internal report, to be published

19) R.H.Hendel, S.S.Pei, R.A.Kiehl, C.W.Tu, M.D.Feuer and R.Dingle : "A 10 GHz Frequency divider using SDHT" in IEEE Elect. Dev. Letters , 10, 406,

20) S.Kurodo, T.Mimura, M.Suzuki, N.Kobahashi : "New device structure for 4Kb HEMT SRAM" in GaAs IC Symposium, 125,

21) R.Katoh, M.Kurata and J.Yoshida : "Numerical CML switching analysis for heterojunction bipolar transistor" in Solid-State Elect., 29, 151,

22) J.Berentz, "HEMT Technology gains on mm-waves" in Microwaves & RF, 121, (1985)

Luminescence and Transport Properties of GaAs Sawtooth Doping Superlattices

E.F. Schubert*, M. Hauser, B. Ullrich, and K. Ploog

Max-Planck-Institut für Festkörperforschung,
D-7000 Stuttgart 80, Fed. Rep. of Germany

Abstract

GaAs sawtooth doping superlattices consisting of a periodic sequence of n- and p-type δ-doped GaAs layers equally spaced by undoped regions were prepared by molecular beam epitaxy. The sheet-charge distribution resulting from the doping profile described by a periodic train of δ-functions leads to a strong sawtooth-shaped modulation of the real-space energy bands. The effective energy gap of the superlattice is smaller than that of GaAs, and for a superlattice period of less than 20 nm it is stable even at high excitation densities. Stable high-intensity luminescence beyond 1 μm is thus observed also at room temperature. In addition, vertical carrier transport due to tunneling becomes important in short-period GaAs sawtooth doping superlattices.

1. Introduction

The technique of molecular beam epitaxy (MBE) allows location of the dopant atoms Si and Be in distinct atomic monolayers of the (100) oriented GaAs host material [1]. Sheet doping densities in excess of $1 \times 10^{13} cm^{-2}$ are easily achieved [2], so that the fractional coverage of the available Ga sites in the (100) plane reaches several percent. The resulting doping profile is mathematically described by the Dirac-δ-function. Due to size quantization, the electrons (or holes) populate quantized energy levels in the V-shaped potential wells of δ-doped GaAs [1]. The new GaAs sawtooth doping superlattice (SDS) is obtained if alternating n- and p-doped GaAs monolayer sheets are equally spaced by thin (typically 5 - 10 nm) undoped GaAs regions. Then the doping profile consists of a periodic train of δ-functions, and the resulting sheet-charge distribution leads to a strong sawtooth-shaped modulation of the conduction and valence band edges [3]. The effective superlattice energy gap is significantly smaller than the gap of the host material, and for a superlattice period of less than 20 nm it is stable even at high excitation densities due to recombination lifetimes in the nanosecond range. In this paper we establish the basic set of equations for the design rules of GaAs SDS, and we discuss the effect of period length on the vertical transport in TYPE A and TYPE B SDS. We then present some details on the preparation of GaAs SDS by MBE, and we conclude with experimental results on the luminescence and vertical transport properties of the superlattice.

---

* present address: AT & T Bell Laboratories, Holmdel, N.J. 07733, USA

2. Electronic Properties of Sawtooth Doping Superlattices

The doping profile $(N_D - N_A)$ of SDS is schematically shown in Fig. 1b. We describe this profile by a periodic train of δ-functions [3] according to

$$(N_D - N_A) = N_D^{2D} \sum_{i=-\infty}^{+\infty} \delta(z-iz_p) - N_A^{2D} \sum_{i=-\infty}^{+\infty} \delta(z - \frac{1}{2}z_p - iz_p) \qquad (1)$$

where z_p is the superlattice period and N_D^{2D} and N_A^{2D} are the two-dimensional (2D) donor and acceptor concentrations, respectively. The charge distribution derived from Eq. (1) via Poisson's equation leads to a sawtooth-shaped modulation of the conduction and valence band edges, which is schematically depicted in Fig. 1c for the case of $N_D^{2D} = N_A^{2D} = N^{2D}$. The amplitude of the band-edge modulation V_{zz} (zigzag potential) is then given by

$$V_{zz} = \frac{1}{4} \frac{q}{\varepsilon} N^{2D} z_p \qquad (2)$$

where ε is the permittivity of the semiconductor and q is the elementary charge. Due to size quantization, electrons and holes populate quantized energy levels in the narrow V-shaped potential wells. Using a simple approximation to calculate the subband energies of electrons (holes) in isolated potential wells (no coupling) [4] we obtain the relation

$$E_n = 128^{-1/3} (n + 1)^{2/3} \left[\frac{2\pi \hbar q^2 N^{2D}}{\varepsilon (m^*)^{1/2}}\right]^{2/3} \qquad n = 0, 1, 2 \ldots \qquad (3)$$

where m^* is the electron or heavy-hole effective mass.

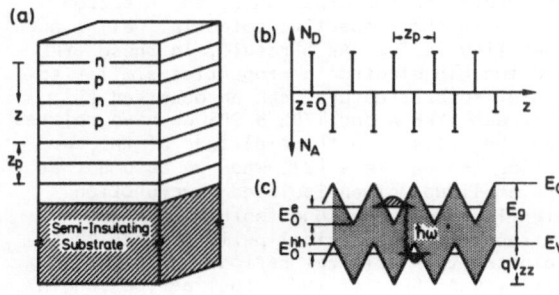

Fig. 1 Schematic illustration of (a) layer sequence, (b) doping profile, and (c) modulation of conduction and valence band edges in a GaAs sawtooth doping superlattice.

In SDS the impurities with discrete charge q are randomly distributed in the respective dopant planes having a mean distance of $(N^{2D})^{-1/2}$. To keep statistical fluctuations of the depth and width of the potential wells low, the mean impurity distance should be less than half the superlattice period, i.e. $\frac{1}{2}z_p > (N^{2D})^{-1/2}$, and the spatial extent of the lowest subband z_0 should be larger than the mean impurity distance, i.e. $z_0 > (N^{2D})^{-1/2}$, where the spatial extent of the subbands z_n is given by $z_n = \frac{1}{2}(n + 1) 2\pi \hbar (2m^* E_n)^{1/2}$

Radiative electron-hole recombination in GaAs SDS involves the lowest electron and heavy-hole subbands. As a result, the optical gap of the superlattice E_g^{eff} given by

261

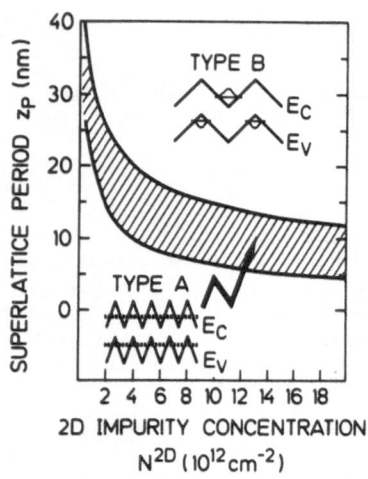

Fig. 2 Classification of TYPE A and TYPE B GaAs sawtooth doping super-lattices in terms of quantum mechanical coupling. The upper curve denotes the critical period length versus doping density. The lower curve indicates the limit where the period length becomes less than the mean impurity distance.

$$E_g^{eff} = E_g - q\,V_{zz} + E_o^e + E_o^{hh} \tag{4}$$

is smaller than the gap of the host material E_g.

At a given doping density N^{2D} the period length z_p determines the quantum mechanical coupling between the V-shaped wells of SDS. Small period lengths imply a strong overlap of electron and hole wavefunctions from adjacent potential wells, which in turn results in short recombination lifetimes in the ns range [3]. This so-called TYPE A SDS thus exhibits a stable energy gap smaller than the GaAs gap which cannot be tuned by carrier injection. In SDS with long period length, on the other hand, the electron and hole wavefunctions are localized in the respective potential wells, and the radiative recombination probability is low. As a result, in these TYPE B SDS, long carrier lifetimes and tunable electronic properties similar to those in semiconductors with a "nipi superstructure" can be observed [5]. For a quantitative distinction between TYPE A and TYPE B SDS we have related the tunneling probability between the wells with the amplitude of the corresponding wavefunctions penetrating the barriers [2]. When we assume that coupling becomes essential if the amplitude of an isolated wavefunction reaching the center of the barrier is larger than one tenth of the wavefunction at the well-to-barrier transition, a critical period length z_p^* is obtained. In Fig. 2 we have plotted this critical period length versus doping density, in order to classify GaAs SDS. If the actual period length z_p is larger than z^*, a TYPE B SDS with distinct potential wells for electrons and for holes ("multi quantum well structure") results. If z_p is smaller than the critical period length, a TYPE A SDS results where coupling between the wells becomes essential. However, the period length of the TYPE A SDS has always to be larger than the mean impurity distance $(N^{2D})^{-1/2}$ in the respective dopant planes. Therefore, a TYPE A SDS is achievable with material design parameters depicted by the shaded area of Fig. 2.

In TYPE A SDS with strong coupling, electron and hole subbands of considerable width are formed, and carrier transport perpendicular to the layers becomes important. In TYPE B SDS we have to apply a voltage perpendicular to the layers, and two tunneling mechanisms are possible depending on the magnitude of the electric field $\underline{E}$. If the energy drop within one period qEz_p, the so-called Stark-ladder energy, exceeds the subband width ΔE but is smaller

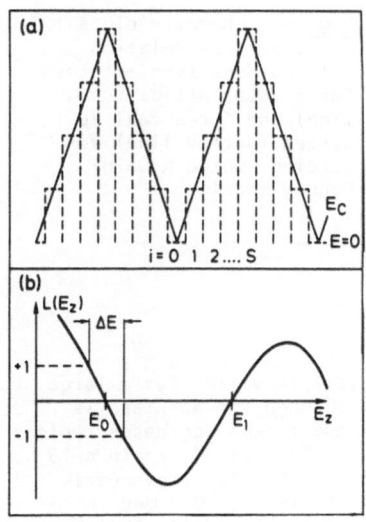

Fig. 3 (a) Approximation of saw-tooth-shaped modulation of band edges by a step-like shape. The height of each step is qV_{zz}/S, where S is the number of steps. (b) Schematic plot of the characteristic function $L(E_z)$ yielding the width ΔE and the energy of the lowest subband $(E_o - \frac{1}{2} \Delta E) \leq E_z \leq (E_o + \frac{1}{2} \Delta E)$.

than the separation of the lowest and first excited subband (E_1-E_o), i.e. $\Delta E < qEz_p < (E_1-E_o)$, phonon-assisted tunneling occurs [6]. Acoustic or optical phonons are required to provide the momentum conservation during tunneling. If on the other hand the Stark-ladder energy is comparable to the subband separation, i.e. $qEz_p \simeq (E_1-E_o)$, resonant tunneling occurs [7]. In this case, electrons from the lowest subband tunnel into the first excited subband of the adjacent quantum well. Resonant tunneling is mainly observed at low temperatures, where phonon-assisted tunneling is reduced.

For a quantitative estimate of vertical transport in SDS we have approximated the sawtooth-shaped modulation of the band edges by a series of square-shaped barriers, as indicated in Fig. 3a, and we have used the Kronig-Penney model to describe the formation of electron and hole subbands. In this way we can calculate the transmission factor of electrons (holes) through arbitrarily shaped barriers. The energy of the lowest subband E_o and its bandwidth ΔE is obtained by solving the characteristic equation given in standard textbooks of quantum mechanics [8], which is schematically plotted in Fig. 3b. For $-1 \leq L(E_z) < 1$ this equation has the solution $E_z = E(k_z)$. Assuming small subband widths ΔE, the function $L(E_z)$ can be linearized in the vicinity of E_o, as indicated in Fig. 3b.

Using the dispersion relation for carrier motion in z-direction, $E_z = E_o - \frac{1}{2} \Delta E \cos (k_z z_p)$, we obtain the total energy for carrier motion in parallel and perpendicular direction, $E = [(\hbar^2 k_{xy}^2)/ 2m^*]+ E_z (k_z)$, which is depicted in Fig. 4a (solid line) and compared with the flat dispersion relation for a real 2D system (dotted curve). In Fig. 4b we compare the plots of the density-of-states for a real 2D system (dotted curve) with that of a superlattice (solid curve) which represents a quasi-3D system with a small extension in z-direction, using the equations

$$D^{2D}(E) = m^*_{xy}/ \pi\hbar^2 \quad \text{and} \quad D(E) = [m^*_{xy}/(\pi^2\hbar^2)](1/z_p)\arccos[(-2/\Delta E)(E-E_o)] \quad (5)$$

The effective mass for carrier motion in z-direction, also called the "confinement mass", is given by $m^*_z = (2\hbar^2)/(\Delta Ez_p^2)$. Since ΔE depends expo-

263

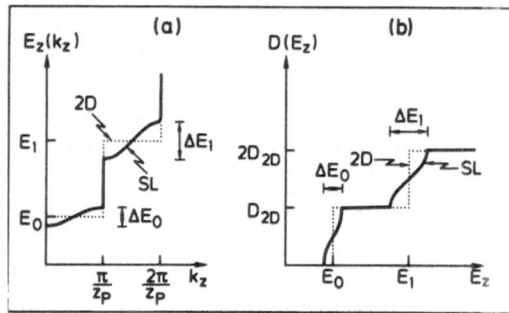

Fig. 4 Schematic plots of dispersion relation (a) and density-of-states for a superlattice (solid line) and for a real 2D system (dotted line) according to the Kronig-Penney model.

nentially on $(-z_p)$, also the relation $m_z^* \sim \exp(z_p)$ is valid. For a large superlattice period, the tunneling probability through the barriers is negligibly small or, in other words, the confinement mass increases rapidly. As a typical example, the confinement mass is 8.7×10^{-31} kg in a 15 nm period superlattice ($m_{xy}^* = 0.067 \, m_o$, $V_{zz} = 0.3$ V, S = 1), i.e. nearly equal to the free-electron mass. The subband width is $\Delta E = 0.7$ meV in this case. In contrast, a superlattice with a 30 nm period length ($m_{xy}^* = 0.067 \, m_o$, $V_{zz} = 0.3$ V, S = 1) has a confinement mass of 4.3×10^{-27} kg, which is larger by a factor of 5×10^3 than the free-electron mass. The mobility for motion in z-direction also decreases rapidly according to Drude's classical transport model for the free electron gas. As a result, at a given electric field and temperature, the current transport in SDS perpendicular to the layers depends exponentially on the barrier thickness or period length, according to $j \sim \exp(-z_p)$.

3. Sample Preparation

The GaAs SDS samples were grown by molecular beam epitaxy (MBE) on (100) o-riented GaAs substrates using a growth rate of 1 μm/hr and a growth tempe-rature of 500 - 550 °C. The δ-function like Si and Be doping profiles were obtained by interrupting the growth of the GaAs host crystal by closing the Ga shutter and leaving the As shutter open. The As-stabilized (2 x 4) surface reconstruction was thus maintained, while the shutter of the res-pective dopant effusion cell was opened for a certain time (up to a few mi-nutes, depending on the desired doping density). In this impurity growth

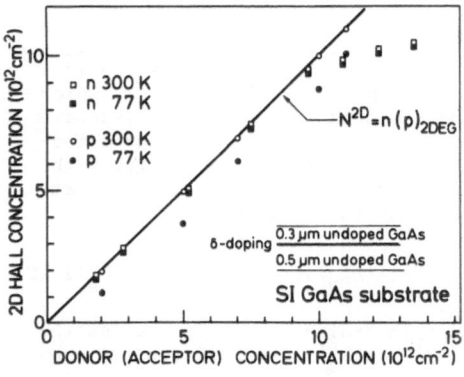

Fig. 5 2D Hall concentration versus supplied dopant atoms in single n- and p-type δ-doped GaAs

mode the host crystal does not grow. To continue GaAs growth, the dopant shutter is closed and the GaAs is opened again. For a single-type δ-doping layer in GaAs we show in Fig. 5 that the measured doping density is in good agreement with the number of supplied dopant atoms up to very high densities [2]. The fractional coverage of the available Ga sites in the (100) plane reaches about 5% if we assume that the GaAs (100) surface contains 1.25 x 10^{15} atoms cm^{-2}, i.e. 6.25 x 10^{14} Ga atoms cm^{-2}. For Si-δ-doping we observe a deviation of measured electron density from the supplied dopant atoms beyond 8 x 10^{12} cm^{-2}. At these high doping densities the electrons in the V-shaped potential well populate high-index subbands so that the Fermi energy is increased by more than 300 meV. This energy shift is comparable with the energy separation between Γ and L conduction band minima in GaAs. Electrons may thus populate the L minima at high Si doping levels.

4. Room-Temperature Photoluminescence

Photo- and electroluminescence measurements on TYPE A GaAs SDS of different material design parameters clearly reveal a stable energy gap which is considerably smaller than the gap of the host material and which does not depend on the excitation density [3]. As an example for the significant red-shift we display in Fig. 6a the room-temperature luminescence spectrum of a represenatative GaAs SDS which exhibits a peak-wavelength of 1.035 μm. The observed linewidth of the luminescence is due to potential fluctuations caused by the random impurity distribution and due to thermal broadening. The red-shift of the luminescence depends critically on the material design parameters of the GaAs SDS. In Fig. 6b we show the variation of the effective energy gap and of the emission wavelength in TYPE A GaAs SDS as a function

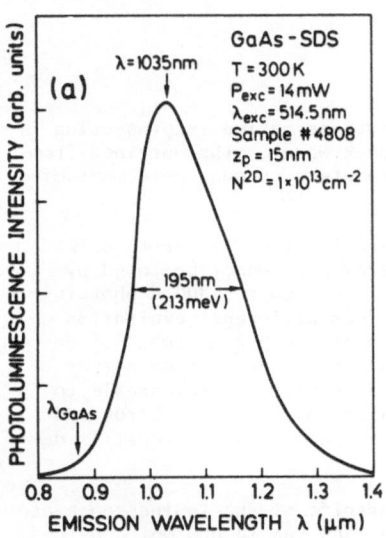
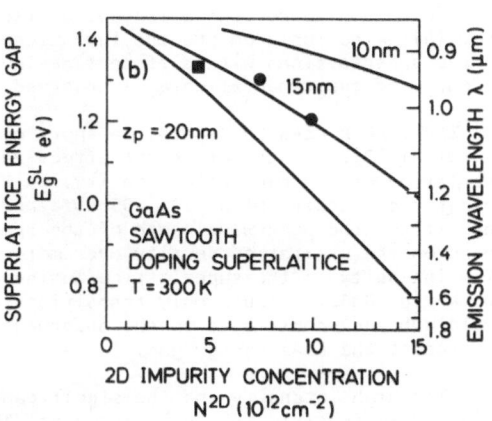

Fig. 6 (a) Room-temperature photoluminescence of a TYPE A GaAs SDS showing a peak wavelength of 1.035 μm. (b) Variation of energy gap and emission wavelength in TYPE A GaAs SDS as a function of 2D doping concentration calculated for three period lengths. Also indicated are the peak wavelengths obtained from SDS samples with z_p = 20 nm (square) and z_p = 15 nm (dots).

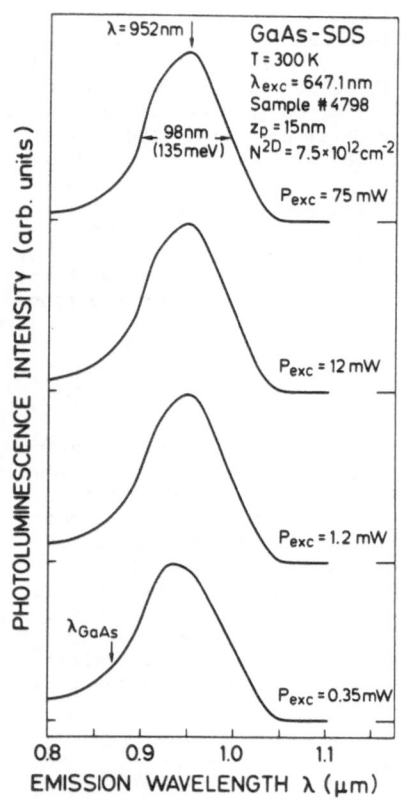

Fig. 7 Room-temperature photoluminescence spectra of a TYPE A GaAs SDS taken at different excitation densities. The arrow indicates the emission wavelength of bulk GaAs at 870 nm

of the 2D doping concentration calculated for three period lengths using Eq. (4). Also included are the luminescence peak wavelengths obtained from three superlattices with different design parameters. A good agreement of theory and experimental data is observed.

The TYPE A GaAs SDS exhibit a short carrier lifetime similar to bulk material [3], and therefore the effective energy gap cannot be tuned by variation of the excitation density. In Fig. 7 we show the 300 K photoluminescence spectra of a GaAs SDS obtained at four differenct excitation densities. The peak-wavelength of the spectrum at $\lambda = 952$ nm does not depend on the excitation density over more than three orders of magnitude. The intensity of the superlattice luminescence is high and comparable to bulk-type GaAs, because only transitions between the lowest electron and hole subbands contribute to the observed luminescence and no signal is detected at the GaAs energy gap.

The high intensity and the significant redshift of the luminescence at 300 K make feasible the application of TYPE A GaAs SDS in photonic devices. We have fabricated light emitting diodes (LED) and injection lasers with the superlattice active region sandwiched by n- and p-type $Al_{0.3}Ga_{0.7}As$ layers for confinement [9, 10]. Edge emitting diodes with different SDS design parameters emit monochromatic light at wavelength of $0.9 < \lambda < 1.0$ um. The threshold current density of broad area lasers emitting at 905 nm was found to be 2.2 kA cm^{-2}.

5. Vertical Carrier Transport

We have used current-voltage (I - V) measurements to study in particular the vertical transport of holes in GaAs SDS in the temperature range 4 - 300 K (due to the smaller effective mass vertical electron transport is not the rate limiting process). The superlattices were grown on heavily p-doped GaAs substrates. To facilitate ohmic contact formation, a thin p^+-GaAs layer forms the cap layer. Mesa-etched circular Cr/Au dots of 500 μm diameter were used as ohmic contact. The real-space energy band diagram of a TYPE A GaAs SDS (z_p = 15 nm) embedded between two p^+-GaAs layers is schematically depicted in the upper part of Fig. 8. The I - V characteristics measured at 300 K and 77 K clearly show ohmic behaviour. The efficient transport mechanisms through the thin triangular barriers are most probably tunneling (field emission) and thermally assisted tunneling (thermionic field emission) of heavy and light holes. For the latter process we assume thermal population of light-hole subbands. Due to the mixing of heavy- and light hole [11] states in the valence band of the superlattice, a quantitative analysis of hole tunneling is difficult. However, if we neglect valence-band mixing, the calculation of the heavy- and light-hole currents yields a resistance of less than 1 Ω cm^{-2} for one triangular barrier. This value is in good agreement with our experiment. The large zigzag energy of qV_{zz} = 517 meV for the present SDS, which is much larger than the thermal energy kT at room temperature, implies that pure thermionic emission of holes over the triangular barrier gives only a minor contribution to the vertical transport. The observed variation of the I - V characteristics with temperature, i.e. enhanced resistance at 77 K and 4 K, implies that thermally assisted tunneling dominates the carrier transport at 300 K. The significant change of the shape of the I - V characteristics at 4 K indicates that an electric field normal to the layers is established for efficient carrier transport.

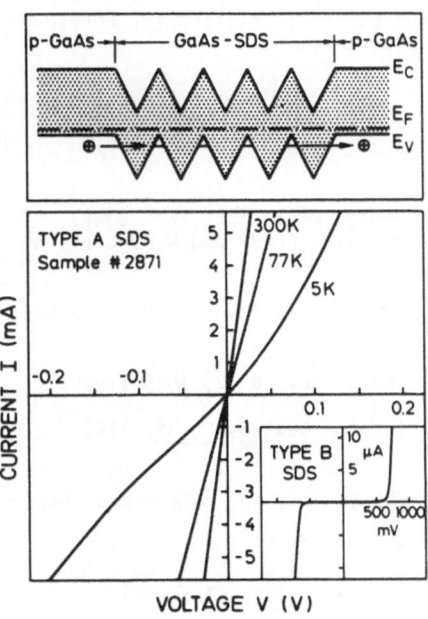

Fig. 8 Current-voltage characteristics of a TYPE A GaAs SDS at different temperatures. In the upper part we show the real-space energy band diagram of the superlattice embedded in two p-GaAs layers. The current-voltage characteristic of a TYPE B SDS depicted in the inset exhibits an extremely small current of less than 10^{-6} A for voltages < 0.5 V.

This means that the horizontal energy band diagram depicted in Fig. 8 (top) is slightly tilted by the applied bias.

Current-voltage measurements on a TYPE B GaAs SDS with a period length of $z_p = 60$ nm reveal a drastically reduced current by more than a factor of 10^3 perpendicular to the layers. The inset of Fig. 8 shows the high resistance of this multiquantum well structure for voltages below 0.5 V. When the voltage is increased beyond 0.6 V, the electric field perpendicular to the layers becomes sufficiently large for a rapid increase of the current. The observed current-voltage characteristic is symmetric for positive and negative voltages.

6. Conclusion

Using molecular beam epitaxy we have prepared GaAs sawtooth doping super-lattices which are composed of a periodic sequence of n- and p-type δ-doped GaAs layers equally spaced by undoped regions. The periodic variation of the sheet-charge distribution resulting from the specific doping profile leads to a distinct sawtooth-shaped modulation of the conduction and valence band edges. The resulting effective energy gap is considerably smaller than the gap of GaAs, and for superlattice periods below 20 nm it is stable also at high excitation densities. We observed a stable red-shift of high-intensity luminescence also at room temperature. In addition, vertical carrier transport due to tunneling of holes was investigated by current-voltage measurements in the temperature range 4 - 300 K.

Acknowledgement

The authors would like to thank A. Fischer for expert help with the sample preparation. This work was sponsored by the Bundesministerium für Forschung und Technologie of the Federal Republic of Germany.

References

1   A. Zrenner, H. Reisinger, F. Koch, K. Ploog: Proc. 17th Int. Conf. Phys. Semicond., San Francisco 1984, Eds. J.P. Chadi and W.A.Harrison (Springer, New York, 1985) p. 325

2   K. Ploog, A. Fischer, E.F. Schubert: Proc. MSS-II (Kyoto 1985), to be published in Surf. Sci. (1986)

3   E.F. Schubert, Y. Horikoshi, K. Ploog: Phys. Rev. B 32, 1085 (1985)

4   E.F. Schubert, K. Ploog: IEEE Trans. Electron Devices ED-32, 1868 (1985)

5   K. Ploog, G.H. Döhler: Adv. Phys. 32, 285 (1983), and references therein

6   R. Tsu, G.H. Döhler: Phys. Rev. B 12, 680 (1975)

7   L.L. Chang, L. Esaki, R. Tsu: Appl. Phys. Lett. 24, 593 (1974)

8   N.W. Ashcroft, N.D. Mermin: Solid State Physics (Molt-Saunders, Tokyo, 1976) p. 146; S. Flügge: Rechenmethoden der Quantentheorie (Springer, Berlin-Heidelberg, 1976) p. 40

9   E.F. Schubert, A. Fischer, K. Ploog: Electron. Lett. 21, 411 (1985)

10 E.F. Schubert, A. Fischer, Y. Horikoshi, K. Ploog: Appl. Phys. Lett. 47, 219 (1985)

11 S.S. Nedorezov, Sov. Phys.-Solid State 12, 1814 (1971).

Physics and Applications of Doping Superlattices

G.H. Dohler

Hewlett-Packard Laboratories, 1501 Page Mill Road,
Palo Alto, CA 94304, USA

Semiconductor superlattices can be fabricated by priodic variation of either composition ($Al_xGa_{1-x}As$ - GaAs, e.g.) or doping (n-i-p-i, e.g.). During the past decade they have proven as ideal model substances for the study of the electronic properties of two-dimensional quantum systems, but also as an interesting material for new electronic and opto-electronic devices. The special interest in *n-i-p-i doping superlattices* originates from the "tunability" of their electronic structure, a feature which is due to an "indirect band gap in real space" and which is not shown by any uniform semiconductor nor by compositional superlattices. In this lecture the basic theoretical concept of n-i-p-i doping superlattices is briefly reviewed. The emphasis is on a discussion of recent theoretical and experimental investigations of new phenomena in doping superlattices and on their implications for device applications. This includes photo response studies of $GaAs$ n-i-p-i structures, where room temperature values of photoconductive gain in excess of 10^7 have been obtained. Applications for fast high-detectivity detectors, based on a combination of the unusual transport properties of n-i-p-i's with selective n- and p-contacts will be discussed. We report on *room temperature* studies of tunable photo– and electroluminescence in conventional n-i-p-i structures and on temperature-dependent and time-resolved luminescence measurements on *type-II hetero n-i-p-i*'s. In the latter case contributions from different subbands can be clearly resolved as individual lines, due to strongly reduced impurity broadening of the subband levels. We discuss recent experimental and theoretical results on the tunable absorption coefficient of n-i-p-i's and on potential applications for optical modulators and switches. Finally, we will discuss the non-linear optical properties conventional and *type-I hetero* n-i-p-i's which result from the unique combination of long electron-hole recombination lifetimes and of weak, but still significant absorption at photon energies *far below* the band gap of the host material(s) of the n-i-p-i structure. These properties again relate to interesting basic phenomena (carrier–induced band gap shrinkage, excitonic absorption, oversaturation of quantum-well absorption) as well as to device applications as purely optical logic components or optical switches.

1. Introduction

During the past decade the investigation of artificial semiconductor superlattices [1,2] has proven that these novel materials represent ideal model substances for the study of two-dimensional electronic systems. At the same time they have turned out as a very useful basic material for many novel electrical and, in particular, electro-optical devices [3-6]. Originally, most of the research in general and actually all the experimental efforts were devoted to compositional superlattices, a periodic sequence of a few nm thick layers of two different, but lattice-matched semiconductors. At about the same time when these studies started, our first detailed theoretical investigation [7] of semiconductors with a periodic doping profile consisting of n- and p-doped layers with intrinsic zones in between ("n-i-p-i superstructure") revealed that such semiconductors should exibit exotic properties, which are neither shown by any bulk crystal nor by the compositional

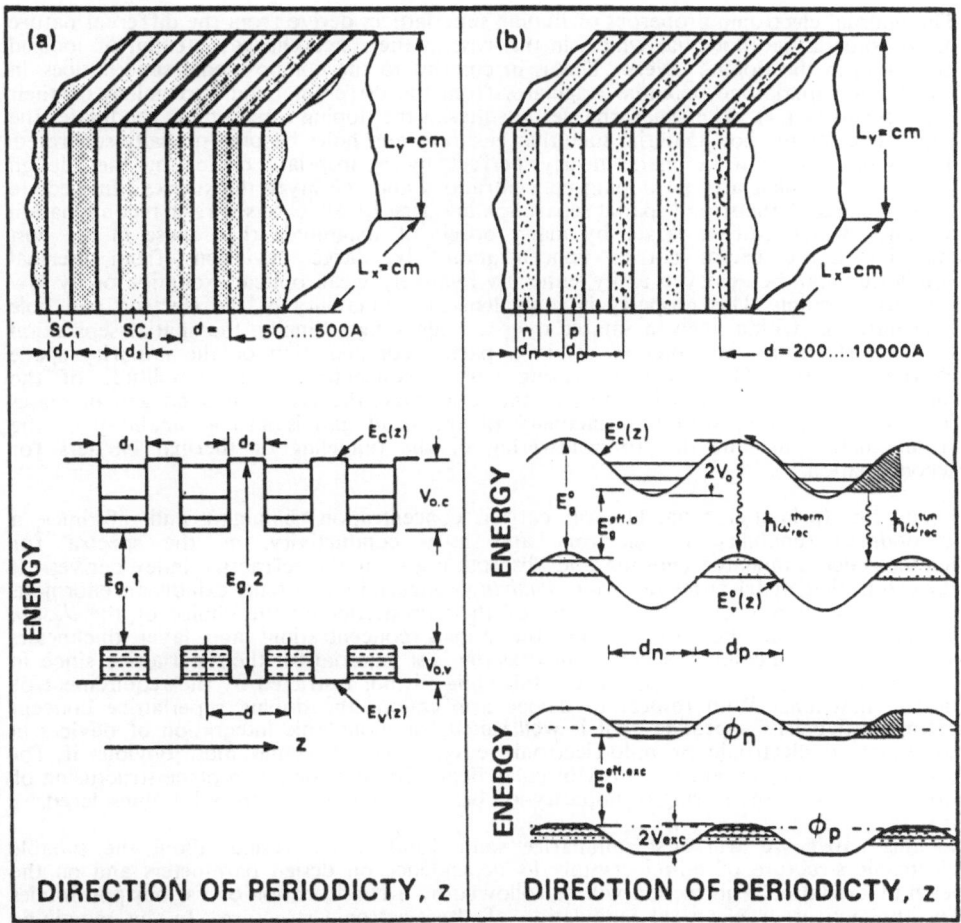

Fig. 1 Schematic real-space and band diagrams of a compositional (a) and a n-i-p-i doping superlattice (b). The superlattice potential is a consequence of varying band gap, $E_{g,1}$ and $E_{g,2}$ respectively, in the compositional superlattice. In the doping superlattice the periodic potential is due to the space charge potential of ionized impurities. Because of the potential barriers spatiallly separating electrons and holes the band gap becomes a tunable quantity in the doping suyperlattices. The shaded areas depict schematically the barriers for electron-hole recombination. The wavy lines indicate the tunneling and the thermally activated recombination processes, respectively.

superlattices. Although the predicted properties appeared quite appealing with respect to fundamental physics as well as with respect to future device applications, their experimental verification [8-15] came much later only, when the first GaAs doping superlattices had been grown by molecular beam epitaxy (MBE) [16]. This is quite surprising, since the requirements for growing high quality doping superlattices are less demanding than those for their compositional counterparts, which where studied much earlier. So far, nearly all of the experimental and most of the later theoretical studies on doping superlattices [17-24] have dealt with doping structures not containing intrinsic regions. The term "n-i-p-i crystal", however, became popular for the whole class of doping superlattices.

271

The unusual electronic properties of doping superlattices derive from the different nature of the superlattice potential which, in this case, is the space charge potential of ionized impurities in the doping layers. This is in contrast to the compositional superlattices, in which the superlattice potential originates from the different band gap values of their components (Fig.1). The space charge potential in the doping superlattices modulates the band edges of the host material such that electrons and holes become spatially separated. This separation can be made nearly perfect by appropriate choice of the "design parameters", which are the doping concentrations and the layer thicknesses. Immediate consequences of the strong spatial separation are, first of all, excess-carrier recombination lifetimes, which can be larger by many orders of magnitude than those in the host material. Large excess carrier concentrations, i.e., large deviations from thermal equilibrium can be achieved easily, either by relatively weak optical excitation or by low injection currents. This property is equivalent to a tunability of the electron and hole concentration within a given sample over a wide range. Finally, the spatial separation between electrons and holes results in a partial compensation of the impurity space charge potential. Thus, with increasing carrier concentration the amplitude of the superlattice potential decreases and, at the same time, the effective band gap increases (Fig.1(b)). Associated with this tunability of the band gap is also a tunability of the recombination lifetime due to a lowering of the tunneling or thermal barriers for recombination.

From the (primary) tunability of carrier concentration, bandgap and lifetimes, a (secondary) tunability of electron- and hole- conductivity, of the spectra for luminescence, stimulated emission and absorption and of the refractive index derives. It turns out that, apart from the wide *tunability* range, n-i-p-i systems exhibit an enormous flexibility with respect to the *tailoring* of their properties by the choice of the *design parameters*. I addition to choosing the doping concentrations and layer thicknesses virtually any semiconductor can be used as the host material for the superlattice, since, in contrast to compositional superlattices, this choice is not restricted by the requirement of lattice matching. With respect to device applications the doping superlattice concept seems particularly appealing as it is well suited for monolithic integration of devices in sophisticated electronic or opto-electronic circuits. This becomes most obvious if, for instance, the specific potential of Molecular Beam Epitaxy for the in-plane structuring of doping layers by shadowing or impurity-ion beam writing during growth is considered.

In this paper we will first summarize some fundamental results about the tunable electronic structure of n-i-p-i crystals, its dependence on design parameters and on the level of excitation. This section will be followed by a consideration of various possibilities for the optical or electrical modulation of the electronic structure. In the remaining sections we will report recent experimental and theoretical results on simple n-i-p-i crystals as well as on *hetero n-i-p-i's*, a system which combines periodic modulation of n- and p-doping and periodic variation of composition. Device applications of the various phenomena will be discussed.

2. Electronic Structure of n-i-p-i Crystals

The lower section of Fig.1 (b) shows schematically the real-space energy diagram of a n-i-p-i crystal for the ground state (upper part) and the excited state (lower part). The modulation of the conduction and valence band edges is due to the positive charge of the donor and the negative charge of the acceptor atoms. The strength of this modulation, $2V_0$, follows from Poisson's equation

$$2V_0 = (2\pi e^2/\kappa_0) [n_D(d_n/2)^2 + n_A(d_p/2)^2]. \tag{1}$$

Here κ_0 stands for the static dielectric constant and e for the elementary charge. We have assumed that there are no undoped layers and that the product of doping concentration and layer thickness for the n-layers (donor concentration n_D and layer width d_n) is the same one as for the p-layers (acceptor concentration n_A and layer width d_p), i.e.

$$n_D d_n = n_A d_p . \tag{2}$$

A term

$$(2\pi e^2/\kappa_o)\, n_D d_n d_i \tag{3}$$

has to be added to the r.h.s. of (1), if undoped layers of thickness d_i exist between the n- and the p-layers.

The motion of the carriers in superlattice direction is quantized by the space charge potential. Just like in the compositional superlattices this leads to the formation of electronic subbands. Since we are dealing with parabolic potentials (instead of rectangular quantum wells) the energies of the low-index subband edges are harmonic oscillator levels for our simple example [20,23]

$$\epsilon_{c,\mu} = \hbar\omega_c(\mu+1/2) \qquad \mu = 0, 1, 2, ... \tag{4}$$

$$\epsilon_{vi,\nu} = \hbar\omega_{vi}(\nu+1/2) \qquad \nu = 0, 1, 2, ...; \, i = l, h. \tag{5}$$

for the conduction and valence subbands, if measured from conduction and valenece band extrema at the center of the n- and the p-layers, respectively. The index i stands for the light and heavy mass valence bands. The (harmonic oscillator) energies $\hbar\omega_c$ and $\hbar\omega_{v,i}$ are formally the same as the plasmoा energies for electrons

$$\omega_c = (4\pi e^2 n_D/m_c\kappa_o)^{1/2} \tag{6}$$

and holes,

$$\omega_{vi} = (4\pi e^2 n_D/m_{vi}\kappa_o)^{1/2} \tag{7}$$

in uniformly doped semiconductors (m_c, and m_{vi} are the electron and light- and heavy-hole effective masses).

The value of the effective band gap, $E_g^{eff,o}$, defined as the energy difference between the uppermost heavy-hole subband and the lowest conduction subband

$$E_g^{eff,o} = E_g^o - 2V_o + \epsilon_{c,o} + \epsilon_{vh,o} \tag{8}$$

depends in most cases mainly on the value of V_o, and therefore, linearly on the doping concentrations, but quadratically on the thickness of the doping layers. It becomes zero if these quantities exceed certain values. A GaAs n-i-p-i crystal, for instance, becomes a "n-i-p-i semimetal" if the layer widths are larger than about $d_n = d_p = 65$ nm at a doping level of $n_D = n_A = 10^{18}$ cm^{-3}. In the semimetal situation, and also for the case where (2) is not fulfilled, the calculation of the electronic structure has to be performed self-consistently [17,18]. This is also true for the excited n-i-p-i crystal, i.e., if there are electrons and holes populating subbands in the n- and p-layers, respectively.

For many cases of interest for device applications, however, it is sufficient to have estimates. Such estimates can be obtained by calculating the superlattice potential as if it was composed of flat and neutral central portions of width

$$d_n{}^o = n^{(2)}/n_D \tag{9}$$

$$d_p{}^o = p^{(2)}/n_A \quad . \tag{10}$$

273

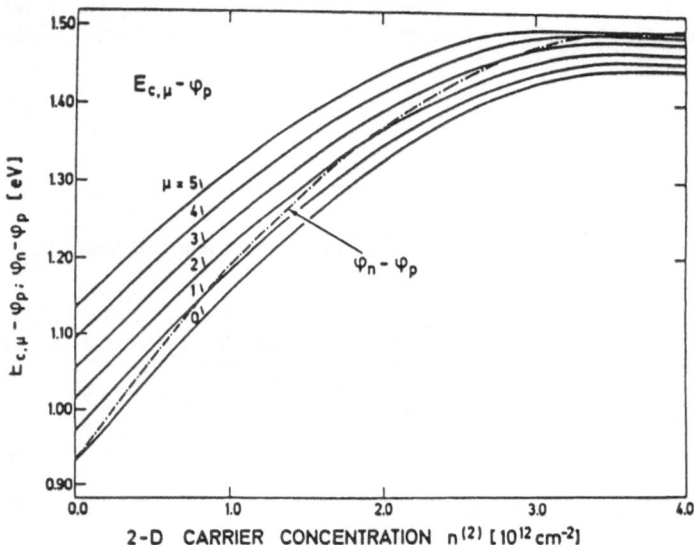

Fig. 2 Relation between (two-dimensional) carrier concentration in the layers and effective band gap and quasi Fermi level difference, calculated selfconsistently for a GaAs doping superlattice with n_D = n_A = 10^{18} cm^{-3} and d_n = d_p = 40 nm. The results for $\phi_n - \phi_p$ vs. $n^{(2)}$ do not differ significantly from those calculated with the approximate Eqs. (11-14).

with parabolic sections in between. $n^{(2)}$ and $p^{(2)}$ in (9) and (10) are the (two-dimensional) carrier densities in the doping layers. The modulation of the band edges, which was given by (2) for the compensated n-i-p-i crystal in its ground state has now to be replaced by

$$2V_{exc} = (2\pi e^2/\kappa_o)\,[n_D(d_n^+)^2 + n_A(d_p^-)^2] \tag{11}$$

with

$$d_n^+ = (d_n - n^{(2)}/n_D)/2 \tag{12}$$

$$d_p^- = (d_p - p^{(2)}/n_A)/2 \tag{13}$$

$$n_D d_n^+ = n_A d_p^- \tag{14}$$

where (14) expresses the reqirement of (macroscopic) charge neutrality of the n-i-p-i structure. The important relation between effective band gap in the excited state, $E_g^{eff,exc}$ and two-dimensional carrier densities $n^{(2)}$ and $p^{(2)}$ is, in fact, rather well described by replacing V_o in (8) by V_n from (11), as apparent from a comparison with the nearly parabolic band gap vs. carrier density relation found in numerical self-consistent calculations (see Fig. 2 for an example).

In the introduction we mentioned, that the tunability of the electronic structure is a consequence of the extremely long excess carrier lifetimes which, again, result from the spatial separation of electrons and holes. In spite of the spatial separation the lifetimes are not infinitely long as the carriers can recombine either by tunneling through, or, at high temperatures, by thermal excitation accross the potential barrier indicated in Fig. 1(b).

A useful analytical expression for estimating the lifetime enhancement for radiative and non-radiative tunneling recombination between electrons and holes as a function of design parameters and degree of excitation, i.e. carrier concentration and effective band gap, $E_g^{eff,exc}$ is the following [20]

$$\tau_{tun}^{nipi}/\tau^{bulk} \simeq \exp[4(E_g^O - E_g^{eff,exc})/\hbar(\omega_c + \omega_{vh})] \tag{15}$$

with ω_c and ω_{vh} given by (6) and (7). A similar expression applies for the lifetime enhancement factor for vertical recombination of thermally excited carriers at finite temperatures:

$$\tau_{th}^{nipi}/\tau^{bulk} \simeq \exp[(E_g^O - E_g^{eff,exc})/kT] . \tag{16}$$

A comparison of (15) and (16) reveals, that tunneling or thermal recombination processes will predominate at a given temperature T, depending on whether

$$kT < \text{or} > kT_O \equiv \hbar(\omega_c + \omega_{vh})/4 \tag{17}$$

but, interestingly, independent of the value of $E_g^O - E_g^{eff,n}$. For a given host material the condition (17) depends only on the doping level. For GaAs, e.g., tunneling will still predominate at room temperature, if $n_D = n_A = 3 \times 10^{18}$ cm^{-3}, or larger. This is important with respect to the question whether the room temperature luminescence is tunable or not, as we will see later.

3. Modulation of the Electronic and Optical Properties by Light and/or External Bias

The most obvious way to modulate the carrier concentration in a n-i-p-i crystal is by the absorption of photons (Fig.3(a)). The photo-excited electrons and holes will relax to the lowest conduction and uppermost valence subbands within very short times (typically within ps or less) by the emission of phonons. The efficiency for this process is very large, since the competing electron-hole recombination processes are much slower, even in

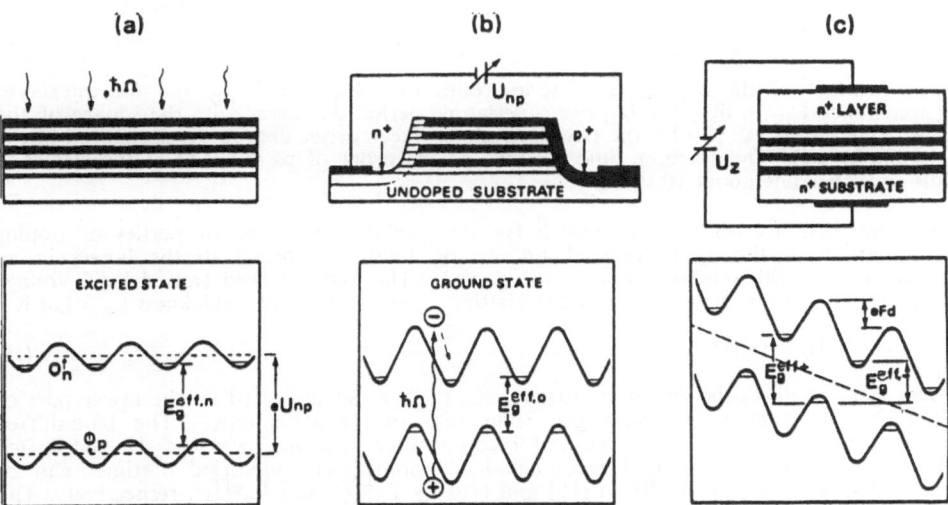

Fig. 3 Optical and electrical modulation of a n-i-p-i crystal. (a) by light; (b) by injection or extraction of electrons and holes using "selective" n- and p-type contacts; (c) by an electrical field normal to the layers.

the unmodulated host material. This is true, *a fortiori*, in n-i-p-i crystals, because of the enhancement of the lifetimes due to spatial separation. The variation of carrier concentration does not only result in a change of the effective band gap and of the electron-hole recombination lifetime as discussed in the previous section. We will show later on, that also the absorption coefficient, the refractive index, the luminescence spectrum and optical gain are strongly affected as well. The characteristic times and the required intensities for this light-induced modulation of properties cover a very wide range, depending on the design of the samples.

The light-induced changes of the optical properties depend on the number of photo-induced carriers. Because of the very long recombination lifetimes the number of photo-induced carriers in many cases depends on the light fluence, i.e. the *product* of intensity and time and *not* on the intensity itself. Consequently,the turn-on time for the photo-induced effects is inversely proportional to the intensity, even at low excitation density. Because of the long recombination lifetimes the turn-off times tend to be particularly long. There are, however, ways to overcome these long turn-off times by quenching of the photo-response by means of electrical contacts as described in the following. It is quite clear, that modulation of the n-i-p-i system by light includes self-induced transient behavior and non-linearities of absorption coefficient and refractive index.

Modulation of a n-i-p-i crystal can also be achieved by an external voltage U_{np} applied through "selective contacts". Such selective contacts, which exhibit ohmic behavior to one type of doping layer, but are blocking with respect to the layers of opposite type of doping are a quite unique feature of doping superlattices. They can be achieved in many different ways. A version which is suitable for application to mesa structures is shown schematically in Fig. 3(b). Carriers will flow through the n-contact into the n-layers and through the p-contact into the p-layers of the n-i-p-i structure until the difference between the electron quasi Fermi level ϕ_n and hole quasi Fermi level ϕ_p equals the external potential eU_{np} applied between the contacts. An obvious advantage of modulating bandgap and carrier concentration in this way is the possibility to inject *or* extract carriers with such electrodes. This improves the frequency response of the device dramatically, if the electrode pattern has a favorable geometry. The time constant for changes induced by variation of eU_{np} is

$$\tau_{np} = (R_{nn} + R_{pp}) \, C_{np} \tag{18}$$

The product of the doping layer series resistances R_{nn} and R_{pp} and the interlayer capacitance $C_{np} = dQ^{(2)}/dU_{np}$ per superlattice period decreases with the square of the electrode distances. ($Q^{(2)}$ is the two-dimensional free carrier charge density stored in the doping layers). The response time becomes of the order of ns in GaAs, if the electrode distances become about 10 μm.

A fundamentally different approach for the modulation of the properties of doping superlattices is the application of an electric field F_z normal to the layers in an arrangement with sandwich contacts (Fig. 3(c)). The average field caused by a voltage U_z accross a n-i-p-i structure of N superlattice periods with a total thickness $L_z = Nd$ is

$$F_z = U_z/L_z = U_z/(Nd) \tag{19}$$

This field modifies the electronic structure in the way as indicated on the lower part of Fig. 3(c). The effective band gap splits into two contributions. The blue-shifted contribution, $E_g^{eff,-}$, exhibits reduced recombination lifetimes, whereas the red-shifted one, $E_g^{eff,+}$ yields increased recombination lifetimes. The modified lifetimes can be estimated by replacing E_g^{eff} in (15) and (16) by $E_g^{eff,-}$ and $E_g^{eff,+}$, respectively. This applies independent of whether the crystal is in its ground state or in an excited state, when the field F_z is applied. The key feature by which the present case differs from the modulation by selectice contacts is the ultrafast reponse time. Independent on whether we are dealing with the variation of absorption, optical gain or luminescence energy or intensity, the response to changes of U_z is quasi-instantaneous, since the modulation is not associated with changes of the carrier concentrations in the layers. Thus, the relevant time constant is no longer given by (18), but, instead, by

$$\tau_z = R_z C_z . \tag{20}$$

R_z is the resistance of the sandwich contacts and the external circuit and can be made very small, if the n-i-p-i structure is sandwiched between strongly doped top and bottom layers. The capacitance C_z also is much smaller than C_{np} in Eq. (18) since the n-i-p-i structure behaves as a dielectric of total thickness L_z with respect to the external voltage U_z.

4. Properties of Doping Superlattices and Device Applications

The tunability of the electron and hole conductivity parallel to the layers by carrier injection and extraction [8], of the photo-[9,10,13] and electro-[12] luminescence spectra, of the optical gain spectra [14], of the subband structure [9,15] and of the absorption coefficient [11] has been discussed previously and has been demonstrated by experiments on GaAs doping superlattices which were grown by molecular beam epitaxy (See Ref. [2] for a review). In the following we will restrict our discussion to recent theoretical and experimental results and on their possible device applications.

4.1 Photoresponse and Photodetectors.

The steady-state photoinduced increase of carrier concentration in a semiconductor is given by the product of carrier generation rate times recombination lifetime

$$\Delta n = (dn/dt) \, \tau_{rec} \tag{21}$$

Because of the long recombination lifetimes the photoresponse of n-i-p-i crystals becomes by many orders of magnitude larger than in bulk semiconductors. Although the experimentally determined lifetimes are significantly shorter than those calculated from (15) and (16) in samples with low doping level and thick layers due to recombination at deep traps and/or at non-perfect selective contacts, we have obtained lifetimes of a few seconds in GaAs n-i-p-i's with doping levels in the 10^{16} cm^{-3} range [25].

The selective contacts have two interesting and useful consequences for the photoresponse of n-i-p-i crystals. In bulk semiconductors with ohmic contacts for the high-mobility carriers (usually the electrons) the effective lifetime to be used in (21) is limited by the "sweep-out" [26] of the low-mobility carriers at these contacts. This means, that the photocurrent, which is proportional to (21) and to the applied voltage saturates if the drift time of the low-mobility carriers decreases below τ_{rec}. Therefore, the photoconductive gain, defined as the number of charges measured per absorbed photon, is limited to $g < \mu_n\mu_p+1$. The selective n-contacts ; however, "repel" the holes in n-i-p-i crystals. Thus, the gain, now being given by

$$g = \tau_{rec}^{nipi}/\tau_{transit}, \text{ with } \quad \tau_{transit} = L_x^2/\mu_n U_{nn}, \tag{22}$$

can become huge.

With a contact distance $L_x = 0.5$ mm and a voltage of $U_{nn} = 1$ V values of $g > 10^7$ have recently been obtained in GaAs n-i-p-i's at room temperature [25].

The second favorable property of selective contacts relates to the dark currents. In order to achieve large signal to noise ratios a low dark conductivity is desirable [26].

Uniform bulk crystals have to be very pure in order to exhibit neither significant electron nor hole conductivity. In n-i-p-i crystals, however, low dark *electron* conductance is most easily obtained by an excess of p-doping. Thus, such n-i-p-i crystals represent photodetectors with extremely high *responsivity* and *detectivity*.

The "independence" of electrons and holes in n-i-p-i crystals has many other interesting implications for photodetector applications [27,25]. Electrons can be swept laterally into small areas by built-in or by external drift fields and recombination lifetimes can be externally adjusted. This allows to make fast low-noise photodetectors with high gain-bandwidth products [27].

Finally we mention, that the absorption at photn energies below the band gap allows to extend the spectral response of n-i-p-i detectors to long wavelengths (See Section 4.3).

4.1 Tunable photo and Electro Luminescence and its Device Applications

The energy of photons emitted in the tunneling recombination process corresponds to the *tunable* effective band gap, E_g^{eff}. The tunneling had been found to represent an efficient light emitting process at low temperatures [9-11]. At room temperature there are also thermally activated vertical recombination processes. Tunable room temperature luminescence becomes impossible, if the latter processes dominate since, in this case, the energy of the emitted photons will correspond to the energy of the band gap of the host material, E_g^0. The theoretical estimate given in section 2 ,in fact, predicts that tunneling recombinations will prevail at room temperature only, if the doping concentrations are

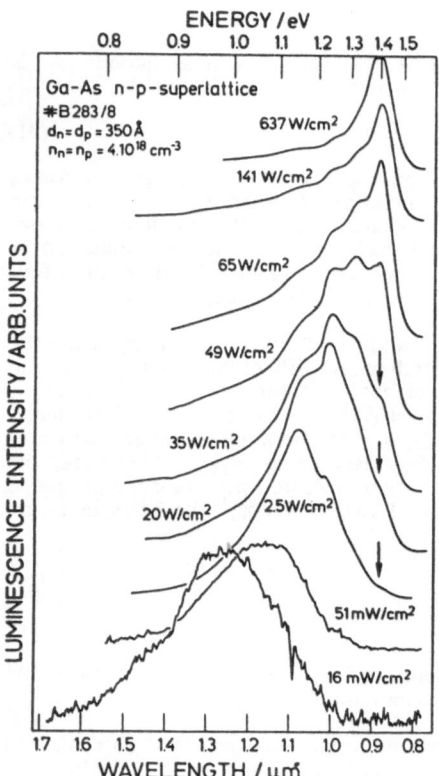

Fig. 4 Room temperature luminescence spectra of a GaAs n-i-p-i doping superlattice with a superlattice period of 70 nm and doping concentrations $n_D = n_A = 4\times10^{18}$ cm^{-3}. At low excitation density the luminescence maximum is red-shifted by about 30% away from its uniform bulk value.

higher than 3×10^{18} cm^{-3} for GaAs, e.g.. This was recently verified by room temperature photo-luminescence experiments [29]. Fig. 4 shows luminescence spectra of a strongly doped GaAs n-i-p-i as a function of the excitation density. The calculated value of $E_g^{eff,0}$ for this sample is zero. With decreasing excitation density the luminescence spectrum, indeed, shifts towards away from the bulk GaAs band gap by about 30%.

With respect tp tunable light sources electrical excitation is, of course, much more attractive than photo-luminescence. By using a new concept for the selective n- and p-contacts [29] we have recently been able to observe tunable electroluminescence with an efficiency of more than 1% at room temperature. This represents a dramatic improvement by several orders of magnitude compared with previous results [12].

The modulation frequency, which is another important property characterizing light emitting devices, is not determined by the long recombination lifetime in n-i-p-i crystals. It is rather more related to the RC time constant τ_{np} from Eq. (18), although it can be substancially faster than its value. The most suitable design, a narrow-stripe edge emitting device, should allow for a modulation bandwidth of about 1 GHz for 10 μm stripe width.

By orders of magnitude faster modulation becomes possible by the use of the sandwich arrangement for the contacts (Fig. 3(c)). Fig. 5 shows the simplest device of this kind, an optically excited and electrically modulated light source. The emitted spectra shift by an amount $\pm eF_z(t)d/2$ around their steady-state position if the modulation frequency of the voltage $U_z(t)$ varies fast compared with the recombination lifetime in the n-i-p-i sample. The design for a purely electrical version of this device with lateral injection and normal modulation is less obvious, but also possible.

Usually it will not be desirable to create simultaneously red- and blue-shifted spectra. The wide flexibility in designing n-i-p-i structures easily allows to resolve this potential disadvantage by choosing an asymmetric configuration, to suppress all the recombination processes of electrons from the n- to their left-hand-side neighboring p-layers, for instance. This can be achieved either by an asymmetric doping profile or by using the idea of a "hetero n-i-p-i" [30,31], with a periodic layer sequence of the type p-Al$_x$Ga$_{1-x}$As / n-Al$_x$Ga$_{1-x}$As / i-GaAs, for instance.

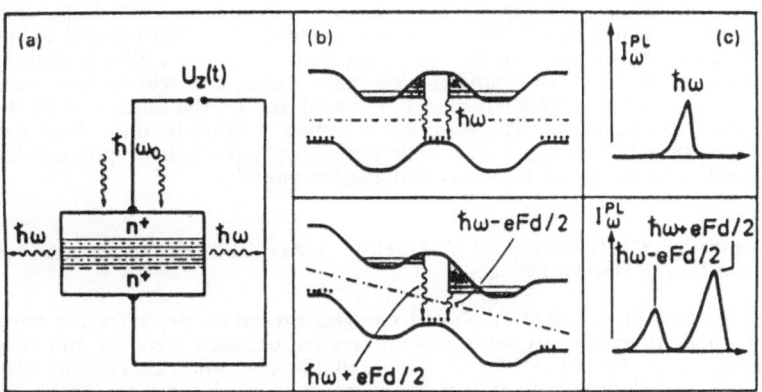

Fig. 5   Device for ultrafast modulation of photoluminescence by an electric field F_z. (a) n-i-p-i sample with low-resistance sandwich contacts. (b) Real space energy diagram and (c) luminescence spectra without (upper part) and with (lower part) external potential U_z applied. The arrows pointing to the dots in part (b) indicate the radiating tunneling recombination processes between electrons in the conduction subbands and holes occupying the accceptor impurity band.

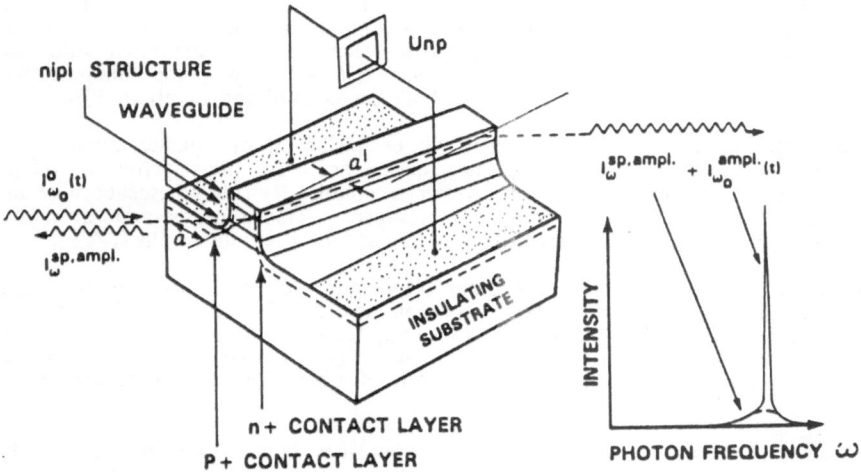

Fig. 6 Optical amplifier, schematically. The angles α and $\alpha' \neq 0$ for the light outside and inside the device should indicate that there is no Fabry-Perot cavity.

Lasers based on n-i-p-i structures have a number of appealing features. They are expected to exhibit low values of threshold current since population inversion can be obtained at much lower injection level than in conventional semiconductor lasers, due to the increased recombination lifetimes. The optical-gain spectrum of n-i-p-i crystals can be tuned over a wide range, as it is directly related to the tunable band gap, $E_g^{eff,exc}$ [18]. Therefore, a n-i-p-i crystal in combination with a grating as a tunable frequency-selective mirror, for instance, seems very appealing for the construction of a tunable laser with wide modulation range.

Low values of gain in n-i-p-i structures occur together with low values of spontaneous luminescence. This combination of properties provides favorable conditions for an optical amplifier, as shown schematically in Fig. 6 . The structure will emit a weak and rather broad spontaneous luminescence signal, which will be amplified for light emitted in the x-direction. If, however, the intensity of the incoming external signal $I_\omega^0(t)$ is already much higher than the amplified background luminescence signal, this will be true even more for the amplified signal $I_\omega^0(t)(1-R)^2 \exp(g_\omega L_z)$ coming out on the other side of the device. The emission at the photon energy hω is not expected to saturate, since there will always be a sufficiently wide energy range with population inversion to prevent the depletion of occupied states by the increased recombination intensity.

4.3 Tunable Absorption Coefficient and Refractive Index and its Application to Modulators and Non-Linear Optical Devices.

Absorption of light occurs in principle at photon energies exceeding the effective band gap, $E_g^{eff,exc}$ (or, more precisely, exceeding the difference between electron and hole quasi Fermi levels $\phi_n - \phi_p$). Near $E_g^{eff,exc}$ the value of the absorption coefficient, $\alpha(\omega)$ may, however, be very small, depending on the design parameters. In fact, the factor by which α is reduced compared with typical values at about 100 meV above the absorption edge of the host material is of the order of the inverse of the lifetime enhancement given by (15). With increasing photon energy the overlap between the wave functions for the initial and the final state of the absorption process increases exponentially. In the semiclassical limit a calculation [11] of $\alpha(\omega)$ gives an exponential tail, similar to the Franz-Keldysh absorption tail in uniform semiconductors with external fields, which extends from the bulk band gap E_g^0 down to $E_g^{eff,exc}$. The slope of this exponential tail varies strongly with the degree of excitation. For n-i-p-i systems of high doping level and short superlattice period a calculation using the real subband energies (4) and (5) and wave functions is necessary. In this case a strong step-like structure of $\alpha(\omega)$ is found [23].

Although large changes in the transmission coefficient of a n-i-p-i structure can be achieved in either case, the latter one appears as the more attractive one for absorptive optical modulators for monochromatic light. We will discuss the concept of an ultrafast modulator using a wave guide n-i-p-i structure with the active material sandwiched between layers of a material with lower refractive index, which also forms sandwich contacts to the sample [4]. In Ref. [23] it was found, that the absorption coefficient of a n-i-p-i crystal with high doping level but short superlattice period drops from values of the order of a few 100 cm^{-1} to a value close to zero near the effective band gap $E_g^{eff,0}$. Thus, a light signal of photon energy $\hbar\omega$ slightly less than $E_g^{eff,0}$ can propagate with low attenuation over rather long distances in the direction parallel to the layers of such a n-i-p-i crystal. The decrease in effective band gap $E_g^{eff,+}$ induced by an external field $F_z(t)$, however changes the attenuation drastically because of the sudden increase of the absorption coefficient once the effective band gap becomes smaller than the photon energy. Due to the large *relative* changes of the absorption coefficient the intensity transmitted over a distance L, $I_0\exp(-\alpha L)$ can be modulated from close to unity down to very low values, if α is varied by a factor of 10 or 20. This means, that the extinction ratio of the device is very high. The frequency response of this modulator, again, is only limited by the short time constant τ_z. A floating of the absorption edge is prevented by the relatively short lifetime of the photo-generated carriers in such a structure. An aspect, which possibly can be of interest is that in principle, the device can be operated in a mode where it switches back and forth between absorption and optical gain as a function of the external bias $U_z(t)$, provided the photon energy is slightly higher than the effective band gap, $E_g^{eff,exc}$.

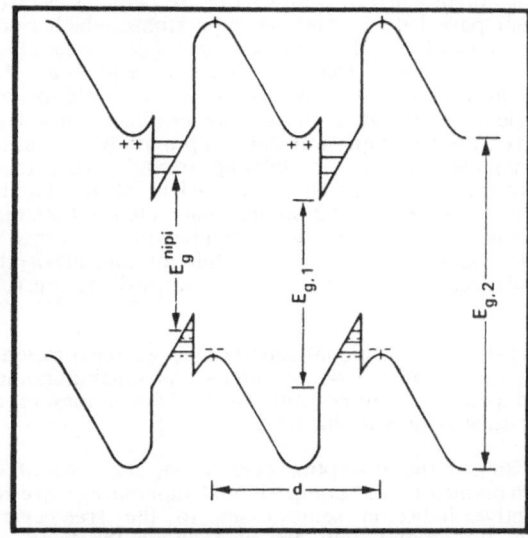

DIRECTION OF PERIODICITY z

Fig. 7 "Hetero n-i-p-i" structure for fast absorptive modulators. A large extinction ratio is achieved if light is travelling through a wave guide structure, with the hetero n-i-p-i sandwiched between a semiconductor material with lower refractive index.

281

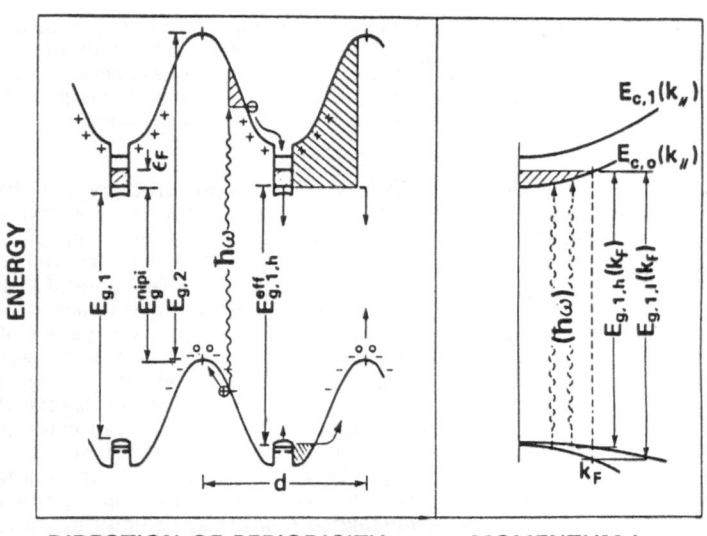

DIRECTION OF PERIODICITY z   MOMENTUM k₍₎

Fig. 8 "Hetero n-i-p-i" for non-linear optical phenomena. Absorption in semiconductor 2 below the gap $E_{g,2}$ is small, but finite. Absorption in the quantum well above $E_{g,1,h}^{eff}$ is large, if it is not forbidden due to band-filling, as shown in the momentum space diagram.

A more sophisticated version of the present modulator is obtained by replacing the simple n-i-p-i structure by an asymmetric hetero n-i-p-i, as discussed in connection with the light emitting structures. The advantage will be a sharper absorption edge and, therefore, very low voltages required for modulation. In this hetero n-i-p-i structure (Fig. 7) electrons and holes are well separated from their parent donor and acceptor atoms, which results in a strong reduction of spatial fluctuations of the effective band gap [32]. It should be pointed out that this version has also some similarity with a fast compositional-superlattice modulator, investigated by Chemla et al. [6]. We believe, however, that our structure has significant advantages. As we are dealing with a large built-in electrical field in the GaAs layer, the modulation of band-gap and absorption is a first order effect, even at low external bias. We are modulating around zero external bias. Dark currents in z-direction are particularly small, as the height of the barriers separating adjacent GaAs layers are increased by the additional space charge potentials. The photo-induced currents are essentially only intra-layer recombination currents, i.e. confined to the GaAs layers. Effective band gap, band gap modulation and absorption coefficient at threshold can be optimized without restrictions imposed by currents becoming too high.

Recently it has been demonstrated [7] that compositional superlattices exhibit properties which make them extremely interesting as bistable optical switches and other non-linear optical devices. The properties of doping superlattices add further dimensions to the possible applications of semiconductor superlattices in this field.

The largest contributions to nonlinearities of the absorption coefficient, $\alpha(\omega)$, and of the refractive index, $n_{opt}(\omega)$, in bulk semiconductors and compositional superlattices are due to band-filling and, for the refractive index in some cases, to the free carrier polarizability [33]. These phenomena are usually only observed above the band gap where the absorption coefficient is generally very high under low illumination intensity.

Because of the relatively short recombination life times, high illumination intensities at photon energies above band gap are required to generate enough free electrons for inducing and maintaining significant changes of $\alpha(\omega)$ and $n_{opt}(\omega)$. The power required in $Al_xGa_{1-x}As$ / GaAs superlattices has been found to be significantly lower than in bulk GaAs [7].

The phenomenology of non-linear optical effects in conventional n-i-p-i doping superlattices differs in two remarkable points from the compositional ones. First, the long recombination lifetime implies very low intensities necessary to create a large increase in free carrier concentration and, consequently, also low power to induce non-linear optical effects. Secondly, the extended absorption tail below E_g^0 and its strong dependence on the photo-induced carrier concentration allows for large light-induced changes of light transmission through wave guide structures of more than a few μm length. The transparent state of "absorption saturation" is characterized by low power dissipation needed to maintain this excited state. The unusual combination of these two properties makes possible the coexistence of weak absorption and yet relatively large changes of the refractive index, even at low light intensities. This is a very appealing combination. It implies, that the optical length of a n-i-p-i crystal for light travelling parallel to its layers in a Fabry-Perot cavity may change by quite a bit more than half a wave length while the light attenuation $I_0[1-\exp(-\alpha L)]$ is still small. This may not only allow for optical bistability but even make possible multi-stability, yet at low light intensities.

Finally, also with respect to optical non-linearities the hetero n-i-p-i's exhibit particularly interesting and for applications desirable properties. We confine our considerations to a version which differs from the previous hetero n-i-p-i by having the undoped lower band gap material at the center of the n-doping layer (Fig. 8). For photon energies above the effective band gap of the quantum well formed by this material, $E_{g,1}^{eff}$, but below the band gap of the wider band gap material, $E_{g,2}$, the absorption coefficient at the beginning will be as high as in a familiar compositional superlattice. The photo-excited holes in the valence band section of the quantum well encounter two channels. Apart from intra-quantum-well recombination with the electrons there is the possibility to escape into the larger band gap material by thermal excitation across the hole quantum well barrier, subsequently being collected at the center of this layer, where the probability for recombination with the spatially separated electrons is very low. With appropriate design the room temperature probability for this escape process is exceeding the recombination probability. Therefore, there will be the most efficient possible built-up of electron density in the quantum well until the absorption finally drops due to band-filling and the associated Burstein-Moss shift [34] of the absorption edge, $E_{g,1,h}(k_F)$ (See momentum space diagram in Fig. 8). In contrast to the familiar case of band filling, where absorption would just saturate at $E_{g,1,h}(k_F) = \hbar\omega$ at a rather high power density level, the typical weak, but finite n-i-p-i absorption for $\hbar\omega < E_{g,2}$ persists and continues to fill-up the quantum wells with electrons, thus shifting the threshold for absorption in the quantum well to even higher values. This new phenomenon of "oversaturation of the quantum well absorption" allows for switching from very high to very low absorption and also for conservation of the transparent state over a longer period of time after switching off the illumination. Also it should be mentioned, that the light induced changes of the refractive index ehibit maximum values at those energies. close to $E_{g,1,h}^{eff}$.

5. Conclusions

In this lecture we have tried to give some idea about the variety of phenomena which occur in n-i-p-i doping superlattices and of the wide range of novel opto-electronic and purely optical devices which can be made by using the quite unique electronic properties of n-i-p-i doping superlattices. Although the practicality of the *device concepts* has been demonstrated only in some cases yet we are quite confident, that this will also happen for the others as well within the near future. This optimism is based on our very encouraging experiences which we have had in the past in observing all the *phenomena* in n-i-p-i structures which had been predicted long time before the first samples had been grown.

References

[1] L. Esaki in: in *Proceedings of the 17th Int. Conf. on the Physics of Semiconductors*, J.D. Chadi and W.A. Harrison, ed., Springer, New York, 473 (1985)
[2] K. Ploog and G.H. Dohler, Advances in Physics 32, 285 (1983)
[3] W.T. Tsang, IEEE QE-20, 1119 (1984)
[4] G.H. Dohler, Superlattices and Microstructures, 1, 279 (1985)
[5] F. Capasso in *Proceedings of the 17th Int. Conf. on the Physics of Semiconductors*, J.D. Chadi and W.A. Harrison, ed., Springer, New York, 1537 (1985)
[6] D.S. Chemla, T.C. Damen, D.A.B. Miller, A.C. Gossard, and W. Wiegmann, Appl. Phys. Lett. 42, 864 (1983); For a review see: N. Peyghambarian and H.M. Gibbs, Optical Engineering, 24, 68 (1985)
[7] G.H. Dohler, Phys. Stat. Sol. (b), 52, 79 and 533 (1972)
[8] K. Ploog, H. Kunzel, J. Knecht, A. Fischer, and G.H. Dohler, Appl. Phys. Lett. 38, 870 (1981)
[9] G.H. Dohler, H. Kunzel, D. Olego, K. Ploog, p. Ruden, H.J. Stolz, and G. Abstreiter, Phys. Rev. Lett. 47, 864 (1981)
[10] H. Jung, G.H. Dohler, H. Kunzel, K. Ploog, P. Ruden, and H.J. Stolz, Solid State Commun. 43, 291 (1982)
[11] G.H. Dohler, H. Kunzel, and K.Ploog, Phys. Rev. B 25, 2616 (1982)
[12] H. Kunzel, G.H. Dohler, P. Ruden, and K. Ploog, Appl. Phys. Lett. 41, 852 (1982)
[13] W. Rehm, H. Kunzel, G.H. Dohler, K. Ploog, and P. Ruden, Physica 117B and 118B, 732 (1983)
[14] H. Jung, G.H. Dohler, E.O. Gobel, and K. Ploog, Appl. Phys. Lett. 43, 40 (1983)
[15] J.C. Maan, Th. Englert, H. Kunzel, A. Fischer, and K. Ploog, J. Vav. Sci. Technol. B 1, 289 (1983)
[16] K. Ploog, A. Fischer, and H. Kunzel, J. Electrochem. Soc. 128, 400 (1981)
[17] G.H. Dohler, Surface Sci. 73, 97 (1978)
[18] P. Ruden and G.H. Dohler, Phys. Rev. B 27, 3538 (1983)
[19] P. Ruden and G.H. Dohler, Phys. Rev. B 27, 3547 (1983)
[20] G.H. Dohler, J. Vac. Sci. Technol. B 1, 278 (1983)
[21] P. Ruden, J. Vac. Sci. Technol. B 1, 285 (1983)
[22] G.H. Dohler and P. Ruden, Surface Science, 142, 474 (1984)
[23] G.H. Dohler and P. Ruden, Phys. Rev. B 30, 5932 (1984)
[24] P.P. Ruden and G.H. Dohler, in *Proceedings of the 17th Int. Conf. on the Physics of Semiconductors*, J.D. Chadi and W.A. Harrison, ed., Springer, New York, 535 (1985)
[25] G.H. Dohler, G. Trott, and J.N. Miller, to be published
[26] For instance: C.T. Elliot in *Handbook on Semiconductors*, T.S. Moss, ed., North Holland, Amsterdam, 1981, Vol. 4, p. 746
[27] G.H. Dohler, Superlattices and Microstructures, 1, 427 (1985)
[28] G.H. Dohler, CRC, Critical Reviews in Solid State and Materials Sciences, to be published
[29] G.H. Dohler, G. Fasol, T.S. Low, J.N. Miller, and K.Ploog, Solid State Commun. 57, 563 (1986)
[30] G.H. Dohler, G. Hasnain, and J.N. Miller, to be published
[31] G.H. Dohler, Physica Scripta 24, 430 (1981); R.A. Street, G.H. Dohler, J.N. Miller, and P.P. Ruden, submitted for publication
[32] H.L. Stormer, R. Dingle, A.C.Gossard, W. Wiegmann, and R.A. Logan, Inst. Phys. Conf. Ser. 43, 557 (1979)
[33/ F. Stern, Phys. Rev. 133, A1653 (1964)
[34] E. Burstein, Phys. Rev. 93, 632 (1954)

Electronic Excitations
in Microstructured Two-Dimensional Systems

D. Heitmann

Institut für Angewandte Physik, Jungiusstr. 11,
D-2000 Hamburg 36, Fed. Rep. of Germany
Present address: Max-Planck-Institut für Festkörperforschung,
D-7000 Stuttgart 80, Fed. Rep. of Germany

Abstract. Lateral structures of submicron periodicity have been fabricated on two-dimensional (2D) electronic systems. In particular,
systems have been realized where the originally homogeneous 2D
charge density is spatially modulated on a submicron scale. We will
discuss the excitation of plasmons and intersubband resonances in
these systems. Novel electronic properties are observed, e.g. minigaps in the plasmon dispersion due to the superlattice effect of the
periodical charge density modulation on the plasmon dispersion.

1. Introduction

The most important technical applications of two-dimensional electronic
systems (2DES) are metal-oxide-semiconductor (MOS) devices [1] based on Si
and, with increasing importance, equivalent AlGaAs-heterostructure devices, e.g. high electron mobility transistors [2]. Higher integration and
performance make it necessary to reduce the lateral dimensions of these
devices which now approach, in particular for experimental samples, characteristic lengths of the 2DES, e.g. elastic or inelastic scattering
lengths, ballistic transport lengths, Fermi wavelengths or cyclotron
radii. This leads to strong modification of the lateral transport. Small
lateral structures make it also possible to realize novel devices based on
an electrically controlled vertical transport, e.g., permeable base transistors [3], static induced transistors [4], hetero-bipolar transistors
[5] or similar devices [2]. Also lateral p-n-superlattices have been prepared [6]. The technical importance, but equally well general interest,
has stimulated an increasing number of fundamental investigations of 2DES
with small lateral confinement (e.g.[7-9]). Here I will report on the
fabrication and investigation of 2DES with lateral structures of submicron
periodicities. Lateral structures are used as grating couplers and, in
particular, to prepare electronic systems with spatially modulated charge
densities. In these systems novel physical effects, e.g. superlattice
effects on the plasmon dispersion, are observed.

2. Fabrication of Lateral Structures by Holographic Lithography

A set-up for holographic lithography is shown schematically in Fig.1. The
superposition of two expanded coherent laser beams creates a sinusoidal
intensity modulation. A photoresist on top of the sample is exposed and
forms after the development a periodic surface relief. The periodicity
$a=\lambda/2\sin\delta$ of the profile depends on the wavelengths λ of the laser and the

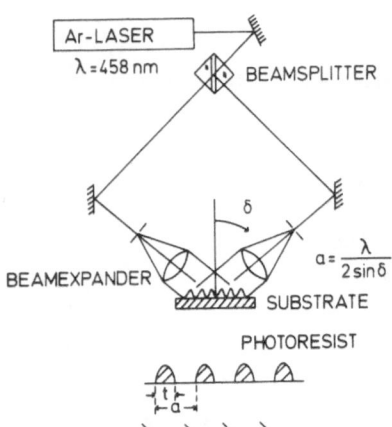

Fig. 1:

Holographic lithography set up for the preparation of small periodic structures (left) and electron micrograph of a photoresist structure with 250 nm periodicity and 50 nm groove width.

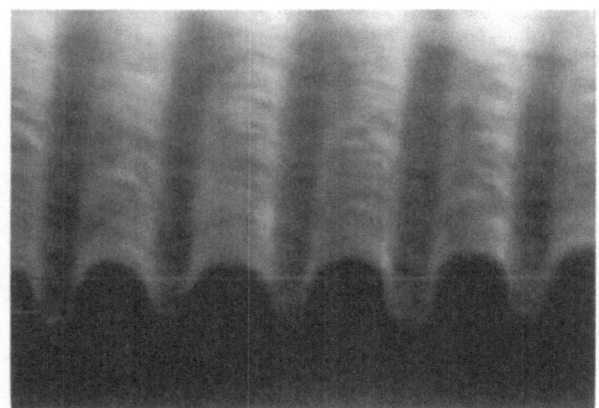

angle of incidence δ. The shape of the photoresist profile can be controlled by the time of the exposure and the development and by additional dry etching processes in an O_2-plasma. In Fig.1 an electron micrograph of a photoresist profile is shown. The periodicity of the profile is $a \approx 250$ nm and the groove width is about 50 nm. Holographic lithography assures an excellent homogeneity of the profile over large areas (1 cm^2 and more). Periodic metal stripes can be prepared by evaporating a metal with a well defined angle of incidence onto the surface relief (shadowing) and then lifting of the photoresist. The photoresist, or if one wants to reverse the structure, shadowed metal stripes, can be used as a mask to etch into the sample. An example is shown in Fig.2a, where the oxide of an Si-MOS capacitor has been etched through the photoresist mask. The thickness of the oxide is then d_1 in the region t_1 and d_2 for the rest $t_2 = a - t_1$ of the period. Via a gate voltage V_g, which is applied to a continuous gate, an electronic system with modulated charge density $N_{si} = \varepsilon \cdot (V_g - V_{ti})/e \cdot d_i$ ($i=1,2$; V_{ti} = threshold voltage) is induced.

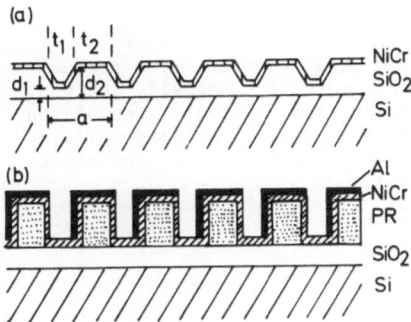

(a)

(b)

Fig. 2:

Microstructured MOS systems with modulated insulator thickness: (a) modulated oxide, (b) modulated photoresist (PR). Via a continuous NiCr gate a modulated charge density can be induced at the $Si-SiO_2$ interface.

A similar structure with a two-layer insulator, SiO_2 and photoresist, is shown in Fig.2b. Here the modulation of the photoresist leads to a high aspect ratio ($d_2/d_1 \approx 7$). The thin thermally grown oxide layer assures a low number of surface states and scatterers. Techniques to prepare microstructured 2DES systems are described in more detail in Refs.15 and 16.

3. Grating Coupler Spectroscopy

Two types of resonances characterize the dynamic excitation spectrum of the 2DES.

a) 2D plasmons [17], the resonant collective intraband excitation of the 2DES representing resonant oscillations of the carriers in the 2D plane (=x-y-plane). 2D plasmons show a characteristic dispersion [1,17]

$$\omega_p^2 = e^2 \cdot N_s \cdot q / 2 \cdot \bar{\varepsilon} \cdot m_p \tag{1}$$

Here q is the plasmon wave vector, $\bar{\varepsilon} = \bar{\varepsilon}(\omega,q)$ an effective dielectric function and m_p the plasmon mass.

b) Intersubband resonances (ISR), the resonant transitions between the different subbands that exist in the 2D system (e.g. [1,18,19]).

Far-infrared (FIR) spectroscopy is a powerful tool to investigate these excitations (e.g., [20] and references therein). In the arrangement of Fig.3 FIR radiation is transmitted normally through the sample which has a semitransparent gate. The relative change of transmission

$$\Delta T/T = (T(V_g) - T(V_t))/T(V_t) \propto Re\ \sigma\ (\omega,q) \tag{2}$$

is measured. For small signals $\Delta T/T$ is proportional to the effective dynamic conductivity of the sample.

In a homogeneous sample (N_s=const) and with an unstructured gate, the incident radiation only has electric field components E_x parallel to the interface. Thus in this case only the non-resonant Drude conductivity $\sigma_p(\omega,q=0)$ is measured. For a periodic structure, e.g., for periodic metal stripes, a modulated oxide or a modulated charge density, the electric field is spatially modulated. It has now field components $\bar{e}_n(\omega,q_n)$ of wave vector $q_n=n \cdot 2\pi/a$. In the experiments the wavelengths λ of the FIR radi-

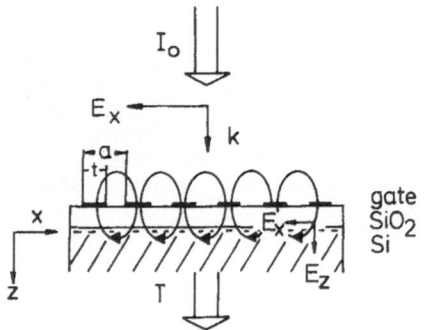

Fig. 3:

Grating coupler effect of a periodical structure. For normally incident
FIR radiation spatially modulated
parallel ($e_x(\omega,q)$) and perpendicular
($e_z(\omega,q)$) electric field components
are induced.

ation is small compared with a. Thus these field components only exist in
the near field. Here the fields can couple with 2D plasmons of the same
wave vectors [21]. We will come back to plasmons in Sect.5. Also of importance is that the grating coupler induces E_z-components of the electric
field perpendicular to the interface, as is shown schematically in Fig.3.
Via these E_z-components intersubband resonances can be excited [22,13].
For a highly symmetric and parabolic surface band structure, e.g.,
Si(100), perpendicular field components are necessary to excite ISR. So
far stripline [18] and prism arrangements [19] have been used to investigate ISR on Si (100). We will show in the next section that grating couplers are a very efficient method to excite ISR.

4. Grating Coupler Induced Intersubband Resonances

In Fig.4 experimental spectra measured on Si(100) samples with homogeneous change density are shown. The grating periodicity is a=1800nm. For
the charge density $N_s=3.3\cdot10^{12}cm^{-2}$ two resonances E_{01} and E_{02} are observed
which correspond, respectively, to resonant transitions from the lowest
subband 0 to the first and second excited subbands. With increasing N_s the
resonances shift to higher frequencies, corresponding to a larger subband
separation in the steeper potential well at larger surface electric
fields. For charge densities $N >8\cdot10^{12}cm^{-2}$ additional resonances E_{01} are
observed which can be attributed to resonant transitions in the primed
subband system [1]. The primed subband system arises from the projection
of four volume energy ellipsoids of Si onto the Si(100) surface and is
separated in k-space by $0.86\cdot2\pi/A$ (A crystal lattice constant) in [001]
and equivalent directions. It is known that this subband system is occupied for $N_s>7.5\cdot10^{12}cm^{-2}$ [23].

The ISR-frequency differs from the subband separation due to depolarization and exciton effects [1]. For Si(111) and Si(110) the surface band
structure results from the projection of volume energy ellipsoids which
are tilted with respect to the surface. Here so-called parallel excitation
of ISR with parallel electric field components E_x is possible. Parallel
excitation is not effected by the depolarization shift. In grating coupler
induced ISR spectroscopy on Si(111) and Si(110) both resonances, directly
parallel excited resonances and perpendicular grating coupler excited -
and thus depolarization shifted - resonances are observed. This allows a
detailed investigation of the depolarization effect.

Grating coupler-induced ISR in electron inversion layers for the three
main surface orientations of Si are treated in detail in Ref.13. There it

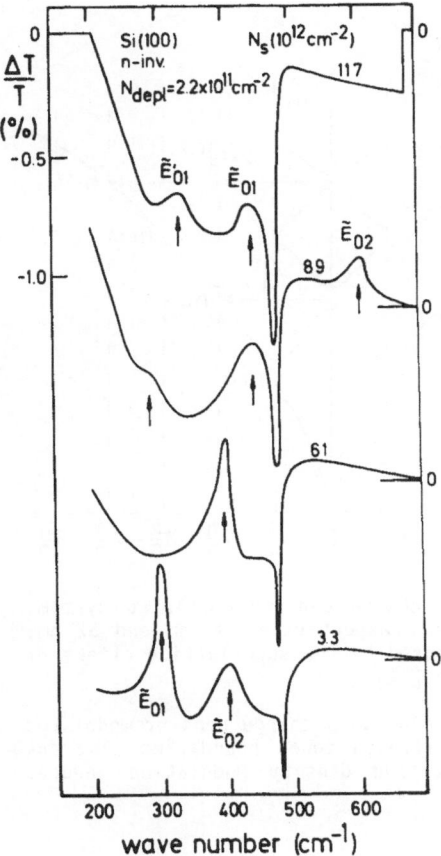

Fig. 4:

Grating coupler induced intersubband resonances (↑) for n-inversion on Si(100) in a homogeneous system at different charge densities N_s. For $N_s > 7 \cdot 10^{12}$ cm^{-2} transitions in the primed subband system are observed. The structure at 480 cm^{-1} is related to polariton type of excitation in the SiO$_2$ [13].

is also discussed that due to the grating coupler effect the transitions are non-vertical ($\Delta k = k_f - k_i = 2\pi/a \gg 2\pi/\lambda$). ($k_f$ and k_i are the wave vectors of final and initial states, respectively.) However, Δk is for the experiments here still small compared with the Fermi-vector k_F, thus effects are small. It is further discussed in Ref.13 that optical phonons in SiO$_2$ at frequencies about 480 cm^{-1} have a significant influence on the ISR (see spectra in Fig.4). In particular, the sharp structure with a negative signal for ($-\Delta T/T$) in Fig.4 is caused by the influence of the 2DES on polariton type of excitations in the SiO$_2$-layer.

5. Superlattice Effects on the Plasmon Dispersion in Charge Density Modulated Systems

In the experiments described in the previous section the microstructure serves in a passive way as an experimental tool which does not change the dynamic excitation spectrum of the 2DES. We will now discuss the insulator modulated systems shown in Fig.2 where via the microstructure the charge density of the system is modulated [10,11].

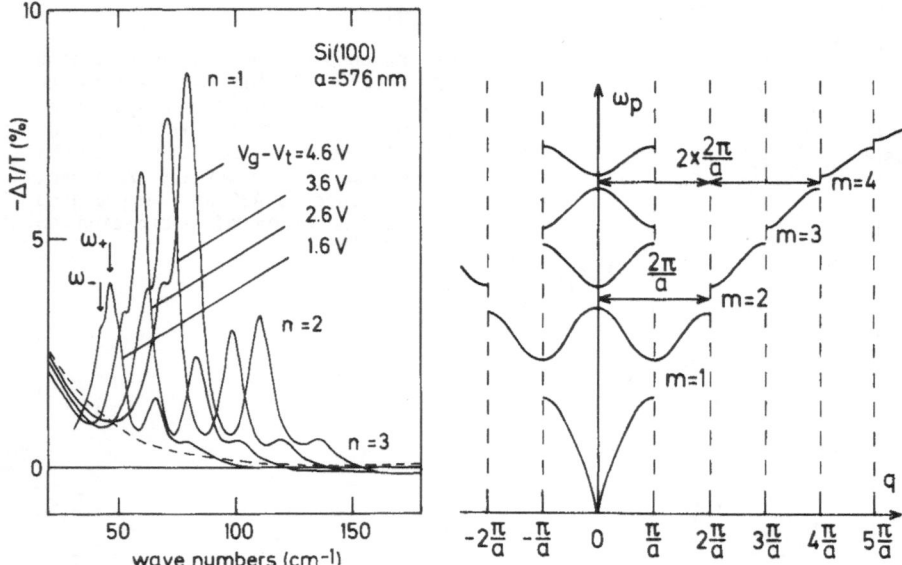

Fig. 5 (left): Plasmon excitation in a charge density modulated system. Periodicity a is 576 nm, d_1 and d_2 are, respectively, 30 nm and 52 nm, t_1/a is 0.25. The splitting ω_-, ω_+ is caused by the superlattice effect of the periodical charge density modulation [11].

Fig. 6 (right): Schematic plasmon dispersion in a charge density modulated electron gas. Dashed lines indicate Brillouin zones boundaries. The superlattice effect of the periodical charge density modulation induces minigaps in the plasmon dispersion.

In Fig.5 experimental spectra in an oxide thickness modulated Si(100) system are shown. For FIR radiation polarized perpendicular to the grating grooves (full lines) several resonances are observed which shift with increasing gate voltage and corresponding increasing charge densities N_{s1} and N_{s2} in the two regions t_1 and t_2 to higher frequencies. The resonances are plasmon excitations with wave vectors $q_n = n \cdot 2\pi/a$ (n=1,2,3). Characteristic for the plasmon excitation in a charge density modulated system is, that resonances are split into two resonances ω_- and ω_+. The amount of the splitting depends strongly on the geometry of sample, in particular for samples with large aspect ratios N_{s2}/N_{s1} splitting is also observed for plasmons with n=2 [10,16]. This splitting arises from the superlattice effect of the periodical charge density modulation on the plasmon dispersion. The physics is illustrated in Fig.6 where we show schematically the plasmon dispersion in a charge density modulated system. The periodic modulation induces Brillouin zones at $\pm m\pi/a$. If we fold the plasmon dispersion back into the first Brillouin zone then we expect gaps in the plasmon dispersion at the center (q=0) and at the boundary π/a of the reduced zone. Since we use the same grating that produces the Brillouin zones also for the coupling process, we can only observe gaps at m=2,4... corresponding to gaps at q=0 in the reduced scheme. The two branches, ω_- and ω_+ at q=0 for m=2 are observed in the spectra of Fig.5 (n=m/2=1).

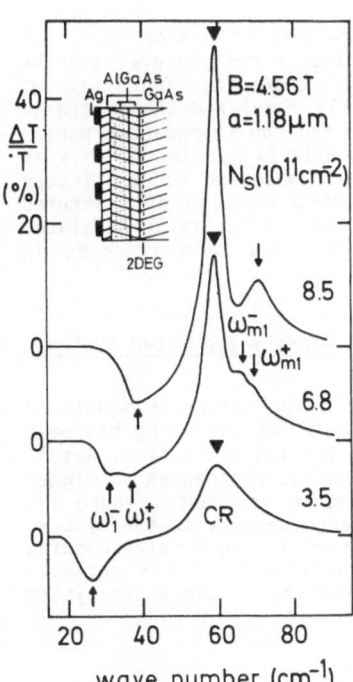

Fig. 7:

Plasmons (↑) for B=0, magneto-plasmons (↓) and cyclotron resonances (CR) at B=4.56T for different charge densities N_s in an AlGaAs-GaAs heterojunction. For N_s= $6.8 \cdot 10^{11} cm^{-2}$ a splitting of the plasmon and magneto plasmon resonance (+,-) is observed [12].

The splitting of the plasmon dispersion in a charge density modulated system has been calculated in the approximation of a small modulation [24]. It is found that the splitting of the mth gap $\Delta\omega^2_{pm}=\omega^2_{+m}-\omega^2_{-m}$ is proportional to $|N_m|/N_s$ where N_m is the mth Fourier coefficient of a charge density Fourier series $N_s(x)=\Sigma^m N_m \exp(i2\pi mx/a)(m=+\infty$ to $-\infty)$. This explains the strong dependence of the splitting on the geometry of the sample that is found in the experiments, in particular, that the splitting is in general smaller for higher m.

The plasmon resonances in a charge density modulated system have the character of standing waves where both branches, ω_+ and ω_-, have a different symmetry. In the symmetric system of Fig.2a the ω_+ branch has a radiative character and is thus stronger excited by FIR radiation. For samples of the arrangement shown in Fig.2b Al stripes are evaporated in a shadowing process onto the photoresist profile. These stripes are asymmetric with respect to the charge density modulation, and it is found that the ω_--branch is excited with higher efficiency [16].

Superlattice effects on the plasmon dispersion have also been observed for 2D plasmon excitation in AlGaAs heterostructure systems [12]. In Fig.7 experimental spectra measured on modulation-doped AlGaAs heterojunctions are shown. The spectra are achieved by dividing the transmission spectra that are measured at a magnetic field B=0 by the spectra measured at B=4.56 T. Thus the resonance ω_1 at low wave numbers corresponds to 2D plasmon excitation at B=0, resonances at higher wave numbers to excitation at a magnetic field B=4.56T, i.e., the cyclotron resonance (CR) of frequency $\omega_c=e\cdot B/m^*$ and the magnetoplasmon resonance $\omega^2_{mp}=\omega^2_c+\omega^2_p(B=0)$ [17]. The charge density in the system has been increased by short light pulses from a near infrared light emitting diode via the persistent photoeffect. For $N_s = 6.8 \cdot 10^{11} \text{cm}^{-2}$ it is observed that both the plasmon resonance ω_1 and the magneto plasmon resonance ω_{m1} are split into two resonances. The origin of this splitting is attributed again to a spatially modulated charge density. Since the illumination is performed through the non-transparent periodic Ag stripes the persistent photoeffect is spatially modulated. This is confirmed by the fact that the splitting is not observed for continuous illumination (spectra at $N_s = 8.5 \cdot 10^{11} \text{cm}^{-2}$) where because of scattered light, donors are more uniformly spatially ionized. A spatially modulated photo-effect induced by holographic illumination has been reported in Ref.25.

6. Intersubband Resonance Excitation in Charge Density Modulated Systems

Figure 8 shows experimental spectra measured in oxide thickness modulated Si(100) systems as shown in Fig.2a [14]. Additional Al shadowing has been used to increase the grating coupler efficiency for ISR excitation. Resonances $\hbar\omega_n$ at low wave numbers are plasmon resonances. Resonances at higher wave numbers are ISR's. In contrast to the plasmon resonance, where the center frequency is governed by the averaged charge density in the system, two ISR resonances $E_{on}(t_1)$ and $E_{on}(t_2)$ are observed in Fig.8 corresponding to ISR excitation in the regions t_1 and t_2, respectively. This reflects the fact that plasmons are the collective excitation of the whole system

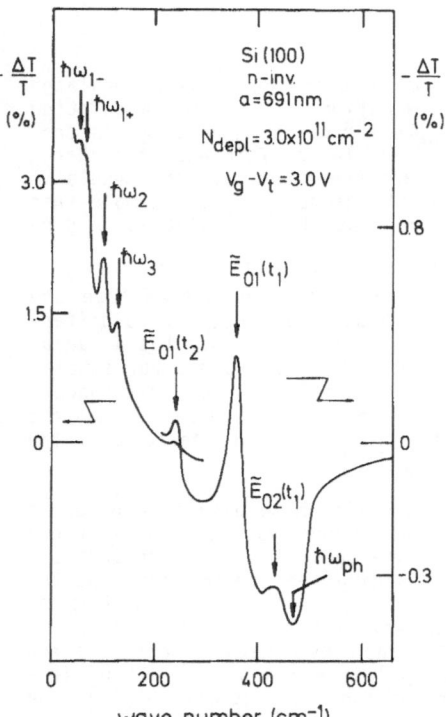

Fig. 8:

Plasmon resonances ($\hbar\omega_i$) and grating coupler induced intersubband resonances E_{ij} for an oxide thickness modulated MOS system ($d_2 = 20\text{nm}$, $d_1 = 47\text{nm}$, $t_1/a \approx 0.7$, $V_g - V_t = 2\text{V}$) [14].

(the mean free path of plasmons is several grating periodicities), whereas the ISR is, except for the depolarisation effect, a single particle excitation which measures the local properties of the system.

Comparing intersubband resonances in the regions t_1 and t_2 of the microstructured system with ISR frequencies in homogeneous systems of equivalent thickness yields for the dimensions here no significant influence of the microstructure on the ISR excitation. Tentatively a small decrease of the depolarisation shift is found [14].

7. Summary

The grating coupler effect of lateral structures with small periodicities can be used as an experimental tool to investigate dynamic excitations of a 2DES, i.e., intersubband resonances and plasmons. Small structures can be used to lower the dimensionality of the system, e.g., to prepare systems with spatially modulated charge density. This leads to novel physical phenomena, e.g., the formation of minigaps in the plasmon dispersion.

Acknowledgement

I would like to thank my colleagues, J.P. Kotthaus, E. Batke, W. Beinvogl, W. Hansen, U. Mackens, K. Ploog and L. Prager, who have contributed in a collaborative effort to the results discussed here. I also acknowledge support from the "Stiftung Volkswagenwerk" and the "Deutsche Forschungsgemeinschaft".

References

1 For a review on 2DES see T. Ando, A.B. Fowler, F. Stern: Rev.Mod. Phys.54, 437 (1982)
2 For a summary on new electronic devices see articles in the proceedings of the 2.Int.Conf. on "Modulated Semiconductor Structures", Kyoto 1985
3 C.O. Bolzer, M.A. Hollis, K.B. Nichols, S. Rabe, A. Vera, C.L. Chen: IEEE Electron Device Lett. EDL-6, 456 (1985)
4 P.M. Campbell, W. Garwacki, A.R. Sears, P. Menditto, B.J. Baliga: IEEE Electron Device Lett. EDL-6, 304 (1985)
5 M.F. Chang, P.M. Asbeck, D.L. Miller, K.C. Wang: IEEE Electron Device Lett. EDL-7, 8 (1986)
6 D.L. Miller: Appl.Phys.Lett.47, 1309 (1985)
7 A.B. Fowler, A. Hartstein, R.A. Webb: Phys.Rev.Lett.48, 196 (1982)
8 W.J. Skocpol, L.D. Jackel, E.L. Hu, R.E. Howard, L.A. Fetter: Phys. Rev.Lett.49, 951 (1982)
9 R.G. Wheeler, K.K. Choi, A. Goel, R. Wisnieff, D.E. Prober: Phys. Rev.Lett.49, 1674 (1982)
10 U. Mackens, D. Heitmann, L. Prager, J.P. Kotthaus, W. Beinvogl: Phys.Rev.Lett.53, 1485 (1984)
11 D. Heitmann, J.P. Kotthaus, U. Mackens, W. Beinvogl: J. Superlattices and Microstructures 1, 35 (1985)
12 E. Batke, D. Heitmann, J.P. Kotthaus, K. Ploog: Phys.Rev.Lett.54, 2367 (1985)
13 D. Heitmann, U. Mackens: Phys.Rev.B, in press
14 U. Mackens, D. Heitmann, J.P. Kotthaus: Surf.Sci. in press

15   U. Mackens, D. Heitmann, J.P. Kotthaus: Proc. of INFOS, Toulouse, 1985
16   E. Batke, W. Hansen, D. Heitmann, J.P. Kotthaus, U. Mackens, L. Prager, K. Ploog: J. Physique, in press
17   For a recent review on 2D plasmons see D. Heitmann: Surf.Sci., in press
18   P. Kneschaurek, A. Kamgar, J.F. Koch: Phys.Rev.B14 1610 (1976)
19   B.D. McCombe, R.T. Holm, D.E. Schafer: Solid State Commun.32, 603 (1979)
20   D. Heitmann: Festkörperprobleme (Advances in Solid State Physics) XXV, p.429, P. Grosse (ed.) Vieweg, Braunschweig 1985
21   S.J. Allen, jr., D.C. Tsui, R.A. Logan: Phys.Rev.Lett.38, 980 (1977)
22   D. Heitmann, J.P. Kotthaus, E.G. Mohr: Solid State Commun.44, 715 (1982)
23   D.C. Tsui, G. Kaminsky: Phys.Rev.Lett.35, 1468 (1975)
24   M.V. Krasheninnikov, A.V. Chaplik: Sov. Phys. Semicond.15, 19 (1981)
25   K. Tsubaki, H. Sakaki, J. Yoshino, Y. Sekiguchi: Appl.Phys.Lett.45, 663 (1984)

Part VII

High Field Transport
and Optical Excitation

Carrier Transport in Semiconductor Devices of Very Small Dimensions

W. Hänsch

Siemens AG, Central Research and Development, Microelectronics, Otto-Hahn-Ring 6, D-8000 München 83, Fed. Rep. of Germany

We present a self-consistent formulation of hot carrier transport in very small devices. Particle balance and energy balance equation as well as the corresponding relaxation times are calculated. As a special case we consider a local solution of the general formulation to derive a local field-dependent mobility and carrier temperature suitable to extend the conventional semiconductor equations. We present results for a realistic two-dimensional device simulation including hot carrier effects.

I. Introduction

An essential guide in miniaturization of modern integrated circuits is the scaling of single components. However, a consequent scaling can not be accomplished. For many reasons the supply voltages have been fixed to 5 V, which give increasing field strengths in small devices. On the other hand there are intrinsic physical properties, as for instance particle interaction strength, which are impossible to scale. All this limits the classical description of device physics. Entering the submicron MOSFET region requires a reconsideration of the physics which gave us the powerful semiconductor equations. This macroscopic formulation proved to be very useful for practical device design /1/. Therefore it is worthwhile to look for an extended macroscopic formulation which meets the requirements of the submicron MOSFET devices. A rigorous treatment should of course start from the microscopic formulation of carrier transport /2/. Its central quantity is the distribution function of particles: f (R, k, t). This function is defined on a seven-dimensional space of time t, momentum k, and space R; if we adopt the semiclassical picture of transport and neglect quantum corrections as well as corrections due to strongly interacting particles. In the following Sec. II we will utilize the equation of motion for the distribution function to derive an extended set of semiconductor equations appropriate for hot carrier transport. This can not be done without any assumptions about the specific form of f and the interactions the carriers are subjected to.

In former analysis a drifted Maxwellian was supposed for f and scattering of carriers was described in a relaxation time approximation, utilizing Boltzmann's equation /3/. This procedure is not self-consistent: Especially in high fields when scattering by optical phonons becomes important no relaxation time exists /4/. To avoid this we will approximate the distribution function in a different way that allows a consistent treatment of the scattering terms. This formulation will provide us with a self-consistent local model of mobility and carrier temperature, for all field strengths.

In Sec. III we will use the local model of Sec II. and show how it can be used in a practical device simulator program for a MOSFET. We will study in detail its implications on carrier and current density distributions in a realistic n-channel MOSFET.

A summary of our findings closes this work in Sec IV.

II The model

For the stationary case the Boltzmann equation reads /2/

$$-e\underline{E}(\underline{R})\underline{\nabla}_k f(\underline{R},\underline{k}) + v(\underline{k})\underline{\nabla}_k f(\underline{R},\underline{k}) = \sum{}^{in}(\underline{R},\underline{k}) - \sum{}^{out}(\underline{R},\underline{k}) \tag{2.1}$$

Here we allow for an inhomogeneous electrical field $\underline{E}$ ($\underline{R}$). For the particle velocity we have $\underline{v}(\underline{k}) = \underline{\nabla}\mathcal{E}(\underline{k})$. For convenience we will assume the effective mass approximation with a parabolic band with effective mass m. This model is sufficient for electron transport in Si, for example. In case of GaAs a two-band model should be employed. More important however are the scattering terms on the right-hand side of Equ. (2.1). Here $\sum{}^{in}$ represents the scattering into the state $\underline{k}$ and $\sum{}^{out}$ that out of $\underline{k}$; their balance equals the kinematic change of f. For our analysis we allow quite general scattering mechanisms: Elastic as well as inelastic scattering. Elastic scattering is represented by the following interaction matrix element /2/.

$$W(\underline{k},\underline{k}') = M(|\underline{k}-\underline{k}'|)\delta(\mathcal{E}(\underline{k}') - \mathcal{E}(\underline{k})) \tag{2.2}$$

We assume isotropic scattering. For the inelastic channel we have /2/

$$W(\underline{k},\underline{k}') = M^{\circ}(|\underline{k}-\underline{k}'|)[g^{\circ}(|\underline{k}'-\underline{k}|)+1]\delta(\mathcal{E}(\underline{k}') - \mathcal{E}(\underline{k}) - \omega_{|\underline{k}-\underline{k}'|})$$

$$+ M^{\circ}(|\underline{k}'-\underline{k}|)\,g^{\circ}(|\underline{k}-\underline{k}'|)\,\delta(\mathcal{E}(\underline{k}') - \mathcal{E}(\underline{k}) + \omega_{|\underline{k}-\underline{k}'|}) \tag{2.3}$$

for electron phonon scattering. We assume an isotropic phonon dispersion relationship $\omega_{|\underline{k}-\underline{k}'|}$

With Equ (2.3) both acoustical and optical phonons are included. Elastic scattering is predominantly provided by electrons scattered off ionized impurities. Once the interaction matrix elements are known $\sum{}^{in}$ and $\sum{}^{out}$ are calculated from

$$\sum{}^{in}(\underline{R},\underline{k}) = \sum_{\underline{k}'} W(\underline{k}',\underline{k})\,f(\underline{R},\underline{k}') \tag{2.4}$$

$$\sum{}^{out}(\underline{R},\underline{k}) = \sum_{\underline{k}'} W(\underline{k},\underline{k}')\,f(\underline{R},\underline{K}) \tag{2.5}$$

With Equ's (2.1) to (2.5) the model is completely specified and the solution of Boltzmann's equation is in principle possible. However, we are not interested in f but rather in its averages or moments. Therefore we do not go to the trouble to solve Equ. (2.1). This in turn will give us relationships between the macroscopic variables: Particle density n, current density $\underline{j}$, energy density $\langle\mathcal{E}\rangle$, and energy current density $\underline{v}\langle\mathcal{E}\rangle$. The former two are the variables of the classical semiconductor equations which determine the particle flow keeping the

carrier system in thermal equilibrium with the lattice. In very high
fields ($\geq 10^3$V/cm) carriers can gain more energy in the field than they
lose by collisions with phonons, therefore they will no longer maintain
thermal equilibrium with the lattice. We have to add an energy balance
for the carrier system.

In former work f was approximated by a drifted Maxwellian with an electronic temperature and a common drift velocity /3/. This form of f is
justified if the electron-electron interaction is strong enough to
maintain the carrier system itself in equilibrium. A critical particle
density n_c has to be exceeded so that collisions among carriers are
much more frequent than collisions between carriers and phonons. A
realistic estimate of n_c is still missing /5/. We use a different Ansatz
which has the advantage that the scattering term in the Boltzmann equation can be treated self-consistently. Separating an isotropic diffusion
term and a drift term weighted in direction of the local field we have

$$f(\underline{R},\underline{k}) = f^0(\underline{R},\epsilon(\underline{k})) + \underline{E}(\underline{R})\cdot\underline{k}\,f^1(\underline{R},\epsilon(\underline{k})) \tag{2.6}$$

where f^0 and f^1 still depend on the electric field. Rigorously f^0 and
f^1 represent the first term of a general expansion of the odd and even
parts of f. As it turns out they are apparently sufficient to describe
the essential features of non-linear response: velocity saturation
and overshoot in a self-consistent way. Furthermore it is assured that
in contrast to the drifted Maxwellian, the heat current contribution
is not identical to zero. With Equ. (2.6) it is straightforward to
calculate the moments of Equ. (2.1). We have to multiply Equ. (2.1)
by k^n, n = 0,1,2,3 and sum over momentum. It turns out that the n-th
moment of Equ. (2.1) contains the (n+1)-th moment of f. Therefore a
truncation is necessary to end up with a finite number of equations.
To include the energy balance we have to approximate the 5-th moment
of f, which corresponds to the heat current. We choose it so that the
heat current near equilibrium is well represented by the Wiedemann
Franz law /2/. As a result we get:

$$\nabla\underline{j} = 0 \tag{2.7}$$

$$\underline{j} = e\mu n\underline{E} + \tfrac{2}{3}\mu\underline{\nabla}<\epsilon> \tag{2.8}$$

$$\nabla\underline{v}_{<\epsilon>} = \underline{j}\,\underline{E} - \frac{<\epsilon>-<\epsilon>_0}{\tau_E} \tag{2.9}$$

$$e\underline{v}_{<\epsilon>} = -\tfrac{5}{3}\underline{j}\cdot\frac{<\epsilon>}{n} - \tfrac{10}{9}\mu<\epsilon>\underline{\nabla}\frac{<\epsilon>}{n} \tag{2.10}$$

with

$$\mu = \frac{e\tau_m}{m} \tag{2.11}$$

the energy and momentum relaxation times τ_E and τ_m, respectively are
calculated from

$$\frac{1}{\tau_m} = e\,\frac{\underline{j}\sum_{\underline{k}\underline{k}'}(\underline{v}(\underline{k}')-\underline{v}(\underline{k}))\,W(\underline{k}',\underline{k})\,f(\underline{R},\underline{k}')}{j^2} \tag{2.12}$$

$$\frac{1}{\tau_\varepsilon} = \frac{\sum\limits_{\underline{k}\underline{k}'} \left(\varepsilon(\underline{k}) - \varepsilon(\underline{k}')\right) W(\underline{k}',\underline{k}) f(\underline{R},\underline{k}')}{<\varepsilon> - <\varepsilon>_0} \tag{2.13}$$

They are usually considered in the relaxation time approximation, which leaves them adjustable for practical calculations. We can remove this ambiguity, utilizing properties of Equ. (2.6) and the matrix elements $W(k,k')$, Equ.'s (2.2) and (2.3). In the first place we expand the distribution function with respect to its moments. This gives

$$f(\underline{R},\underline{k}) = \sum_{n=0}^{\infty} M_{2m}(\underline{R}) \, g_{2m}(k) + \underline{k} \sum_{n=0}^{\infty} \underline{M}_{2m+1}(\underline{R}) \, g_{2m+1}(k) \tag{2.14}$$

The functions g_i depend only on the microscopical variable k. To be consistent within our truncation scheme we only have to retain the first four terms of this expansion. Inserting this in Equ.'s (2.12) and (2.13) gives, using simple symmetry arguments, without further approximation

$$\frac{1}{\tau_m} = A + B \, \frac{\underline{j} \, v <\varepsilon>}{\underline{j}^2} \tag{2.15}$$

$$\frac{1}{\tau_\varepsilon} = \frac{1}{\tau_{\varepsilon_0}} \tag{2.16}$$

where $A, B,$ and τ_{ε_0} do not depend on the macroscopical variables. We can express Equ. (2.15) in a more suitable way by experimentally accessible quantities

$$\frac{1}{\tau_m} = \frac{1}{\tau_{m_0}} \left\{ 1 - \frac{\tau_{m_0}}{\tau_{\varepsilon_0}} \, \frac{\frac{3}{2}\frac{kT}{e} + \frac{3}{2}\frac{\underline{j} \, v <\varepsilon>}{\underline{j}^2}}{\frac{m v_{sat}^2}{e}} \right\} \tag{2.17}$$

Here τ_{m_0} will be provided by the Ohmic low-field mobility $\mu_0 = e\tau_{m_0}/m$. We get a more transparent expression if we expend Equ (2.17) with respect to τ_{ε_0}. This gives for the mobility

$$\mu = \frac{\mu_0}{1 - \frac{2}{3} \, \frac{\mu_0}{\tau_{\varepsilon_0} v_{sat}^2} (u_T^0 - u_T)} \tag{2.18}$$

Here we have used $<\varepsilon> = \frac{3}{2}enu_T$. Both Equ. (2.16) and Equ. (2.18) are the key results of our calculation. A constant energy relaxation time τ_{ε_0} is backed by high-field Monte Carlo calculations at room temperature /6/. The mobility Equ. (2.18) shows in a very elucidate way that velocity saturation is caused by carrier heating. With Equ's (2.7) to (2.11), (2.16) and (2.17) or (2.18) we have a consistent formulation of hot carrier transport, without ambiguities concerning the relaxation times. They are calculated without severe restrictions on the electric field or scattering mechanisms. However, we do not intend to do a microscopic calculation of the still material-dependent quantities τ_{m_0}, v_{sat} and τ_{ε_0}. They can most easily be determined by experiments in simple geometries /7/.

The mobility Equ. (2.18) has been derived without any assumptions about the electrical field. Its form allows an easy discussion of the

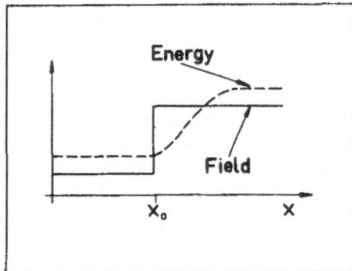

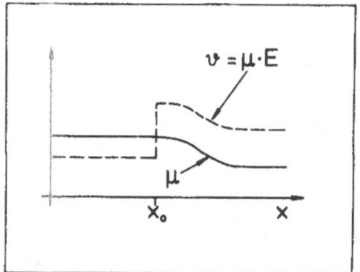

Fig. 1: Energy response
to a step profile of the
electric field

Fig. 2: Velocity over-
shoot at a step profile
of the electric field

overshoot effect. Suppose we have a step-like field profile, as shown
in Fig. 1. The energy per particle u_T will respond in a general non-
local theory with retardation. The corresponding particle velocity
will schematically behave as shown in Fig. 2. There is a short transition region where the velocity exceeds its stationary value considerably.

This phenomenon is called velocity overshoot, and is also present for
rapid changes of the field in time. Overshoot effects drastically alter
the performance of high-speed devices on GaAs basis /8/. For Si they
are less important. Therefore it is justified to employ a local approximation. To this end we solve Equ.'s (2.8) to (2.10), Equ. (2.16),
and Equ. (2.17) disregarding all spatial derivatives. The result is
a field-dependent mobility and carrier temperature.

$$\mu(E) = 2\mu_o / \left(1 + \left(1 + 4\, \tfrac{\mu_o^2}{v_{sat}^2}E^2\right)^{1/2}\right)$$ (2.19)

$$kT_e = kT + e\, \tfrac{2}{3}\, T_{\mathcal{E}_o}\, v_{sat}^2 \left(\tfrac{1}{\mu(E)} - \tfrac{1}{\mu_o}\right)$$ (2.20)

Equ.'s (2.19) and (2.20) resemble very closely the corresponding results from Monte Carlo solutions of high-field transport, in constant
electric field /6/. This gives in turn a justification of our procedure.

III. Local model for electron transport

The local mobility and temperature model developed in the previous
section shall now serve as the starting point for a modification of
the conventional semiconductor equations. Only the current equation
has to be modified, because we will consider the energy balance as lo-
cally fulfilled. The current Equ. (2.8) will read in the local model,
if we use $\langle \mathcal{E} \rangle = \tfrac{3}{2}kT_e n$.

$$\underline{j} = e\mu(F)n\underline{E} + \mu(F)\underline{\nabla}(kT_e n)$$ (3.1)

Here μ and T_e are given by their local expressions Equ.'s (2.20) and
(2.21), respectively. Their field dependence has now been replaced

by a general driving force F which accounts of the heating of the carriers. For the homogeneous case this force is equal to E. The replacement is necessary because only in the homogeneous case carrier heating is related to Joule heating alone. From Equ. (2.9) we see that in the inhomogeneous situation the gradient of the energy current also contributes to the energy balance. The choice of F is however not unique /9/. The classical current relationship emerges from Equ. (3.1) if $\tau_{\varepsilon_0} = 0$, which implies $T_e = T$. This limit has no consequences with regard to the mobility! This is true only in the local approximation. A field-dependent mobility is usually contained in conventional device simulation. It enters however only from an empirical level and was not derived before /10/. With the thermal voltage $u_{T_e} = kT_e/e$ Equ. (3.1) is brought in a more suitable form

$$\underline{j} = e\mu(F)n[-\underline{\nabla}(\varphi - u_{T_e}) + u_{T_e}\tfrac{1}{n}\underline{\nabla}n] \qquad (3.2)$$

Here φ is the electrostatic potential. Utilizing Equ. (3.2) has an important consequence: The quasi Fermi potential whose gradient is the driving force of the current loses its physical meaning /1/. We can however cast Equ. (3.2) in the form

$$\underline{j} = e\mu(F)n\underline{\nabla}\bar{\Phi} \qquad (3.3)$$

to construct a quasi Fermi potential equivalent. However, the particle density is not expressible in a simple way by $\bar{\Phi}$. This potential proved to be a very useful choice for providing the driving force $F = \underline{\nabla}\bar{\Phi}$. The current equation preserves its homogeneous case form. Up to now we do not understand why this choice of F appears to be so convenient.

Hot carrier transport according to the present formulation was realized in the two-dimensional device simulation program MINIMOS 3 /11/. In the following we like to present some results calculated for a realistic n-channel MOSFET with 0.5 µm effective channel length L appropriatly biased. In Fig. 3 we recall the geometrical spedifications of a MOSFET-Device. In Fig.'s 4 and 5 we show the electron density on the drain side. Fig. 4 was calculated by using the conventional semicon-

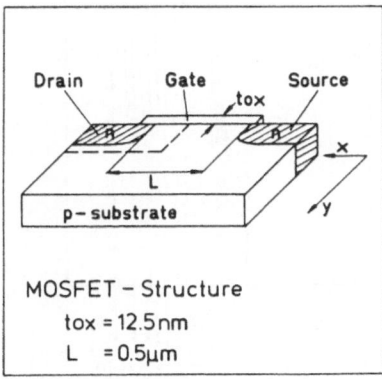

MOSFET – Structure
tox = 12.5nm
L = 0.5µm

Fig. 3: Schematic structure of a MOSFET device

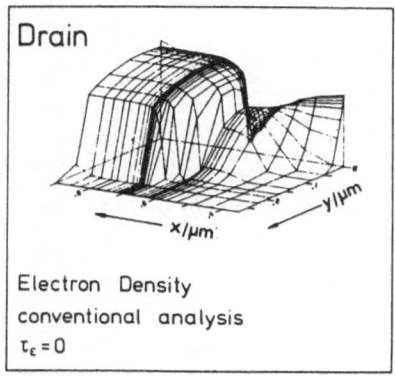

Electron Density
conventional analysis
$\tau_\varepsilon = 0$

Fig. 4: Density profile at the drain side of a MOSFET

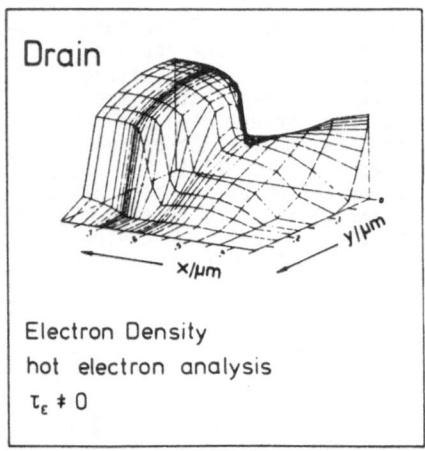

Fig. 5: Density profile
at the drain side of a MOSFET

Electron Density
hot electron analysis
$\tau_\varepsilon \neq 0$

ductor equations and in Fig. 5 the modified current relationship Equ.
(3.2) was used. The electrical potential is almost the same in both
cases.

We observe that due to carrier heating the electron density becomes
smoother. The electrons are pushed towards the bulk. This increases
the average distance of the carriers from the Si/SiO_2 interface. This
modification will effect current models of gate oxid injection or impact
ionisation /12/. In Fig. 7 we show the electronic temperature in the
drain region. A maximum of $T_e = 8T$ is reached. By comparison of Fig.
6 and Fig. 7 we see the expected correlation of the electron temperature with the electric field.

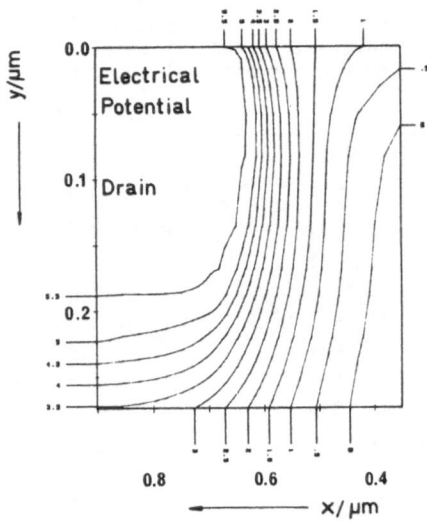

Fig. 6: Electrical potential at the
drain side of a MOSFET

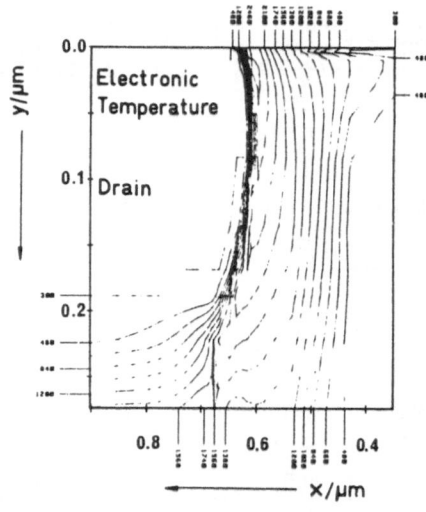

Fig. 7: Electronic temperature distribution at the drain side of a MOSFET

V. Summary

In the previous chapters we discussed a self-consistent treatment for hot electron transport in small devices. From the general balance equations for particles and energy we derived a local solution which provided a local mobility and carrier temperature model. These were used for a first extension of the conventional current equation. This approach is of course only a first step in a more rigorous treatment, which has to solve Equ.'s (2.7) to (2.10) in a two-dimensional geometry.

Our local approach should be a good approximation in Si because here overshoot effects, which are related to non-local effects, are less important. The general approach should hold for not too low temperatures and reasonably large devices so that ballistic transport is negligible /13/.

This work was supported by the Technological Program of the Federal Department for Research and Technology of the Federal Republic of Germany. The authors alone are responsible for the contents.

References

1. S.M. Sze, Physics of Semiconductor Devices (John Wiley & Sons, New York 1981), 2nd Edition

2. O. Madelung, Festkörpertheorie Bd II (Springer Berlin 1972), chap. VIII

3. K. Blotekjaer, IEEE Trans. on Electr. Devices ED 17, 38 (1970)

4. E.M. Conwell, Solid State Physics: Supplement 9, ed. F. Seitz, D. Turnbull, and H. Ehrenreich (Academic Press New York 1967), chap. V

5. R. Stratton, Proc. Poy. Soc. A242, 355 (1957)

6. L. Reggiani, Hot Electron Transport in Semiconductors, Topics in Applied Physics vol. 58, ed. H.J. Güntherodt and H. Beck (Springer Berlin 1985)

7. C. Jacoboni, C. Canali, G. Ottaviani, and A. Albergi Quaranta, Solid State Electr. 20, 77 (1977)

8. R.K. Cook and J. Frey, Compel , 65 (1982)

9. K. Hess and C.T. Sah, IEEE Trans. on Electr. Devices ED 25, 1399 (1978)

10. S. Selberherr, Analysis and Simulation of Semiconductor Devices, (Springer Wien 1984)

11. P. Pichler, H. Pötzl, A. Schütz and S. Selberherr, MINIMOS 3-Users Manual, Technische Universität Wien (Austria), 1986, unpublished

12. K.R. Hoffmann, C. Werner, W. Weber and G. Dorda, IEEE Trans. on Electron. Devices ED-32, 691 (1985)

13. J.P. Barker and D.K. Ferry, Solid State Electr. 23, 519-544 (1980)

Parallel-Transport Experiments in 2D Systems

R.A. Höpfel*

AT & T Bell Laboratories, Holmdel, NJ 07733, USA

Transport experiments in two-dimensional (2D) carrier systems are reviewed, with emphasis on high-field transport relevant for modern device structures (Si-MOSFETs, GaAs-AlGaAs heterostructures and quantum wells). Three methods have been applied to investigate parallel transport in 2D systems: (1) current measurements with simultaneous Hall- or magnetoresistance measurements, (2) direct drift time ("time-of-flight") measurements with picosecond opto-electronic techniques, (3) modeling of dynamical device performance. Hot carrier effects (optical-phonon emission, real-space and valley transfer) play a central role for the understanding of the velocity-field characteristics in these systems.

1. Introduction

Two-dimensional carrier systems in semiconductors are not only wonderful systems for fundamental physics research (see, e.g., the Nobel Prize in Physics 1985), but they are also the basis for important device technologies. The Si-MOS-technology [1] is well-established, the technology of field-effect transistors based on III-V-heterostructures [2] (GaAs-AlGaAs and related compounds) promises to be the technology of future high-speed devices and circuits: The SDHT (selectively doped heterojunction transistor, also called HEMT, MODFET, and TEGFET) is presently the fastest purely electronic switching device. Scientists at AT&T Bell Laboratories recently reported 0.35 μm gate ring oscillators with 5.8 ps propagation delay time per gate at 77 K (10.2 ps at room temperature) [3].

The device speed is closely related to the physics of parallel transport, since in field-effect transistors a fundamental limit for the switching time is given by the transfer time t_r through the gate length x_L, limited by the drift velocity v_d as

$$t_r = x_L/v_d \quad . \tag{1}$$

In most cases, however, the RC time constant of the FET limits the high speed operation, so that the fundamental limit given by the transfer time cannot be reached. Instead the transconductance g_m and the gate capacitance C_{GS} determine the upper frequency limit f_T of the FET [1,4], by

$$f_T = g_m/2\pi C_{GS} \quad , \tag{2}$$

where g_m contains the drift velocity as a linear factor. A recently proposed new concept of "velocity modulation" [5] (instead of carrier density modulation) theoretically would

---

* on leave from: Institut für Experimentalphysik, Universität Innsbruck, Austria.

overcome the transfer time limit (1), since the carriers remain in the conducting channel in both logic states. The RC time constant, however, still limits the switching time, so that we can say that *the drift velocity of the two-dimensional carrier system limits the high-speed performance of all field-effect transistors based on 2D systems*. Therefore, measurements of the velocity-field characteristics of 2D carrier systems, parallel to the planes of confinement, are of central interest for the design of high speed devices. The main question thereby is, whether the transport in 2D systems is qualitatively and quantitatively different from bulk semiconductors. For example: Does the high low-temperature mobility in 2D systems ($\sim 10^6 cm^2/Vs$) lead to higher drift velocities under "device conditions"? Is the shorter switching time in heterostructure FETs (SDHTs) due to higher drift velocities in 2D systems? Three classes of experiments have been performed in order to investigate this problem:

— Current-measurements with simultaneous Hall- or magnetoresistance measurements

— Direct drift-time ("time-of-flight") measurements

— Modeling of device performance.

2. Current-measurements

A "classical" way to determine the drift velocity v_d of charge carriers is to measure the integral sheet current density j_s in a 2D system and to divide by the charge concentration q_s,

$$v_d = j_s/q_s \quad . \tag{3}$$

The current measurement is straightforward (pulsed electric fields avoid lattice heating), the electric field is applied and measured with separate voltage probes (or one relies on the quality of the ohmic contacts). The difficulty of measuring v_d in this way is the charge concentration q_s that must be measured *at high fields*, since the carrier concentration might be changed in high electric fields. The carrier concentration can be monitored by measuring the transverse Hall field E_y at low magnetic fields B_z, given by

$$E_y = r_H j_s B_z/q_s \quad . \tag{4}$$

The "Hall factor" r_H, that takes into account the carrier energy distribution function (the dependence of momentum scattering on carrier energy [6]), has to be known for an exact interpretation of the experiment. This point produced much discussion and ambiguity in some of the experiments reported in the literature.

A second way to eliminate q_s from equation (3) has been applied by Masselink et al [7]: Instead of measuring the Hall voltage, the geometrical magnetoresistance is determined from measuring the sample resistance R with and without (low) magnetic fields. Neglecting the "physical magnetoresistance" ($\mu(B) \sim \mu(0)$ - which is justified at low magnetic fields), the mobility μ can be obtained from

$$R(B) = R(0)(1+r_H^2 \mu^2 B_z^2) \quad . \tag{5}$$

Again, r_H must be known from the distribution function and the dominant momentum scattering mechanism, which is not a trivial problem. The technique, however, has one big advantage over the Hall-measurement: Since only two contacts are required, microstructures similar to real FET devices can be studied, which makes this technique very interesting for transport experiments in microstructures.

305

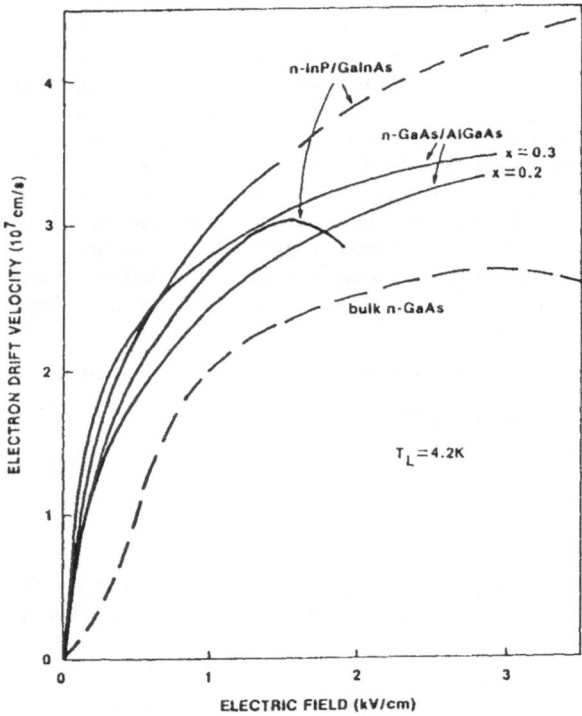

Fig. 1:   Drift velocities as a function of electric field in single heterostructures at
T_L = 4.2 K, as obtained from Hall-measurements, after M. Inoue [8]
(GaAs/AlGaAs) and Tsubaki et al. [10] (InP/InGaAs).

In Fig. 1 experimental results on the velocity-field curve in single heterostructures are
shown, that were obtained from current and Hall-voltage measurements at 4.2 K lattice
temperature. The electron velocities in n-GaAs/AlGaAs single heterostructures (after
M. Inoue [8]) are much higher than in bulk GaAs, not only in the low-field range, where
it is expected due to the high mobilities in modulation-doped structures. It is not clear
why the drift velocity increases up to more than 3×10^7 cm/s, without any indication of
transfer effects. At the highest value the input power per electron exceeds 1.5×10^{-8} W,
which must lead to extremely high carrier temperatures [9], and subsequently dramatic
(real-space and valley) transfer effects. In n-InP/GaInAs single heterostructures
(experiments by Tsubaki et al. [10]), the electron drift velocities even exceed
4×10^7 cm/s in one sample. Clearly, more experimental work is needed to understand
these results. The problems connected with current-measurements have been pointed out
in a paper by Schubert and Ploog [12], who showed that the 2D carrier concentration
can strongly decrease already at electric fields of some 100 V/cm, leading to possible
parallel conduction in AlGaAs.

Figure 2 shows the results obtained at higher lattice temperatures with the geometrical
magnetoresistance technique by Masselink et al. [7]. At 77 K drift velocities
$>1 \times 10^7$ cm/s are reached already at electric fields < 500 V/cm, due to the high

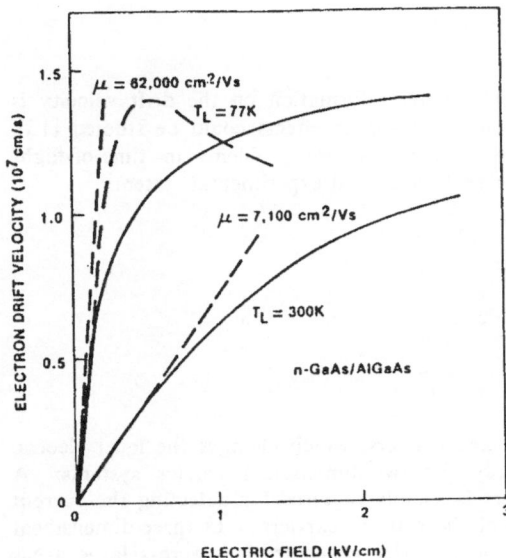

Fig. 2: Drift velocities as a function of electric field in GaAs-AlGaAs single heterostructures, as obtained from geometrical-magnetoresistance measurements, after Masselink et al. [7] (full curves). Dashed curve: Hall-measurements by Tsubaki et al. [13].

electron mobility. At 300 K, the remarkable feature of these experiments is the mobility of 7,100 cm^2/Vs, that is present at electron concentrations of 1×10^{12} cm^{-2}. In bulk GaAs of comparable 3D carrier concentrations ($10^{18} cm^{-2}$) the mobility is reduced by impurity scattering to about 3,000 cm^2/Vs [11]. Thus high drift velocities in selectively-doped heterostructures are reached at lower electric fields than in comparable bulk structures, which might be a key to the understanding of the superior performance of devices based on modulation-doped heterostructures. In the same figure, the results of Tsubaki et al. [13] (Hall-measurements at 77 K) are shown. Negative differential conductivity is observed at electric fields > 500 V/cm, and is interpreted as being due to the onset of inter-subband scattering. An additional mechanism for negative differential conductivity has been theoretically predicted by Al-Mudares and Ridley [14]: The abrupt threshold for the emission of optical phonons should lead to a "scattering-induced" negative differential conductivity in 2D systems.

3. Time-of-flight-measurements

The direct way to determine carrier drift velocities is to resolve the drift motion of injected carriers spatially and in time, i.e., to measure the drift time along a given distance. Several methods have been applied in bulk semiconductors [15], they differ in respect to:

— minority or majority carrier injection

- optical or electrical injection

- optical or electrical detection

From time-of-flight measurements not only direct information on the drift velocity is obtained, but also diffusion, recombination and trapping effects could be studied [15]. The following physical processes, however, create severe problems in time-of-flight experiments and have to be understood for each individual experimental system:

- space charge effects of injected carriers

- dielectric relaxation of injected carriers

- diffusion, recombination, trapping

- injection from the contacts

- inhomogeneity of the electric field

The problem of the space charge of injected carriers, which changes the local electric field, is the most serious one, especially for two-dimensional carrier systems: A minimum total number of injected carriers is usually required for detecting the current pulse (in the case of electrical detection of the drifting carriers). In three-dimensional systems this can be achieved with low sheet carrier concentrations across large areas (several mm^2 !). This leads to low electric fields from space charges (given by Poisson's equation), and has allowed time-of-flight experiments with majority carriers in, e.g., bulk-GaAs by Ruch & Kino [16]. In two-dimensional systems, however, this is not possible by principle, and therefore, other ways to prevent space charge effects have to be found.

Two groups of time-of-flight experiments have been reported so far in 2D carrier systems. Cooper and Nelson [17, 18] have successfully performed time-resolved drift measurements of electrons and holes in Si-MOSFET structures. The authors overcame the problem of space charge effects by applying the parallel electric field along a resistive (polysilicon) gate, which screens the space charge field of the injected carriers within the dielectric relaxation time. The results of these experiments strongly differ from earlier not time-resolved works (40% higher drift velocity for electrons, 60% higher for holes), which demonstrates the necessity and strength of time-resolved experiments.

In the following, the time-of-flight experiments in GaAs/AlGaAs heterostructures, reported by Höpfel, Shah and coworkers [19-21] are described. These experiments for the first time directly investigated the parallel transport problem in III-V-heterostructures by time-resolved techniques. The drift velocity of *minority* electrons in p-modulation-doped multiple quantum wells is measured as a function of the applied field. The majority hole plasma provides a homogeneous electric field, screening the space charges of the injected and separated electron-hole-pairs. In addition, the presence of the hole plasma will lead to some very interesting physical processes.

The principle of the experiments is shown in Fig. 3a: Electron-hole pairs are injected by short laser pulses (from a mode-locked synchronously pumped dye laser) into p-modulation-doped quantum well structures. The high hole concentration causes very rapid dielectric relaxation of the hole plasma, screening the excess charge of the electrons. The dynamics of the electron transport is given by the ambipolar transport equations [6], which reduce to the electron transport equations, if $nu_h << pu_p$ (minority carrier condition [6]). The current in the circuit is increased during the time of the

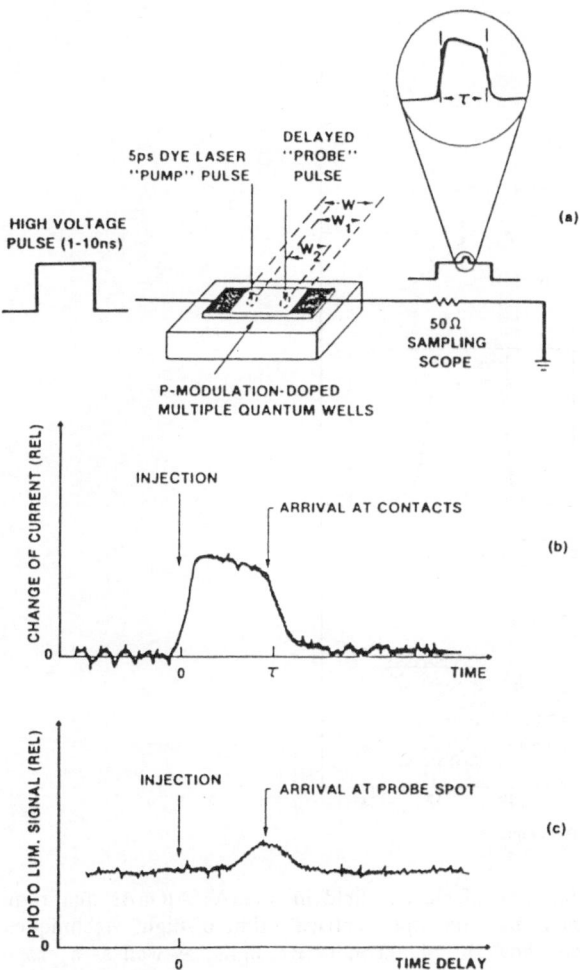

Fig. 3: Schematic illustration of opto-electrical and all-optical time-of-flight measurements (a), relative current signal in opto-electrical experiment (b), luminescence signal as a function of time-delay of the probe pulse in all-optical experiment (c) (after [19]).

minority electron drift by [15] $\Delta I = \Delta Q v_d / W$, where ΔQ is the total charge of the electrons and W is the distance between the contacts (Fig. 3a). The drift velocity is obtained from the duration of the additional current signal τ, as $v_d = W_1/\tau$.

In Fig. 3 also the principle of an all-optical technique is shown, where the drift time to a different point on the sample is detected by an optical laser pulse. The physical process used for detection is the nonlinear photoluminescence [22], leading to an increased time-averaged luminescence signal if the local carrier concentration is increased by the simultaneous arrival of the drifting carriers and the injection at the distance W_2. All-

309

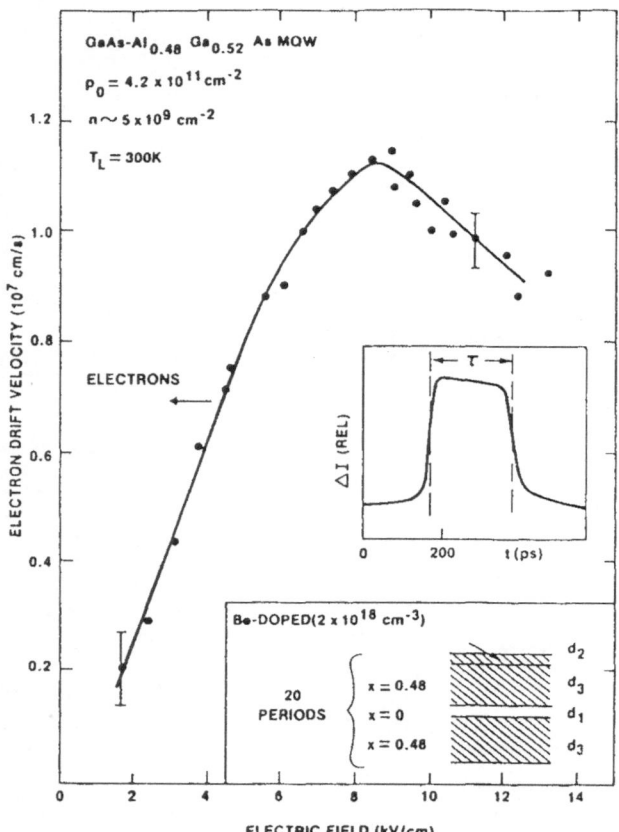

Fig. 4: Electron velocity as a function of electric field in p-GaAs/AlGaAs quantum well structures, obtained by the optoelectrical time-of-flight technique. $T_L = 300\ K$. The insets show the photoconductive signal as well as a cross section of the sample structure. (After [20]).

optical techniques [23] have one main advantage: The time resolution is not limited by the speed of circuit and instrumentation (25 ps for the Tektronix S-4 sampling head), but instead by the laser pulse length. With femtosecond techniques [24] 2D transport experiments should be possible also in the subpicosecond range for studying non-equilibrium transport phenomena.

In Fig. 4 the results of the picosecond time-of-flight measurements, as obtained from the picosecond current measurements [20], are plotted: The drift velocity of the electrons increases nearly linearly with a low field mobility of about 1,500 cm^2/Vs. This low value, as compared to the intrinsic mobility of $\sim 8,000\ cm^2/Vs$ in GaAs at 300 K, indicates that the presence of the high density hole plasma causes strong momentum relaxation by electron-hole Coulomb scattering. After a peak velocity of $1.1 \times 10^7\ cm/s$

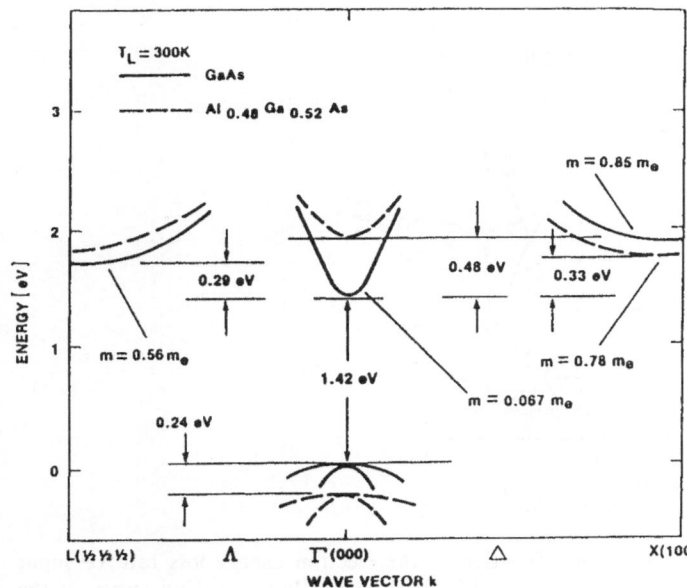

Fig. 5: Band extrema of GaAs and AlGaAs relevant for the discussion of real-space transfer effects, according to [20] (m ... density-of-states-mass).

the velocity *decreases* at fields higher than 8 kV/cm. This negative differential mobility is due to real-space transfer of hot electrons into the low mobility X-minimum of AlGaAs, as it has been shown in [20]. Measurements of the electron temperature by analyzing the band-to-band luminescence spectra show that at the electric fields where negative differential mobility is observed, the electron temperature is around ~650 K. This temperature is sufficient to transfer enough electrons over the confining conduction band barrier into AlGaAs. This effect is referred to as "real-space transfer" [25, 26].

In the case of this experiment, the transfer effect is even more interesting, since at the high Al-concentration of x=0.48 AlGaAs is an *indirect* semiconductor with the absolute conduction band minimum in the X-valley. Figure 5 shows the two bandstructures of GaAs and AlGaAs, with band-offsets of $\Delta E_C/\Delta E_V$ = 60:40 for the Γ-minima up to x=0.43 and a linear dependence of ΔE_V on x for the entire range of $0 \leqslant x \leqslant 1$ [27]. A calculation of the carrier distribution in the different band minima based on total thermalization of the electron distribution [20] shows that at the measured electron temperature of 650 K about 25% of the conduction band electrons are in the X-valley of AlGaAs. Thus the observed negative differential mobility can be well understood as being due to a *combined real-space and valley transfer* effect. This effect occurs at lower electron temperatures than the bulk valley transfer due to the lower energy barrier to the X-minimum of AlGaAs and the high density-of-states there.

The carrier heating thus leads to several transport effects in 2D carrier systems. Ref. [9] represents an excellent review on the physical processes that determine the carrier heating in polar 2D systems. The system of minority electrons in a 2D hole plasma is a

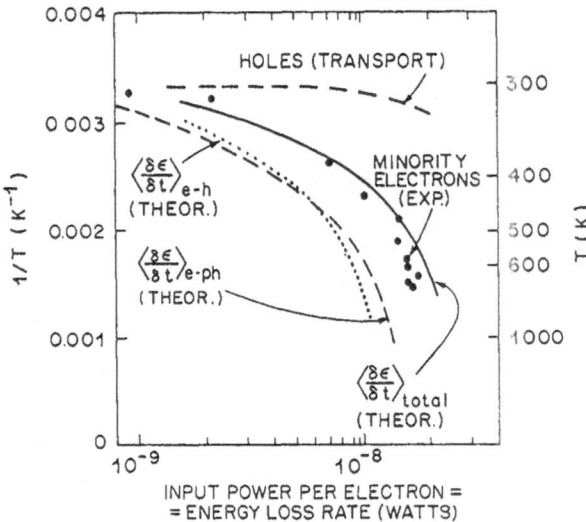

Fig. 6:   Carrier temperatures as a function of the electron energy loss rate (= input power per electron), after [21]. The total energy loss rate (full curve) is the sum of the energy loss rates by phonon scattering and by electron-hole scattering, $<\delta\epsilon/\delta t>_{total} = <\delta\epsilon/\delta t>_{e-ph} + <\delta\epsilon/\delta t>_{e-h}$.

very special case in this respect, electrons and holes - due to the different mobilities - acquire different energy in the electric field, and also have different energy loss rates to the lattice (via optical-phonon emission). Shah et al. [28] showed that this leads to a much stronger heating of electrons in n-modulation-doped quantum wells than of holes in p-GaAs quantum wells. For the situation in the minority injection experiments this would mean that the electrons should be driven to much "hotter" distribution functions than the holes. In fact, in Ref. [21] it could be experimentally shown that the energy distributions of electrons and holes can be characterized by *different* temperatures. Figure 6 illustrates this situation: Electrons (experimental points) are heated up to ~650 K, whereas at the same electric field (same input power per electron) the holes are not heated to more than ~350 K (measured from the change of mobility with electric field). In this *nonequilibrium* situation net energy transfer by electron-hole Coulomb collisions takes place, which acts as an additional cooling (energy loss) mechanism for the minority electrons. The total energy loss rate of the minority electrons therefore is higher (or: the minority electron temperature is lower) than in the case of majority carriers, where only electron-phonon interaction is present. Taking the sum of the two energy loss rates (Fig. 6), the total energy loss rate of the minority electrons is obtained, in agreement with the experimental results.

4. *Modeling of Device Performance*

In several publications, statements on the drift velocity in 2D carrier systems of GaAs/AlGaAs single heterostructures have been made on the basis of the transconductance [29] or the transit frequency of FETs [30]. The quantity obtained is the so-called "saturation drift velocity" [1] which is an average velocity over the varying electric field present in the device. Since under microwave performance conditions the electric field is not homogeneous along the 2D carrier system and not constant in time, the obtained values for the "saturation drift velocity" - although enormously important for device applications - only indirectly reflect the physical processes that determine the 2D-carrier transport in these systems. However, quantitative trends can be seen from these experiments indicating higher saturation velocities of heterostructure FETs as compared to MESFETs, as well as strong effects from the doping ("pulse" doping vs. homogeneous doping) [30]. Furthermore, information on the hot-electron transfer effects in 2D systems can be obtained from modeling the several types of hot-carrier devices [31] that use the real-space transfer effects for transistor applications.

5. Conclusion

The physics of transport in 2D systems is "enriched" as compared to 3D systems by the additional effects of (not perfect) confinement, by the high mobilities at low temperatures and low fields due to modulation-doping, by the possibilities of confining or separating electrons and holes, by the variety of different materials and material combinations available, and by the several structure parameters, that determine the shape of the potential wells. Several highly interesting experiments have been performed; the field, however, can be still regarded as a widely open area for ambitious experimentalists. Many questions concerning the physical basis of device performance are yet to be answered, especially the question of transfer effects under hot-carrier conditions. Negative differential mobility is observed in time-of-flight experiments at room temperature, and should be present also in FET-like single heterostructures. Exciting experiments can be expected for the near future.

Acknowledgment

The author thanks J. Shah for permanent collaboration and many stimulating discussions.

REFERENCES

[1] S. M. Sze, "*Physics of Semiconductor Devices*", 2nd ed., Wiley & Sons, New York 1981.

[2] For recent reviews, see: S. Hiyamizu, Proc. 2nd Int. Conf. on "Modulated Semiconductor Structures", Kyoto, Japan 1985, Surface Sci. (1986, in print); T. P. Pearsall, Surface Sci. *142*, 529 (1984).

[3] N. J. Shah, S. S. Pei, C. W. Tu, IEEE Trans. Electron. Devices (May 1986, in print).

[4] H. Dämbkes, K. Heime, Springer Series in Solid State Sciences *53*, 125 (1984).

[5] H. Sakaki, Jpn. J. Appl. Phys. *21*, L381 (1982).

[6] K. Seeger, "*Semiconductor Physics*" (2nd ed.), Ch. 4, Springer-Verlag, Berlin Heidelberg 1982.

[7] W. T. Masselink, W. Kopp, T. Henderson, H. Morkoc, IEEE Electron. Device Lett. *6*, 539 (1985).

[8] M. Inoue, Superlattices and Microstructures *1*, 433 (1985).

[9] J. Shah, IEEE J. Quantum Electron. (special issue on Quantum Well structures, 1986, in print).

[10]   K. Tsubaki, T. Fukui, H. Saito, Appl. Phys. Lett. *46*, 875 (1985).

[11]   *"Landolt-Börnstein,"* Vol. III/17a (ed. O. Madelung), p. 531, Springer-Verlag Berlin 1982.

[12]   E. F. Schubert, K. Ploog, H. Dämbkes, K. Heime, Appl. Phys. *A33*, 183 (1984).

[13]   K. Tsubaki, A. Livingstone, M. Kawashima, H. Okamoto, K. Kumabe, Solid State Comm. *46*, 517 (1983).

[14]   M. Al-Mudares, B. K. Ridley, Physica *134B*, 526 (1985).

[15]   For a review on the time-of-flight technique, see: L. Reggiani, in *"Physics of Nonlinear Transport in Semiconductors,"* p. 243, Plenum Press, New York 1980, and references therein.

[16]   J. G. Ruch, G. S. Kino, Phys. Rev. B&, 2693 (1973).

[17]   J. A. Cooper, Jr., D. F. Nelson, J. Appl. Phys. *54*, 1445 (1983).

[18]   D. F. Nelson, J. A. Cooper, Jr., A. R. Tretola, Appl. Phys. Lett. *41*, 857 (1982).

[19]   R. A. Höpfel, J. Shah, A. C. Gossard, W. Wiegmann, Physica *134B*, 509 (1985).

[20]   R. A. Höpfel, J. Shah, D. Block, A. C. Gossard, Appl. Phys. Lett. *48*, 148 (1986).

[21]   R. A. Höpfel, J. Shah, A. C. Gossard, Phys. Rev. Lett. *56*, 765 (1986).

[22]   A. Von Lehmen, J. M. Ballantyne, Appl. Phys. Lett. *44*, 87 (1984).

[23]   B. F. Levine, W. T. Tsang, C. G. Bethea, F. Capasso, Appl. Phys. Lett. *41*, 470 (1982).

[24]   R. L. Fork, Physica *134B*, 381 (1985).

[25]   K. Hess, J. Physique *C7*, 3 (1981).

[26]   For a review, see: K. Hess, G. J. Iafrate, Springer Topics in Appl. Physics *58*, 201 (1985).

[27]   H. Kroemer, Proc. 2nd Int. Conf. on "Modulated Semiconductor Structures," Kyoto, Japan 1985, Surface Sci. (1986, in print).

[28]   J. Shah, A. Pinczuk, A. C. Gossard, W. Wiegmann, Phys. Rev. Lett. *54*, 2045 (1985).

[29]   Y. Takanashi, N. Kobayashi, IEEE Electron Device Lett. *6* 154, (1985).

[30]   T. Mimura, M. Abe, A. Shibatomi, M. Kobayashi, Proc. 6th Int. Conf. on "Electronic Properties of Two-Dimensional Systems," Kyoto, Japan, 1985, Surface Sci. (1986, in print).

[31]   S. Luryi, A. Kastalsky, Physica *134B*, 453 (1985), and S. Luryi, this volume.

Time-Resolved Spectroscopy
of Hot Carriers in Quantum Wells

J.F. Ryan

Clarendon Laboratory, University of Oxford, Oxford, U.K.

The energy relaxation of hot carriers confined in quantum well structures has been measured directly by time-resolved photoluminescence spectroscopy. Both p-type and n-type modulation-doped GaAs/GaAlAs quantum wells show a much reduced energy-loss rate compared to bulk GaAs. The theoretical interpretation of these results is still somewhat controversial: the effects of degeneracy and screening of the electron-phonon interaction are now believed to produce only a small decrease in the electron-phonon scattering rate, but it has been proposed that hot-phonon effects can produce a dramatic reduction. However, recent calculations of electron-2D phonon scattering suggest that reduced dimensionality can also cause a substantial reduction in the energy-loss rate.

1. Introduction

Nonequilibrium phenomena continue to be a source of great interest and fascination for a wide variety of reasons, not least of which is the considerable challenge they present to measure and understand their basic mechanisms. For twenty years or so the problem of nonequilibrium carriers in semiconductors has recieved great attention, with the result that the properties of bulk materials are now fairly well understood. The importance of the various scattering processes including carrier-carrier and carrier-phonon interactions was appreciated very early on, and they are of course utilized now in many common devices. However, the fabrication of quantum wells and superlattices has given rise to new forms of nonequilibrium behaviour, in some cases due to the confinement of carriers and phonons, and so interest in the problem has reawakened. Furthermore, with the advent of ultrafast time-resolved spectroscopy it has become possible to make real-time measurements of the dynamics of relaxing hot carriers, and so new information has become available which is leading to the development and refinement of new theoretical models. In this paper I will describe some of the progress achieved in this area over the past year or so, placing special emphasis on the properties of GaAs quantum wells.

The essence of the optical experiments described here is that hot electrons and holes are created within a very short time interval (~ 1 ps), and that the subsequent measurement of the changing populations of carriers in excited states gives direct information about the nature of the relaxation processes. Initially the carriers are nonthermal, but carrier-carrier interactions, which have a characteristic time of $\leqslant 100$ fs, cause rapid thermalization to a temperature that is very much in excess of the lattice temperature. In bulk GaAs the primary mechanism by which hot electrons lose energy is LO phonon emission via the Frohlich interaction; hot holes also lose energy by this process, but there is an additional interaction with TO phonons via the deformation potential interaction. These relaxation processes are effective for about a hundred picoseconds or so, and reduce the carrier temperature to a value close to the lattice temperature. An example of this behaviour will be described below. In low-dimensional structures the carrier-phonon scattering will be quite different because of the modified electron energy band structure and the 2D nature of the phonons, and so the energy-loss rate is expected to differ.

2. Optical Measurement of the Energy-Loss Rate

Photoluminescence experiments can provide an estimate of the energy-loss rate of hot carriers because the spectral lineshape yields a relatively direct measurement of the carrier temperature T_c. The luminescence intensity arising from the band-to-band recombination of e-h pairs is [1]:

$$I(E) = AG(E)\rho_{eh}(E)f_e f_h |M_{eh}|^2 \tag{1}$$

where A is a constant, $G(E)$ is the photon density of states, $\rho_{eh}(E)$ is the joint density of e-h states, and M_{eh} is the matrix element for the radiative transition. The quantities f_e and f_h are the electron and hole distribution functions respectively, and so contain information about the carrier temperature. In general f_e and f_h have the full Fermi-Dirac form, but we can see that at sufficiently high energy the product varies as $\exp(-(E-E_G)/k_B T_c)$, so that the luminescence spectrum will display an exponentially decreasing high-energy wing. This statement neglects of course any energy dependence of the other terms in (1); this is approximately valid for M_{eh} and for $G(E)$ (which varies slowly as E^2), but for bulk materials $\rho_{eh} \sim (E-E_G)^{\frac{1}{2}}$ and so must be taken into account. For 2D systems ρ_{eh} is approximately constant, and so the statement is reasonably valid. A time-resolved photoluminescence measurement will therefore yield a cooling curve $T_c(t)$ for given experimental conditions of incident photon energy and photoexcited carrier density. The mean loss rate per particle is then obtained from:

$$\frac{d\epsilon}{dt} = -C \frac{dT_c(t)}{dt} \tag{2}$$

where C is the mean specific heat of the carrier distribution.

This analysis may be taken a stage further in order to obtain an estimate of the mean carrier-phonon scattering rate. The transition rates for absorption (+) and emission (−) of LO phonons are:

$$R = \frac{2\pi}{\hbar} \sum_{q,k} f(\underline{k})(1-f(\underline{k}\pm\underline{q})) \left| \frac{M_{ep}(\underline{q})}{\epsilon(\underline{q},\omega_{LO})} \right|^2$$
$$\times \delta(E_{\underline{q}} - E_{\underline{k}+\underline{q}} \pm \hbar\omega_{LO}) \tag{3}$$

$M_{ep}(\underline{q})$ is the matrix element for the Frolich interaction:

$$\left| M_{ep}(\underline{q}) \right|^2 = \frac{2\pi e^2 \hbar\omega_{LO}}{Vq^2} \left[\frac{1}{\epsilon_\infty} - \frac{1}{\epsilon_0} \right] \left[N(\underline{q}) + \frac{1}{2} \pm \frac{1}{2} \right] \tag{4}$$

$\epsilon(\underline{q},\omega)$ is the wavevector and frequency-dependent dielectric function, and ϵ_∞ and ϵ_0 are the high- and low-frequency dielectric constants. $N(\underline{q})$ is the Planck function giving the occupation of the phonon of mode the wavevector $\underline{q}$. The energy-loss rate per particle can be obtained from (3) by the appropriate summation over $\underline{q}$ and $\underline{k}$. For nondegenerate carriers and low phonon mode populations (i.e. $N(\underline{q}) \sim 0$, so that phonon absorption is neglible) the result is the familiar expression [2]:

$$\frac{d\epsilon}{dt} = -\frac{\hbar\omega_{LO}}{\tau_0} \exp - \left[\frac{\hbar\omega_{LO}}{k_B T_c(t)} \right] \tag{5}$$

317

where τ_0^{-1} is the mean (energy-averaged) scattering rate given by:

$$\tau_0 = \frac{4\pi\hbar^2\kappa_0}{m^*e^2} \left[\frac{m^*}{2\hbar\omega_{LO}}\right]^{\frac{1}{2}} \left[\frac{1}{\epsilon_\infty} - \frac{1}{\epsilon_0}\right] \qquad (6)$$

For bulk GaAs the theoretical value is $\tau_0 \simeq 0.15$ps. Experimentally τ_0 can be estimated by fitting the cooling curve $T_c(t)$ using eqs. (2) and (5). This approach has been widely used in the literature for both bulk semiconductors [3] and more recently, for quantum well structures [4].

3. Experimental Results

Before discussing the results on 2D systems I will describe briefly the behaviour of bulk GaAs. The photoluminescence spectra obtained from an undoped MBE GaAs sample is shown in Figure 1. The laser photon energy is ~ 250 meV above the bandgap, and the estimated photoexcited carrier density is $\simeq 10^{18}$cm^{-3}. A high-energy exponential wing is clearly evident, and the increasing slope with increasing time is evidence of cooling of the electron-hole plasma.

Figure 2 shows the measured cooling curve, together with a fitted curve obtained from eqs. (2) and (5) with $\tau_0^{3D} \simeq 0.6$ ps, which is roughly four times the theoretical value obtained above. Similar results were obtained in early transient absorption experiments

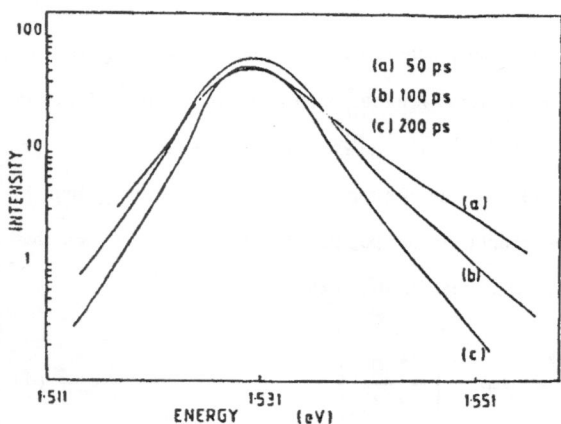

Figure 1   Time-resolved luminescence spectra of bulk GaAs (hν_L=1.78eV, T_L=4K). Photoexcited carrier density $n\approx 10^{18}$ cm^{-3}.

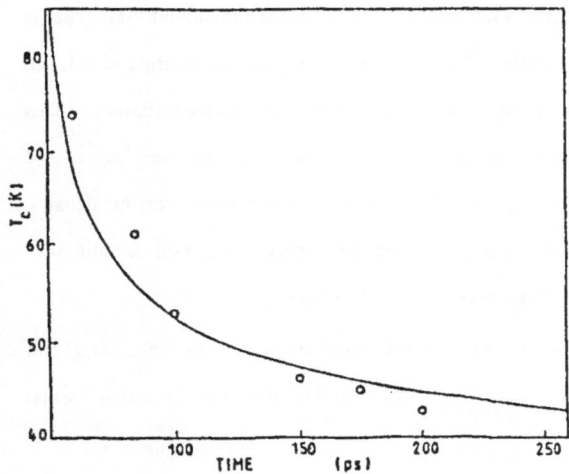

Figure 2 Cooling of hot carriers in bulk GaAs (condtions as in Fig. 1). The fitted curve gives τ^{3D}=0.6 ps.

[3,5] and also in photoluminescence measurements [6,7,8]. Von der Linde and Lambrich [3] measured $\tau_0^{3D} \simeq 0.1$ ps for low carrier density, whereas Leheny et al. [5] found a substantial increase in τ_0^{3D} with increasing carrier density, a factor of 5 increase being measured for n~ 5×10^{17} cm^{-3}. Similar values of τ_0^{3D} were obtained in the luminescence experiments and a factor of 10 increase was observed at rather higher densities 10^{18} - 10^{19} cm^{-3}. Many of these authors assigned this reduction of the scattering rate to screening of the electron-LO phonon interaction: this effect can be seen in eq (1) arising from the density-dependent dielectric function $\epsilon(q,\omega)$. However, Potz and Kocevar [9] and others [10,11] have argued on the basis of detailed numerical calculations that a large nonequilibrium LO phonon population is produced by the relaxing hot carriers which gives rise to substantial phonon reabsorption and so to a reduced energy-loss rate. So far this effect has not been confirmed experimentally. It should also be noted that the analysis of the experimental data considers both types of carriers to contribute equally to the mean energy-loss rate; the approximation is usually justified by invoking fast and efficient energy transfer through electron-hole scattering.

The experiments on the QW structures are technically similar to those on bulk samples, but there are several important new considerations. First, the excitation pulse can be chosen to have a spectral energy which lies either above or below the bandgap of the GaAlAs barriers; in principle this could give information about the charge transfer from the barriers into the wells. Second, the wells can be modulation-doped n-type or p-type so as to provide a cold carrier gas prior to photoexcitation. This enables the heating and subsequent relaxation of a single carrier species to be measured, provided the dopant density greatly exceeds the photoexcited carrier density. Finally, if low-dimensional effects are important then the energy–loss rate should vary with well width, and so appropriate structures should be chosen.

Result on n-type GaAs/GaAlAs MQWs are the most extensive to date, and can best be illustrated using the cooling curves obtained under the various experimental conditions.

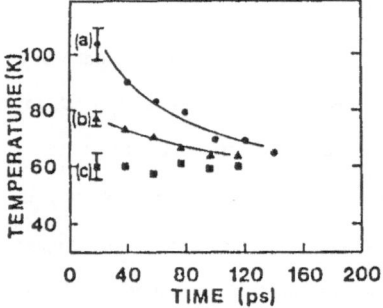

Figure 3 Time-dependence of T_C in a GaAs/ GaAlAs MQW, $d_{GaAs}=258A$. The photoexcited carrier density are: a) $n\approx5\times10^{18}$ cm^{-3} b) 6×10^{16} cm^{-3} and c) 9×10^{14} cm^{-3}. $\tau_0^{2D}=7$ ps. [4]

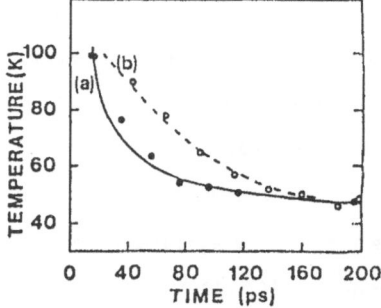

Figure 4 Time-dependence of T_C in MQW sample: a) $d_{GaAs}=258A$, b) $d_{GaAs}=150A$. The solid curve gives a value $\tau_0^{2D}=1.5$ ps. [12].

Figure 3 shows the cooling curves obtained for a sample with $d_{GaAs} = 258$ Å and a dopant density $n_0/d_{GaAs} = 2\times10^{17}$ cm^{-3} when the incident photons are absorbed in both the GaAlAs and GaAs layers [4]. For $n\sim5\times10^{18}$ cm^{-3} the temperature measured at 20 ps after photoexcitation is 105K and this reduces to 65K after 100 ps. For lower pump intensities ($\sim5\times10^{16}$ cm^{-3}) the initial temperature is lower but the same base temperature is reached. At the lowest intensity ($\sim10^{15}$ cm^{-3}) very little cooling of the carriers is detected. The solid lines are obtained from the analysis described above in §2. This amounts to assuming a single electric subband and bulk-like phonons which is obviously a great over-simplification, but it nevertheless allows a convenient comparison to be made with bulk GaAs. A value of $\tau_0^{2D} \simeq 7$ ps is obtained [4] which is dramatically larger than the theoretical value τ_0^{3D}. On the other hand, when the laser photon has energy lying below the GaAlAs bandgap, a substantially more rapid cooling is observed and a lower base temperature is obtained. (See Figure 4(a)). In this case $\tau_0 \simeq 1.5$ps, which is still roughly an order of magnitude larger than τ_0^{3D} [12].

The variation of energy-loss rate with well width has not been examined very closely to date. Ridley [13] has calculated that the intersubband scattering rate is likely to greatly reduce when the confinement energy is greater than $\hbar\omega_{LO}$. For GaAs quantum wells this condition obtains for $d_{GaAs} \leqslant 150\text{Å}$. Figure 4 compares the cooling curves obtained for two modulation-doped samples, (a) $d_{GaAs} = 258\text{Å}$, and (b) $d_{GaAs} = 150\text{Å}$, under approximately the same experimental conditions *viz.* $n \sim 10^{18}$ cm^{-3} and $h\nu_L = 1.78$ eV. The dashed line is not a fitted curve, only a guide to the eye; although the initial and final temperatures in (b) are approximately equal to those in (a) the shape of the cooling is quite different and the simplified model of §2 clearly fails.

The results obtained from p-type quantum wells also show a substantially reduced energy-loss rate. Kash *et al* [14] have measured the energy-loss rates in two samples with different dopant densities but the same well width, $d_{GaAs} = 90\text{Å}$, and have made a number of important observations. They find, firstly, that the cooling curve is independent of the dopant density, whereas it changes substantially with photoexcited carrier density. In addition, they observe energy-loss rates which differ greatly from those obtained from the same sample using steady state photoluminescence from carriers heated by an applied electric field [19].

This result is shown in Figure 5: at short times (10 ps) the time-resolved measurement (solid line) is about a factor of 25 less than the steady-state value (broken line). The latter, it should be said, is in close agreement with a theoretical estimate for 3D holes [19]. At much longer times the two curves come into agreement with each other and

321

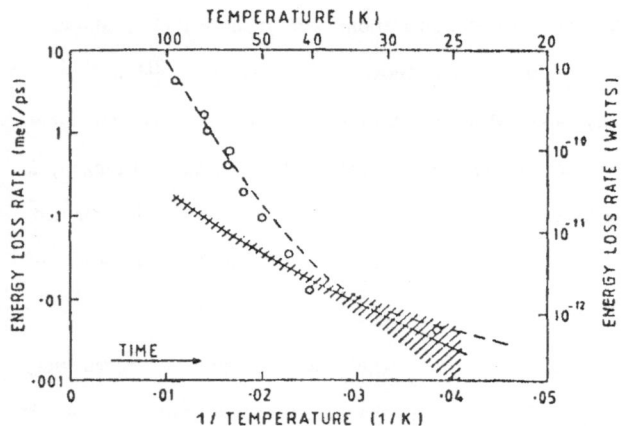

Figure 5 The energy-loss rate in a p-type MQW with d_{GaAs}=90A, p_0=4x10^{17} cm^{-3} Time-resolved data (solid line), steady state data (open circles) and model calculation (broken curve). [13]

with the theory. Kash *et al* interpret these results as follows. The difference between the time-resolved and the steady-state measurements seems to rule out dimensional effects. Also, the fact that the energy-loss mechanism for holes includes deformation potential scattering, which cannot be screened out and which is possibly as large as that due to the bare Frohlich interaction, together with the measured independence of cooling of dopant density, leads to the exclusion of screening as the mechanism responsible for reduced energy-loss rates. The eventual agreement at long times between both experimental values and the theory leads Kash *et al* to propose the hot-phonon mechanism.

4. Discussion

The experiments described above raise a number of important issues and have stimulated several theoretical calculations of the electron-phonon interaction in 2D systems. These calculations have considered the various constituent parts of the problem more or less separately, and an understanding of their relative importance is now beginning to emerge. Firstly, there is the question of screening of the electron-phonon interaction in 2D: the data on the p-type samples seem to suggest that it is a weak effect. Second, there is the question of slow intersubband transitions due to reduced dimensionality of the electrons: the behaviour of the n-type samples

322

indicates a lower scattering rate. Finally, there is the question of the effects of hot phonons and phonon confinement.

The initial calculations of Ridley [13] considered intrasubband scattering of 2D electrons by 3D phonons via an unscreened Frohlich interaction and found that $\tau_0^{2D} < \tau_0^{3D}$, the reduction being due to the finite density of states at zero energy for 2D carriers. Recent calculations by das Sarma and Mason [15] and by Yang and Lyon [16] have considered the effects of screening; the former used static screening and degenerate carriers, whereas the latter authors used dynamic screening and nondegenerate carriers. In both cases 3D phonons were used. Yang and Lyon find that while the scattering is reduced at higher temperatures and densities due to screening, high-energy electrons scatter more efficiently due to coupling with the 2D plasmon [16]. They conclude that the mean scattering rate is reduced by a factor of 2 for the temperatures and densities typical of the above experiments. This result is not expected in the static screening approximation; however das Sarma and Mason found a large increase in τ_0^{2D} due to an unexpectedly high probability for electrons with $\epsilon < \hbar\omega_{LO}$ to reabsorb phonons when the carriers are degenerate - in other words a hot phonon effect.

Price [18] has estimated the magnitude of the hot phonon reduction of the scattering rate for n-type quantum wells, and finds good agreement with steady-state luminescence measurements [19]. At high excitation density he finds that $\tau_0 \to \tau_p$, the phonon lifetime, which for GaAs is 7 ps [20]. However, the agreement of this value with that obtained from the cooling curves in Figure 3 should be regarded as fortuitous.

More recent calculations have addressed the issue of treating the phonons as 2D .excitations. Riddoch and Ridley [21] considered the interaction of electrons with the confined or slab modes of a single layer, and as expected the scattering rate is greatly reduced: the reason for this is that long wavelength modes for $q//z$ do not exist, and since the Frohlich interaction favours $q \approx 0$ modes the overall interaction is weaker. However, in extending their calculation to a MQW structure Riddoch and Ridley found that added interactions with remote layers produced a resultant scattering rate not too different from that of the bulk. Sawaki [22] on the other hand has considered a MQW structure which has a modulation of the elastic constants (as opposed to the modulation of the dielectric constant as considered by Riddoch and Ridley); he finds that the e-2D phonon scattering is again smaller, but now the remote interaction is very weak.

5. Conclusions

Measurements of the hot carrier relaxation in quantum wells using pulsed laser techniques show reduced energy – loss rates. Theoretical models show that nonequilibrium phonon effects are likely to be important, and also that confinement of phonons might produce a significant effect. In order to resolve these issues it is clear that the phonons generated in the relaxation process must be investigated more thoroughly in future experiments.

Acknowledgements

I wish to thank Alzamir da Costa who provided Figures 1 and 2, and also Andrew Turberfield, Robert Taylor and John Worlock who have collaborated with me in this work.

References

1. E.Barry Bebb and E.W.Williams, Semiconductors and Semimetals Vol.8, p181 (Academic Press, New York, 1971)
2. E.M.Conwell, Solid State Physics, Supplement 9, Ed. F.Seitz, D.Turnbull and H. Ehrenreich (Academic Press, New York, 1969)
3. D. von der Linde and R.Lambrich, Phys. Rev. Lett. 42, 1090(1979)
4. J.F.Ryan, R.A.Taylor, A.J.Turberfield, A.Maciel, J.M.Worlock,A.C.Gossard and W.Wiegmann, Phys. Rev. Lett. 53,1841(1984)
5. R.F.Leheny, Jagdeep Shah, R.L.Fork, C.V.Shank and A.Migus, Solid State Commun. 31,809(1979)
6. S.Tanaka, H.Kobayashi, H.Saito and H.Shionoya, J.Phys.Soc.Japan 49,1051(1980)
7. W.Graudszus and E.Göbel, Journal de Physique Colloq 42, C7,445(1981)
8. J.Shah, Journal de Physique Colloq 42, C7,445(1981)
9. W.Potz and P.Kocevar, Phys Rev B28,7040(1983)
10. J.Collet,A.Garnett, M.Riguet and T.Amand, Solid State Commun.42,883(1982)
11. A.C.S.Algarte and R.Luzzi, Phys. Rev. B27,7563(1983)

12. J.F.Ryan, R.A.Taylor, A.J.Turberfield and J.M.Worlock, Surface Science (1986) (to be published)
13. B.K.Ridley,J.Phys.C Solid State Physics 15,5899(1982)
14. K. Kash, J.Shah, D.Block, A.C.Gossard and W.Wiegmann, Physica 134B+C,189(1985)
15. S. Das Sarma and B.A.Mason, Pysica 134B+C,2,301(1985)
16. C.H.Yang and S.A.Lyon, Physica 134B+C,309(1985)
17. P.J.Price, J.Vac.Sci.Technol. 19,599(1981)
18. P.J.Price, Physica 134B+C,164(1985)
19. J.Shah, A.Pinczuk, A.C.Gossard and W.Wiegmann, Phys. Rev. Lett. 54,2045 (1985)
20. D. von der Linde, J.Kuhl and H.Klingenberg, Phys. Rev. Lett. 44,1505 (1980)
21. F.A. Riddoch and B.K.Riddley, Physica 134B+C,342(1985)
22. N.Sawaki, Surface Science (1986) (to be published)

Index of Contributors

F. R. Aussenegg, A. Leitner, M. E. Lippitsch (Eds.)

Surface Studies with Lasers

Proceedings of the International Conference
Mauterndorf, Austria, March 9–11, 1983

1983. 146 figures. IX, 241 pages. (Springer Series in Chemical Physics, Volume 33)
ISBN 3-540-12598-1

Contents: General Surface Spectroscopy. – Surface Enhanced Optical Processes. – Laser Surface Spectroscopy. – Laser Induced Surface Processes. – Index of Contributors.

D. Bäuerle (Ed.)

Laser Processing and Diagnostics

Proceedings of an International Conference,
Linz, Austria, July 15–19, 1984

1984. 399 figures. XI, 551 pages. (Springer Series in Chemical Physics, Volume 39)
Hard cover. ISBN 3-540-13843-9

Contents: Laser – Solid Interactions: Fundamentals and Applications. – Photophysics and Chemistry of Molecule – Surface Interactions. – Photo-assisted Chemical Processing. – Diagnostics of Laser Processing, Materials and Devices. – Laser Diagnostics in Reactive Gaseous Systems. – Index of Contributors. – Subject Index.

A. Rosenfeld (Ed.)

Multiresolution Image Processing and Analysis

1984. 198 figures. VIII, 385 pages.
(Springer Series in Information Sciences, Volume 12). ISBN 3-540-13006-3

Contents: Image Pyramids and Their Uses. – Architectures and Systems. – Modelling, Processing, and Segmentation. – Features and Shape Analysis. – Region Representation and Surface Interpolation. – Time-Varying Analysis. Applications. – Index of Contributors. – Subject Index.

Y. Tarui (Ed.)

VLSI Technology

Fundamentals and Applications

1986. 377 figures. XIV, 450 pages.
(Springer Series in Electrophysics, Volume 12). Hard cover.
ISBN 3-540-12558-2

Contents: Introduction. – Electron Beam Lithography. – Pattern Replication Technology. – Mask Inspection Technology. – Crystal Technology. – Process Technology. – Fundamentals of Test and Evaluation. – Basic Device Technology. – References. – Subject Index.

Springer-Verlag
Berlin Heidelberg New York Tokyo

Springer

H.-J. Queisser (Ed.)

X-Ray Optics

Applications to Solids

With contributions by A. Authier,
U. Bonse, R. Feder, W. Graeff, W. Hart-
mann, S. Kozaki, H.-J. Queisser, E. Spiller,
M. Yoshimatsu

1977. 133 figures, 14 tables. XI, 227 pages
(Topics in Applied Physics, Volume 22)
ISBN 3-540-08462-2

Contents: Introduction: Structure and
Structuring of Solids. – High Brilliance
X-Ray Sources. – X-Ray Lithography. –
X-Ray and Neutron Interferometry. –
Section Topography. – Live Topography.

G. Bauer, F. Kuchar, H. Heinrich (Eds.)

Two-Dimensional Systems, Hetero-structures, and Superlattices

Proceedings of the International Winter
School, Mauterndorf, Austria, February
26–March 2, 1984

1984. 231 figures, IX, 293 pages. (Springer
Series in Solid-State Sciences, Volume 53)
ISBN 3-540-13584-7

Contents: Physics of Heterostructures and
Inversion Layers. – Growth and Devices. –
Multi Quantum Wells and Superlattices. –
Doping Superlattices. – Quantum Hall
Effect. – Index of Contributors.

G. A. Mourou, D. M. Bloom, C.-H. Lee
(Eds.)

Picosecond Electronics and Optoelectronics

Proceedings of the Topical Meeting, Lake
Tahoe, Nevada, March 13–15, 1985

1985. 202 figures. X, 258 pages. (Springer
Series in Electrophysics, Volume 21). Hard
cover. ISBN 3-540-15884-7

Contents: Ultrafast Optics and Electronics.
– High-Speed Phenomena in Bulk Semi-
conductors. – Quantum Structures and
Applications. – Picosecond Diode Lasers.
– Optoelectronics and Photoconductive
Switching. – Cryoelectronics. – Index of
Contributors.

G. Allan, G. Bastard, N. Boccara,
M. Lannoo, M. Voos (Eds.)

Heterojunctions and Semiconductor Superlattices

Proceedings of the Winter School
Les Houches, France, March 12–21, 1985

1986. Approx. 168 figures. Approx. 280
pages. Hard cover. ISBN 3-540-16259-3

Springer-Verlag
Berlin Heidelberg New York Tokyo